Fractals, Quasicrystals, Chaos, Knots and Algebraic Quantum Mechanics

NATO ASI Series

Advanced Science Institutes Series

A Series presenting the results of activities sponsored by the NATO Science Committee, which aims at the dissemination of advanced scientific and technological knowledge, with a view to strengthening links between scientific communities.

The Series is published by an international board of publishers in conjunction with the NATO Scientific Affairs Division

A	Life Sciences	Plenum Publishing Corporation
B	Physics	London and New York
C	Mathematical and Physical Sciences	Kluwer Academic Publishers
D	Behavioural and Social Sciences	Dordrecht, Boston and London
E	Applied Sciences	
F	Computer and Systems Sciences	Springer-Verlag
G	Ecological Sciences	Berlin, Heidelberg, New York, London,
H	Cell Biology	Paris and Tokyo

Series C: Mathematical and Physical Sciences - Vol. 235

Fractals, Quasicrystals, Chaos, Knots and Algebraic Quantum Mechanics

edited by

A. Amann
Laboratory of Physical Chemistry,
ETH-Zentrum, Zürich, Switzerland

L. Cederbaum
Physical Chemistry Institute,
University of Heidelberg, Heidelberg, Germany

and

W. Gans
Institute of Physical Chemistry,
Freie Universität Berlin, Berlin, Germany

Kluwer Academic Publishers

Dordrecht / Boston / London

Published in cooperation with NATO Scientific Affairs Division

Proceedings of the NATO Advanced Research Workshop on
New Theoretical Concepts in Physical Chemistry
Acquafredda di Maratea, Italy
October 4–8, 1987

Library of Congress Cataloging in Publication Data

Fractals, quasicrystals, chaos, knots, and algebraic quantum mechanics
/ edited by A. Amann, L. Cederbaum, W. Gans.
      p.   cm. -- (NATO ASI series. Series C, Mathematical and
physical sciences ; vol. 235)
   "Published in cooperation with NATO Scientific Affairs Division."
   Includes index.
   ISBN 9027727503
   1. Chemistry, Physical and theoretical--Mathematics--Congresses.
2. Fractals--Congresses. 3. Metal crystals--Congresses. 4. Chaotic
behavior in systems--Congresses. 5. Knot theory--Congresses.
6. Quantum theory--Congresses.. I. Amann, A., 1956-   .
II. Cederbaum. L., 1946-   . III. Gans. W., 1949-   . IV. North
Atlantic Treaty Organization. Scientific Affairs Division.
V. Series: NATO ASI series. Series C, Mathematical and physical
sciences ; no. 235.
QD455.3.M3F73 1988
541--dc19   88-12648
                                                         CIP

ISBN-13: 978-94-010-7850-4   e-ISBN-13: 978-94-009-3005-6
DOI: 10.1007/ 978-94-009-3005-6

Published by Kluwer Academic Publishers,
P.O. Box 17, 3300 AA Dordrecht, The Netherlands.

Kluwer Academic Publishers incorporates the publishing programmes of
D. Reidel, Martinus Nijhoff, Dr W. Junk, and MTP Press.

Sold and distributed in the U.S.A. and Canada
by Kluwer Academic Publishers,
101 Philip Drive, Norwell, MA 02061, U.S.A.

In all other countries, sold and distributed
by Kluwer Academic Publishers Group,
P.O. Box 322, 3300 AH Dordrecht, The Netherlands.

TABLE OF CONTENTS

vi

FOREWORD

At the end of the workshop on "New Theoretical Concepts in Physical Chemistry", one of the participants made an attempt to present a first impression of its achievements from his own personal standpoint. Apparently his views reflected a general feeling, so that the organizers thought they would be suitable as a presentation of the proceedings for future readers. That is the background from which this foreword was born.

The scope of the workshop is a very broad one. There are contributions from mathematics, physics, crystallography, chemistry and biology; the problems are approached either by means of axiomatic and rigorous methods, or at an empirical phenomenological level. This same diversification can be found in the new basic concepts presented. Some arise from pure theoretical investigation in C*-algebra or in quantum probability theory; others from an analysis of very complex experimental data like nuclear energy levels, or processes on the frontier between classical and quantum physics; others again have their origin in the discovery of new ordered structures like the icosahedral crystal phases, or the knots of DNA molecules; others follow from the application of ideas like fractals or chaos to new fields like spectral theory or chemical reactions.

It is to be expected that readers will have to face the same sort of difficulties as did the participants in understanding such diverse languages, in applying themselves to subjects possibly far from their own experience, and in grasping highly sophisticated new concepts.

The risk inherent to a workshop like the present one is to be too broad, and to stick at the level of piecing problems together without intercommunication and synthesis. The organizers were fully aware of these dangers, but they nevertheless took up the challenge.

In my view, the goals of this workshop have been well realized. This despite the fact that I really could follow only parts of the contributions. I expect that this will also be the case for most of the readers of these proceedings, and which part of the book they find the more accessible will of course depend on their scientific background.

The reason why I consider the workshop successful is because it has revealed a number of guiding principles at the roots of the new concepts presented, which are fruitful in other and very different areas. Let me give some examples :

1. *Data analysis*. A fairly complex data set (like spectral energy levels, a geometrical atomic arrangement, the time/energy dependency of a dynamical system, or a non-trivial set of classical/quantum mechanical observables, or a family of chemical reactions, or a set of knots, and so on) permits, in general, more than one meaningful approach. Each different way of analyzing the data allows one to extract some of the rich infor-

mation contained in the system. The theoretical (and experimental) data are often so sophisticated that only by adopting different points of view can their structure be made explicit.

2. *Higher/lower dimensional description*. A concretisation of the previous principle is realized in some cases by plotting the data in higher dimensional spaces (in terms either of suitable parameters, and/or coordinates). This is an "unfolding principle". Typical examples are 3--dimensional crystal structures described in a 6-dimensional space. Other examples are ultrametric spaces in random walk problems.

3. *Self-similarity*. The deep-rooted experience that one has of euclidean space had as consequence that metrical relations (and symmetries) were the first to be developed. More recently the relevance of natural and mathematical structures with a fractal character has become clearer. Self-similarity seems to be a general invariance principle of these structures. Interesting is the fact that self-similarity also appears in non-fractal structures (like quasi-crystals).

4. *Multilevel approach*. A more fundamental approach does not eliminate the value of a phenomenological, or empirical one, and vice versa. One example can be found in spontaneous symmetry breaking, another in chirality, another in chaotic behaviour, and so on. Communication between these different theoretical levels appears often to be extremely difficult, especially when dealing with the concepts involved (which can be of very different natures). More effective than a communication based directly on the concepts themselves seems to be one focussing attention rather on the consequences they have.

These ideas reflect personal views and are not exhaustive. One will certainly recognize in the papers of this workshop other frontiercrossing principles as well. Searching for these will frequently lead to deeper insight, even on the basis of a partial understanding. And this will be by no means the only fruitful approach to this text. Other readers can also simply enjoy the beauty of the mathematical theory of knots, and admire the results of genetic engineering where knots are tied and undone or learn a lot about chaos and fractals in natural phenomena.

The heartfelt thanks of all participants are due to the organizers of this workshop and the editors of the present proceedings. Thanks are also due to the NATO Scientific Affairs Division for their financial support.

Nijmegen, December 1987   A. Janner

PREFACE

The Advanced Research Workshop "New Theoretical Concepts in Physical Chemistry" was actually an experiment: instead of dealing with a single field of research, it was devoted to five different and at first sight disconnected topics which we consider to play an important role in the future. Our motivation for the workshop was twofold. First, researchers are usually specialists in their individual fields and a meeting of workshop character provides a good opportunity to learn about methods and problems in relevant neighbouring fields. Second, the workshop provided an excellent forum for discussing similarities and common features present in the five topics chosen:

- *Fractals:* relevant for heterogeneous catalysis on rough surfaces, transport and chemical reactions in amorphous materials.

- *Quasicrystals:* connected with the study of alloys.

- *Chaotic motion:* relevant for the description of spectral properties of complicated nuclei, atoms and molecules.

- *Knot theory:* is important for the investigation of macromolecules such as coiled DNA.

- *Algebraic quantum mechanics:* permits to describe systems with quantum *and* classical properties and relates classical properties to broken symmetries of a system.

The articles presented here contain a general introduction, not at textbook level, but rather appealing to researchers with a keen interest in natural sciences. So the reader should get an idea of the subject even if he/she is lacking part of the background. During the workshop connections among the various fields showed up in particular due to the presence of scientists with broad interests. Two round table discussions on 'Fractals in Chemistry' and 'Chaotic Motion in Quantum Systems' were also helpful in this respect.

We thank the NATO Scientific Affairs Division for providing a generous grant to an unusual interdisciplinary workshop. It is also a pleasure to thank our secretary Mrs.M.Schiessl whose help in preparing the workshop and the proceedings volume was indispensible.

Zürich, December 1987

Anton Amann   Lenz Cederbaum   Werner Gans

LIST OF PARTICIPANTS

Amann, A., Laboratorium für Physikalische Chemie, ETH-Zentrum,
 CH-8092 Zürich, Switzerland
Argyrakis, P., University of Thessaloniki, Department of Physics,
 GR-54006 Thessaloniki, Greece
Blumen, A., Universität Bayreuth, Universitätsstrasse 30,
 D-8580 Bayreuth, Germany
Bohigas, O., Division de Physique Théorique, Institut de Physique
 Nucléaire, F-91406 Orsay Cedex, France
Casati, G., Istituto di Fisica, Università di Milano, I-20133 Milano,
 Italy
Cederbaum, L.S., Physikalisch-Chemisches Institut der Universität
 Heidelberg, Im Neuenheimer Feld 253, D-6900 Heidelberg, Germany
Elser, V., AT & T Bell Laboratories, 600, Mountain Avenue, Murray Hill,
 NJ 07974/2070, USA
Farantos, St., University of Crete, Department of Chemistry, P.O.Box
 1470, GR-71110 Iraklion, Greece
Founargiotakis, M., University of Crete, Department of Chemistry,
 P.O.Box 1470, GR-71110 Iraklion, Greece
Gans, W., Institut für Quantenchemie, Freie Universität Berlin,
 Holbeinstrasse 48, D-1000 Berlin 45, Germany
Gotsis, H., University of Thessaloniki, Department of Physics,
 GR-54006 Thessaloniki, Greece
Gratias, D., Centre d'Etude de Chimie Metallurgique, CNRS,
 15, Rue G. Urbain, F-94400 Vitry, France
Guarneri, I., Dipartimento di Fisica Teorica e Nucleare, Università
 di Pavia, I-27100 Pavia, Italy
Haase, R.W., Universität Tübingen, Institut für Theoretische Physik,
 Auf der Morgenstelle 14, D-7400 Tübingen 1, Germany
de la Harpe, P., Section de Mathématiques, Université de Genève,
 C.P.240, CH-1211 Genève 24, Switzerland
Hässelbarth, W., Institut für Quantenchemie, Freie Universität Berlin,
 Holbeinstrasse 48, D-1000 Berlin 45, Germany
Janner, A., Institute of Theoretical Physics, Catholic University of
 Nijmegen, NL-6525 ED Nijmegen, The Netherlands
Jost, R., Service National des Champs Intenses, CNRS, B.P. 166 X,
 F-38042 Grenoble Cedex, France
Köppel, H., Physikalisch-Chemisches Institut der Universität Heidel-
 berg, Im Neuenheimer Feld 253, D-6900 Heidelberg, Germany
Kramer, P., Institut für Theoretische Physik der Universität Tübingen,
 Auf der Morgenstelle 14, D-7400 Tübingen 1, Germany
Krutzen, B., Institute of Theoretical Physics, Catholic University of
 Nijmegen, NL-6525 ED Nijmegen, The Netherlands

Lewis, J.T., Dublin Institute for Advanced Studies, School of
    Theoretical Physics, 10, Burlington Road, Dublin 4, Ireland
Leyvraz, F., UNAM, Laboratorio de Cuernavaca, Instituto de Fisica,
    01000 Mexico DF, Mexico
Maassen, H., Institute of Theoretical Physics, Catholic University
    of Nijmegen, NL-6525 ED Nijmegen, The Netherlands
Meyer, H.-D., Physikalisch-Chemisches Institut der Universität Heidel-
    berg, Im Neuenheimer Feld 253, D-6900 Heidelberg, Germany
Molinari, L., Istituto di Fisica, Università di Milano, I-20133 Milano,
    Italy
Stasiak, A., Institut für Zellbiologie, ETH-Hönggerberg, CH-8093 Zürich,
    Switzerland
Strocchi, F., International School for Advanced Studies (ISAS),
    I-Trieste, Italy
Sumners, D., Department of Mathematics, Florida State University,
    Tallahassee, FL 32306, USA
Thomas, U., Institut für Theoretische Physik der Universität Tübingen,
    Auf der Morgenstelle 14, D-7400 Tübingen, Germany
Zimmermann, Th., Physikalisch-Chemisches Institut der Universität
    Heidelberg, Im Neuenheimer Feld 253, D-6900 Heidelberg, Germany
Zumofen, G., Laboratorium für Physikalische Chemie, ETH-Zentrum,
    CH-8092 Zürich, Switzerland

PROGRAMME OF THE WORKSHOP

4th October

Blumen   "Fractal concepts in reaction kinetics"
Meyer   "Chaotic behavior of classical Hamiltonian systems as
                indicated by non-vanishing Liapunov exponents"
Kramer   "From periodic to non-periodic order in solids I, II"
de la Harpe   "Introduction to knot and link polynomials I"
Stasiak   "Knots in DNA, how they arise and what we can learn
                from them"
Lewis   "The large deviation principle in statistical
                mechanics I"

5th October

Lewis   "The large deviation principle in statistical
                mechanics II"
Sumners   "Knot theory and DNA"
Guarneri   "Microwave ionization of highly excited H-atoms"
Strocchi   "Long range dynamics and spontaneous symmetry break-
                ing in many-body quantum systems I, II"
Cederbaum   "Statistical properties of energy levels"
Casati   "Exponential photonic localization and chaos in the
                hydrogen atom in a monochromatic field"
Bohigas   "Chaotic motion and energy level fluctuations"

6th October

de la Harpe   "Introduction to knot and link polynomials II"
Amann   "Chirality as a classical observable in algebraic
                quantum mechanics"
Maassen   "Theoretical concepts in quantum probability: Quantum
                Markov processes and Langevin equations I, II"
Jost   "Correlations of energy levels by Fourier transform"
                Round Table Discussion
                "Chaotic motion in quantum systems"

7th October

de la Harpe	"Introduction to knot and link polynomials III"
Elser	"A model of the structure and growth of quasi-crystals"
Krutzen	"On the electronic structure of calaverite"
Gratias	"About the structural determination of quasi-periodic structures: a 6-dimensional Fourier analysis of the $Al_{73}Mn_{21}Si_6$ icosahedral phase"

8th October

Janner	"Crystallography of quasicrystals"
Argyrakis	"Fractal character of chemical reactions in disordered media"
Zumofen	"Fractal patterns in chemistry"
Leyvraz	"Reaction kinetics for diffusion-limited coagulation"
Farantos	"A ro-vibrational study for the regular/irregular behaviour of the CO-Ar vdW system"
	Round Table Discussion
	"Fractals in chemistry"

FRACTAL PATTERNS IN CHEMISTRY

G. Zumofen[$], A. Blumen[&] and J. Klafter[*]

[$]Laboratorium für Physikalische Chemie, ETH-Zentrum,
CH-8092 Zürich, Switzerland

[&]Physics Institute and BIMF, University of Bayreuth,
D-8580 Bayreuth, West Germany

[*]School of Chemistry, Tel-Aviv University, Ramat-Aviv,
69978 Israel

ABSTRACT. Many structures in nature are fractal, i.e. are similar to themselves on different length-scales of observation. This geometrical property has been studied for a great variety of irregular shapes, many of which result from growth processes. In this paper various growth models are considered and discussed in terms of their scaling properties. Deterministic fractals are introduced to demonstrate the exact relationship between geometry and scaling. Several experimental techniques applied to measure the fractal dimension on various scales are reviewed. The method based on the energy-transfer process is described in more detail.

1. INTRODUCTION

In a now classical work B.B. Mandelbrot pointed out to the general public that many structures in nature ranging from coastlines, trees, rivers, clouds, to Brownian motion exhibit self similar behavior [1]. He also coined the term fractal which he took from the Latin term 'fractus' and which accounts for two interpretations, fragmented and irregular. Discussions of such objects, now called fractals, were given already long before by mathematicians who realized that wide classes of patterns display scale-invariance; their structures remain unchanged under dila(ta)tion operations. The fractal approach has now developed in many scientific disciplines such as physics, chemistry and biology.

Scaling properties have been studied extensively for percolation models [2-5]. Percolation theory deals with clusters formed when the probability of randomly occupying a lattice site with a particle is p. Increasing the probability p there is a threshold where a cluster spanning the entire lattice (infinite cluster for an infinite lattice) appears. This threshold constitutes a sharp transition in the concentration dependence of various properties which are related to the long-

1

A. Amann et al. (eds.),
Fractals, Quasicrystals, Chaos, Knots and Algebraic Quantum Mechanics, 1–20.
© 1988 by Kluwer Academic Publishers.

range connectivity. One of the important results is that the infinite
cluster is a fractal object [4,5]. Percolation ideas have been applied
to different physical and chemical situations for which transitions have
been observed, for instance: the flow of liquids in porous media, the
electric conductivity in mixed conductor–insulator materials, exciton
transport in mixed molecular crystals, polymer gelation and vulcanization processes. In this paper we do not discuss further the percolation model, the reader being referred to another contribution in this
book.

In chemistry the analysis of chemical properties often relies on
the spatial configuration of atoms and molecules. For large assemblies,
however, the arrangements are often complicated and irregular, such that
the configurational data are not anymore meaningful. Therefore, other
methods are needed to quantify the properties related to irregular
structures. The fractal concept provides an elegant tool to describe the
degree of irregularity: irregularity then turns out to be a symmetry
property and structure functions can often be identified by simple power
laws [6,7].

Gelation processes are very important in chemistry. They are
related to the percolation problem because they show a sol–gel transition which marks the situation where in a polymerization process for
the first time a macromolecule appears which spans the whole space
available [6,8,9]. This transition is accompanied by an abrupt change of
material properties. For instance, below the transition point the
material behaves like a liquid, at the transition elastic properties
become observable. Gels and aerogels [10] attract the attention of
material scientists, the hope being to invent new materials with
entirely new properties.

Glasses show the full complexity of disordered systems [11]. They
not only lack long-range spatial order but they are also characterized
by non–equilibrium states. These give rise to relaxation processes which
occur on different time scales, even at low temperatures. Hence, geometrical disorder also implies temporal and energetic disorder and hence,
fractal concepts have to be combined with those of continuous time
[12,13] and of (hierarchical) energy structures [13].

Many irregular patterns observed in nature are the result of some
kind of growth process [8,9,14,15]. Flocculation, aggregation, polymerization, and gelation may produce complicated patterns. In order to
obtain a basic understanding of the relationship between growth and
irregular structures, growth models have been developed. Most of the
growth models are very simple and can be easily implemented on a
computer. In several cases the patterns obtained by simulation calculations are in striking agreement with those observed experimentally.

In Section 2 we report on random and deterministic fractals. In
Section 3 experimental techniques are reviewed, the direct energy
transfer is outlined and its experimental relevance is demonstrated. The
paper terminates with conclusions in Section 4.

2. MODELLING OF STRUCTURES

In this section we present several ways of obtaining geometrical
patterns which scale with distance. Common to all objects is their
underlying selfsimilarity, which is most easily visualized in terms of
the fractal concepts. We start thus by introducing fractals and continue
by describing the stochastic and deterministic means of creating fractal
shapes.

Perhaps the simplest fractal is the Sierpinski gasket [1,16]
embedded in the two-dimensional space, as displayed in Figure 1. The
structure can be generated by a prescription which renders clear the
underlying symmetry. One starts from a triangle of side-length 2, which
includes three smaller, upwards-pointing triangles of side-length 1.
This basic pattern is called generator in the nomenclature introduced by
Mandelbrot [1]. A dilatation by a factor of 2 from the upper corner

Figure 1. The Sierpinski gasket at the 6th iteration stage.

transforms the upper small triangle into the large one, and creates two additional, larger triangles. The procedure is then iterated n times, and leads to the nth stage structure. Thus, the portion of the Sierpinski gasket depicted in Figure 1 is at the 6th stage.

Sierpinski gaskets can be generated in embedding spaces of arbitrary dimension d, by starting with the corresponding hypertetrahedrons [17,18]. Also more general patterns can be constructed along the same lines of reasoning, as we will demonstrate towards the end of the Section.

The properties of fractal objects, related to physical quantities, such as mass distribution, density of vibrational states, conductivity and elasticity are describable through several non-integer parameters. These parameters play roles similar to that of the spatial dimension. Here we mainly consider the fractal (Hausdorff) dimension $\bar{d}$, being the parameter of the mass-density distribution. Denoting by N the number of lattice points inside a sphere of radius R, N scales as

$$N \sim R^{\bar{d}}. \tag{1}$$

where $\bar{d}$ is defined by $\bar{d} = \lim_{R \to \infty} \ell n N / \ell n R$. Hence, for a Sierpinski gasket embedded in a d-dimensional space one has [1,19]

$$\bar{d}_S = \frac{\ell n(d+1)}{\ell n(2)} \tag{2}$$

The spectral (fracton) dimension $\tilde{d}$ is another important parameter appearing in connection with dynamical properties such as diffusion controlled reactions and heat conduction. Using a scaling argument Alexander and Orbach [19] have found that for the Sierpinski gasket

$$\tilde{d}_S = \frac{2\ell n(d+1)}{\ell n(d+3)} \tag{3}$$

Therefore, for the Sierpinski gasket of Figure 1, $\bar{d}_S = 1.584$ and $\tilde{d}_S = 1.365$. In general, the relation $\tilde{d} \leqslant \bar{d} \leqslant d$ holds; furthermore, for Sierpinski gaskets $1 < \bar{d}_S < 2$, as may be seen from Eq. (2).

Not all fractal systems display the strong symmetry of the Sierpinski gaskets, which are 'deterministic' fractals [17,18], designated as such since for each point in space it is unambiguously clear whether it belongs to the structure or not. The distribution of points in a fractal pattern may be random, as exemplified in percolation clusters or diffusion-limited aggregates. Nevertheless, even for such structures one has scaling with distance according to Eq. (2), $N \sim R^{\bar{d}}$. One calls such random, scaling patterns 'stochastic' or 'statistic' fractals [17,18, 20]. As pointed out above, stochastic fractals are often obtained through growth processes, on which we now focus.

The interest in growth and form has developed rapidly in the last few years because of the increase in computer ability and because of the recent realization that the shape of such structures can be understood in terms of scaling ideas.

A large number of growth models have been studied [8,14,15] so that

Figure 2. The trail of a random walk on a square lattice. The different shading shows the situations at different times.

we must restrict ourselves to a few representative examples. Here one of the simplest is that of the random walk. We consider a two dimensional lattice and a walker which at each step chooses randomly one of the nearest neighbor sites. After many steps the set of points visited by the walker becomes plane-filling and hence has the fractal dimension $\bar{d} = 2$. Interestingly, the boundary of the set of visited sites is also found to be fractal [1], in fact these related patterns can be used to mimic the shape of landscapes and islands as demonstrated in Figure 2.

A famous variant of random walk models are self avoiding random walks (SAW) [6,8]. Here one imposes the restriction that each lattice site may be visited only once during the walk. SAW are important, since they have been successfully used to depict the geometrical properties of linear polymers [6]. For SAW as for simple random walks the mean squared end-do-end distance $\langle R^2 \rangle$ scales as a function of the number of steps N:

$$\langle R^2 \rangle \sim N^\alpha \qquad\qquad (4)$$

The parameter α plays the role of a critical exponent. One has $\alpha = 1$ for simple random walks, whereas for SAW α is in general different from unity, and depends on the dimension d of the embedding space for the walk.

A further model is that of epidemic growth [8,21]. Here one regards the lattice sites as cells, which may get infected or become immune. The infection starts with one cell (seed). The growth proceeds by choosing

6

at random at each growth step a fresh cell adjacent to the boundary of
the cluster. This cell either gets infected with probability p (and
hence joins the cluster) or becomes immune. The problem is identical to
percolation [4]; for p smaller than the critical concentration p_c the
growth of the cluster always terminates, while for $p > p_c$ infinite
clusters may appear. In Figure 3 an example is shown for a square
lattice and for p = 0.60, which is close to the critical concentration
p_c = 0.593. Models similar to that of epidemics have been developed to
study the spread of forest fires [4,22,23].

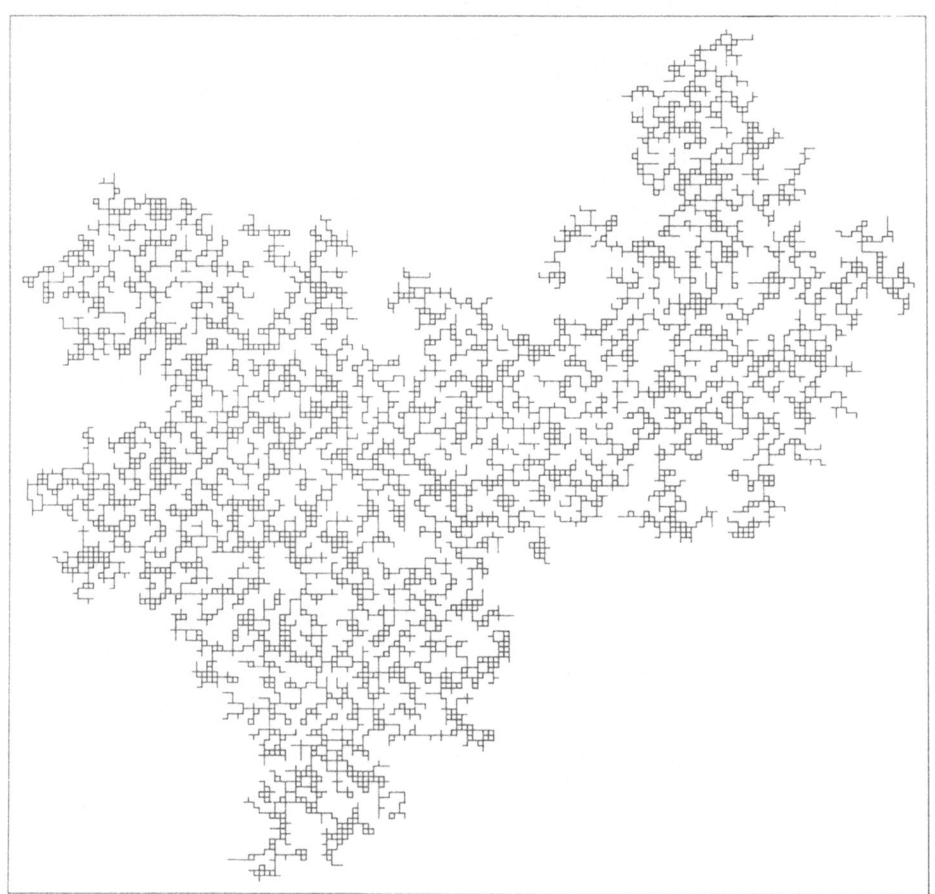

Figure 3. Cluster on a square lattice grown by the epidemic model for
p = 0.6. The situation after 2×10^4 steps is shown.

We proceed our discussion of growth by considering a new element:
the growth should result from having particles diffuse to a cluster and
getting attached to it. This new element leads to diffusion limited

aggregation.

Starting with the pioneering work of Smoluchowski [24], the role of diffusion (Brownian motion) in colloidal aggregation has been a subject of continuous interest. However, only recently large scale computer simulations háve allowed in-depth investigations of growth processes. Thus Witten and Sander [25] showed that a simple diffusion limited growth model leads to complex patterns for the aggregates. In their model a particle seed is placed at the origin. Then further particles are added, one by one; these are released far from the seed, and join the aggregate by performing a random walk. The walk terminates when the particle reaches a site nearest neighbor to the already-formed, diffusion-limited aggregate (DLA). The number of particles in such DLAs scales with the distance from the seed. Thus, for medium-sized DLA (10^3-10^4) one finds as fractal dimensions $d = 1.7$ for $d = 2$ and $d = 2.5$ for $d = 3$ [26]. Typical is the appearance of a tree-like shape which is due to the fact that later arriving particles are less prone to reach the central region and thus attach themselves rather at the periphery of the cluster as may be seen from Figure 4. Only for very large scale patterns (more than 10^5 particles) deviations from universality (e.g. a dependence of d on the lattice type) appear [27,28].

Several different versions of DLA models have been considered. One extension consists in introducing a sticking probability p, i.e. a new

Figure 4. Diffusion limited aggregate (DLA) on a 600×600 lattice. The different shading shows stages in the evolution of the DLA.

particle, in a nearest-neighbor position to the aggregate, joins it only
with probability p (p < 1). With decreasing p the DLAs become more
compact [26]. In the limiting case, p → 0, all surface sites get
occupied with uniform probability. This is then akin to the Eden model
[29] in which the aggregate grows by adding randomly any of the empty
neighboring sites to it. The fractal dimension of the Eden model is
equal to the Euclidean dimension, $\bar{d} = d$.

Experimentally, DLAs are created in several ways: electrodeposition
[30,31], discharge patterns (Lichtenberg figures) in dielectric breakdown [32,33], and branched fingers formed when a low viscosity fluid
penetrates a high viscosity one (viscous fingering) [34]. Such patterns
can be obtained from a construction based on the Laplace equation. The
corresponding potential Φ is taken to be zero at the surface of the
growing object and one at infinity. In between, Φ obeys the Laplace
equation $\Delta\Phi = 0$. The growth occurs at the surface as a random process
weighted by $(\nabla\Phi)^{\eta}$, η being a parameter. Simulation calculations show
that the fractal dimension depends on η [33]; for $\eta = 1$ the ordinary DLA
is recovered.

Up to now we considered growth of one cluster only. However, many
aggregation processes rely on cluster-cluster aggregation. Here the
modelling is more involved because the structure of the aggregates
depends on how the diffusion coefficient is related to the mass of the
clusters. Generally one takes an algebraic dependence $D_j \sim j^{\alpha}$, where D_j
is the diffusion coefficient for a cluster consisting of j unit masses
and α is a parameter. However, for $\alpha < 0$, i.e. when large clusters diffuse
more slowly than small clusters (as is to be expected intuitively) the
fractal dimension of the aggregate is found not to depend on α. The
results are $\bar{d} = 1.42$ (d = 2) and $\bar{d} = 1.78$ (d = 3) [35]; the latter value
is in good agreement with the experimental finding for metal particle
aggregates [36,37].

Finally we mention kinetic gelation [8,9,38]. Here one has for
instance the following model, which is similar to percolation: tetra-
functional, bifunctional and zerofunctional monomers are randomly
distributed on a lattice. The functionality gives the number of bonds a
monomer can form with its nearest neighbors. The growth process is
initiated by activating several monomers. Activated monomers can link
themselves to their nearest neighbors provided that both monomers can
functionally undergo a bond. The process continues by having the newly
bonded monomers activated. Here we remark that clusters formed by
gelation may scale, but in general with other exponents than those found
in percolation.

After this short review of stochastic fractals we now turn to the
deterministic ones. In these the distribution of sites is determined by
an unambiguous, non-random prescription. Examples for such structures
are the Sierpinski gaskets, as shown above, and extensions [17,18,39]
some of which will be discussed in the following. Evidently, deterministic prescriptions standardize the fractals and make them very
useful as model systems. Hence many recent analyses have centered on
Sierpinski type fractals [13,40].

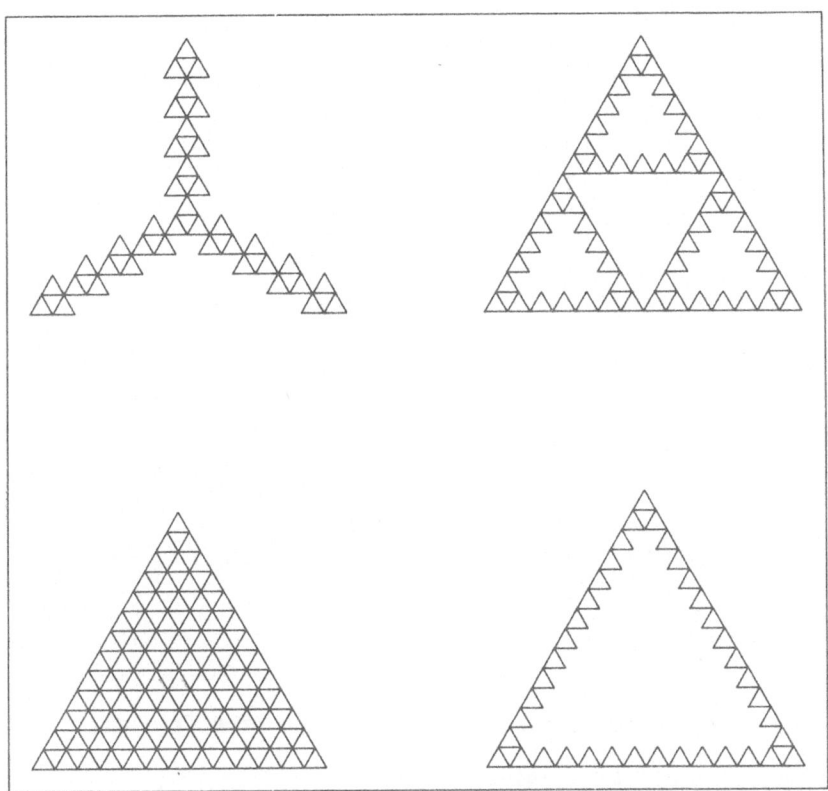

Figure 5. Various symmetric tetrahedral generators for d = 2.

As already mentioned, fractals are described by (at least) three distinct dimensions: the spatial dimension d of the embedding Euclidean space, the fractal dimension d̄ and the spectral dimension d̃. It should be emphasized that d̄ and d̃ are amenable to experimental observation [13,19,39-41]. For modelling purposes it is therefore desirable to construct deterministic fractals with prescribed d̄ and d̃ values. Let us thus consider Sierpinski-type structures. The spectral dimensions d̃ for these structures lie between that of a 2-d Sierpinski gasket, d_S = 1.36 and the value 2. In Refs. [17,18] one took as generators d-dimensional hypertetrahedrons (HT) of side length b. A particular generator $G = G(b,d)$ is obtained by filling such a HT with smaller HTs of unit side length. From G the fractal is iteratively constructed: the structure at stage n+1 is obtained by enlarging G by b^n and then filling all upward pointing HTs with the stage-n structure. More general symmetrical fractals are produced through generators which are only partially filled with HTs. Several connected graphs with tetrahedral symmetry were used for this purpose [17,18]. Examples are displayed in Figure 5 while in Figure 6 the iterative construction of a fractal is demonstrated. Let N

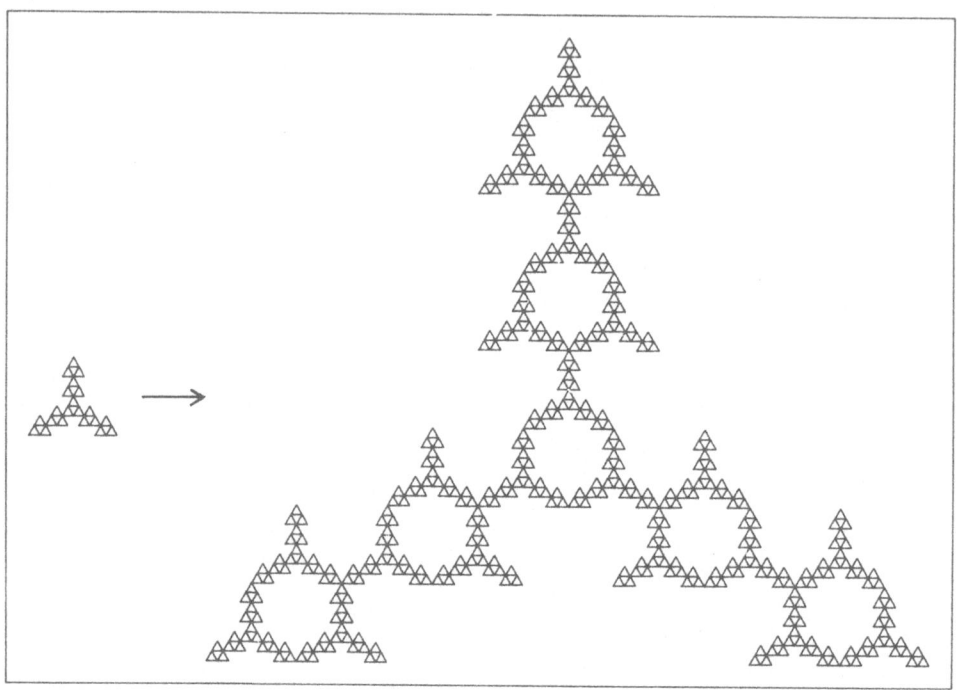

Figure 6. One step in the iterative construction of a fractal.

be the number of small HTs inside G. Then

$$\bar{d} = \ln N / \ln b \qquad (5)$$

On the other hand $\tilde{d}$ can be obtained from the probability to be at the origin at long times (see Refs. [17,18] for details) and lies between 1 and 2. Hence such symmetric fractals are a wide class of lattices which generalize the Sierpinski gaskets.

Finally one can extend such fractals by direct set multiplication. As an example we show in Figure 7 the stage 2 result of multiplying a Sierpinski-gasket with a one-dimensional lattice. We call this following Ref. [13] and [18] the "Toblerone"-lattice. Its spectral dimension can be obtained from the low-frequency behavior of its eigen-modes: $\tilde{d} = 1 + d_S = 2.36.$ On the other hand one realizes by direct inspection that here $\bar{d} = 1 + \bar{d}_S = 2.58.$

3. MEASUREMENT TECHNIQUES FOR DISORDER

According to the previous Section, the main idea which allows to mathematically describe disordered systems as objects <u>sui generis</u> is their scale invariance. Most of the techniques probe the geometry of the samples; it should be clear that dynamical processes also lead to scaling, but this aspect has received much less attention, probably because geometry is more easily visualized than the temporal development. We have in previous works [13,42] also indicated means to investigate dynamics, but we will concentrate here mainly on the geometrical aspect. Several experimental techniques have been applied to monitor the geometry of complex random objects in order to assert whether they are self-similar.

Consider first the transmission electron microscopy (TEM). It has been used by Forrest and Witten [43] to analyze the structures of aerosol particles which have complicated structures. These investigations show that the density of the particles scales with a power law of the distance. Then TEM has been employed in the study of various silica [44] and gold [36,45-47] aggregates. These aggregates, up to several hundred nanometers in extent, consist of spherical particles of almost constant diameters, e.g. around 15 nm in the case of gold colloids [45]. Again scaling is found.

Figure 7. Toblerone fractal, the direct (set-) product of a two-dimensional Sierpinski gasket with a linear chain.

The standard TEM techniques run into difficulties when the individual particles overlap or are of different sizes. A more sophisticated method has been chosen by Tence et al. [37] by which they have investigated aggregates of iron particles of variable size.

We note that more flexible ways to determine scaling in small objects are the scattering techniques. Here either light, X-ray or neutron scattering are applied, the choice being determined by the characteristic length-scale to be observed and by the material which forms the aggregate. Because of its wave length, light scattering monitors distances of 0.2-100 μm. Most aggregates investigated are, however, smaller, and thus neutron and X-ray scattering (which cover the range 0.05-500 nm) are used preferentially [48]. The basic idea is that the scattering intensity I is proportional to the Fourier transform of the density-density correlation function. For large scattering angles one sees the short-range structure. For small angles the scattering intensity I is determined by the density-density correlation function, which for fractals depends on d̄. If the object itself is fractal, then, according to Refs. [44,49-51]:

$$I \sim \theta^{-\bar{d}} \tag{6}$$

where θ is the scattering angle. On the other hand, when the object is three dimensional, but its surface is fractal, the scattering intensity should follow [52]

$$I \sim \theta^{\bar{d}_s - 6} \tag{7}$$

where $\bar{d}_s$ is now the fractal dimension of the surface.

Another method to probe irregular surfaces is the adsorption technique. It consists in determining the minimal number N(R) of adsorbed molecules of radius R required to cover the entire surface. If their radius is varied, one should have for a fractal surface [53]:

$$N(R) \sim R^{-\bar{d}_s} \tag{8}$$

This behavior was reported by Avnir an Pfeifer [54] in their study of crashed glass, carbon black, charcoal and of other porous materials. The adsorbates which they use include N_2, alkanes and polycyclic aromatics (which probe lengths of around 4 to 8 Å). For polymer adsorbates the lengths to be measured extend from 25 to 300 Å [54-56].

We also mention the electrodeposition of metal aggregates as another technique. This is a growth process limited by the diffusion of metal ions. Following Brady and Ball [30] one records the electric current, which is related to the deposited mass. In this way copper aggregates were found fractal over the range of 20-300 μm [30].

Another method to determine the fractal dimension of self-similar objects centers on the direct transfer of electronic excitations. This method, proposed by us [40,42], is now of widespread use. We consider the energy transfer to take place from donor molecules to acceptor molecules and assume that donors and acceptors occupy substitutionally sites on a regular lattice. In this classical problem [57,58] one considers an exited donor molecule, at position r_0, surrounded by acceptors which occupy some of the sites r_1 of a given structure. The transfer rates $w(r)$ depend on the mutual distance r between each donor-acceptor pair. Neglecting back-transfer, the probability of the decay of the donor due to the presence of an acceptor at r_1 is thus

$$f(t, r_1, r_0) = \exp[-tw(r_1-r_0)] \qquad (9)$$

One assumes the acceptors to act independently, which means that they contribute multiplicatively to the decay. Let $g(j)$ be the probability of having j acceptors at one site. The decay of the donor, averaged over all possible distributions of acceptors around it, is given by

$$\Phi(t, r_0) = \prod_{r_1}' \left\{ \sum_j g(j) \, [f(t, r_1, r_0)]^j \right\} \qquad (10)$$

where the product extends over all structure sites with the exception of r_0. Thus, for a binomial distribution, $g(j) = (1-p)\delta_{0,j} + p\delta_{1,j}$ ($\delta_{i,j}$ being the Kronecker delta), one has

$$\Phi(t, r_0) = \prod_{r_1}' \left\{ 1 - p + p \exp[-tw(r_1-r_0)] \right\} \qquad (11)$$

Eq. (11) is exact, and thus theoretically very valuable. For a small acceptor concentration, $p \ll 1$, one may replace Eq. (11) by

$$\tilde{\Phi}(t, r_0) = \exp\left[-p \sum_{r_1}' \left\{ 1 - \exp[-tw(r_1-r_0)] \right\} \right] \qquad (12)$$

which is, in fact, the exact decay of Eq. (10) under the Poisson law, $g(j) = e^{-p}p^j/j!$.

Usually, in treating energy transfer on infinite, regular lattices the dependence of $\Phi(t, r_0)$ on r_0 is irrelevant, since there all sites are equivalent. The situation is different in confined geometries and for irregular objects such as fractals, in which cases the dependence on r_0 may be important. Staying in the general case and introducing a site-density function $\rho(r)$ we can transform the sum in Eq. (12) to the usual integral form. We set

$$\rho_0(r) = \sum_{r_1}' \delta(r-r_1), \qquad (13)$$

where the index 0 in $\rho_0(r)$ acts as a reminder that r_0 is excluded from the sum of the right-hand side. With $\rho_0(r)$ one obtains

$$\tilde{\Phi}(t, r_0) = \exp\left[-p \int dr \, \rho_0(r) \left\{ 1 - \exp[-tw(r_1-r_0)] \right\} \right] \qquad (14)$$

For regular geometries in Euclidean spaces one arrives at the Förster-type decays by taking $\rho_{\bullet}(r) = \rho = $ const and extending the integration over the whole space:

$$\tilde{\Phi}(t) = \quad \exp\left[-p\,\rho \int dr\left\{1 - \exp[-tw(r)]\right\}\right] \tag{15}$$

To obtain decay patterns it is now only necessary to specify the interaction $w(r)$ and to perform the integration in Eq. (15). An often encountered form for $w(r)$ is

$$w(r) = ar^{-s} \tag{16}$$

which holds for isotropic multipolar interactions [57]. The parameter equals 6 for dipole, 8 for dipole–quadrupole and 10 for quadrupole–quadrupole interactions. Inserting Eq. (16) into Eq. (15), one obtains

$$\tilde{\Phi}(t) = \exp(-At^{d/s}) \tag{17}$$

Here d is the spatial dimension of the underlying lattice and A is a time independent constant

$$A = V_d\, p\, \rho\, \Gamma(1-d/s)\, a^{d/s} \tag{18}$$

where V_d is the volume of the unit sphere in d dimensions and $\Gamma(z)$ is the Euler–gamma function.

Let us now focus on optical properties related to fractals. From Eq. (17) we see that the Förster-type decays are determined by the spatial dimension d of the region accessible to acceptors. The natural starting point for the extension is Eq. (11) since it is exact and depends only on combinatorial arguments. For small acceptor concentrations the continuum approximation again leads to Eq. (14) for which the density $\rho_{\bullet}(r)$ has to be specified. Now, in determining the density we should remark that: first, a fractal structure is not translationally symmetric, so that the decay may depend on the location of the donor. For relatively homogenous fractals, such as the Sierpinski gaskets, this is not a very serious matter, since on such objects the local densities around the sites are quite similar. Second, $\rho_{\bullet}(r)$, being a density, must fulfill

$$N = \int_V \rho(r)\, dr \tag{19}$$

where N is the number of lattice sites in the volume V. Combining Eq. (19) with Eq. (1), one sees that for homogeneous fractals one has to a good approximation

$$\rho_{\bullet}(r) \simeq \beta\, r^{\bar{d}-d}, \tag{20}$$

where $\bar{d}$ is the fractal dimension and β is a proportionality constant. Now Eq. (20) no longer depends on the origin. Inserting it into Eq. (14) one obtains

$$\tilde{\Phi}(t) = \exp\left[-p \, \beta \int d\mathbf{r} \, r^{\bar{d}-d} \left\{ 1 - \exp[-tw(\mathbf{r})] \right\} \right]$$ (21)

For isotropic multipolar interactions, the integration is immediate [42]. One obtains as long-time behavior

$$\tilde{\Phi}(t) = \exp(-\hat{A}t^{\beta})$$ (22)

with $\hat{A}$ time independent and $\beta = \bar{d}/s$. This simple form, Eq. (22), is very useful in the interpretation of observed data and it enables for fractals an efficient determination of $\bar{d}$. Experimentally this is usually achieved as follows: the donor molecules are excited with a short light pulse and the radiative emission is used to measure the excitation decay. From this procedure we learn that not only the energy transfer (ET) limits the excitation's lifetime; inherently, radiation also shortens the excitation's lifetime. Since the two processes, radiation and ET, are independent, they contribute multiplicatively to the survival probability

$$\tilde{\Phi}(t) = \exp(-\hat{A}t^{\beta} - t/\tau)$$ (23)

with τ for the radiative lifetime. Since $\beta < 1$ the decay is governed by radiation at long times. This, on the other hand, also limits the spatial range for which the ET method can probe the structure of the underlying material. Typically, this range is of the order of 100 Å, which is also the region amenable to the small angle X-ray scattering technique. Hence, in this region these two methods give complementary information.

The ET technique, proposed in 1984 [42], has first been applied by Even et al. [59] in the investigation of the porosity of Vycor glass. They studied the kinetics of the ET between rhodamine B and malachite green in solution inside the glass. Using Eq. (23) the authors of Ref. [59] found $\bar{d} = 1.74$. This result has recently been interrogated with respect to its compatibility with light and X-ray scattering data [60,61]. These data and those of neutron scattering [62] give indications for $\bar{d} \simeq 2$, but the experiments do not allow to make definitive statements about the roughness of the internal surfaces. This roughness has also been invoked in the discussion of the spectral and photo-physical properties of $Ru(bpy)_3^{2+}$ adsorbed onto Vycor glass [63].

Rojanski et al. [64] have measured the fractal dimension of mesoporous silica gel by three independent different techniques: adsorption of probe molecules, X-ray scattering and ET. All techniques indicate that the internal surface is very irregular with a fractal dimension of $\bar{d} \simeq 3$. This is an important achievement since these independent measurements show that the techniques are equivalent when it comes to determine $\bar{d}$. In a more recent paper Pines-Rojanski et al. [65] have reestablished their result by an improved ET method. A fractal structure has also been envisaged by Brundage and Yen [66] to explain the small β exponent, they have obtained for the ET among Yb^{3+} ions diluted in silicate glass. However, for spin recoveries in Yb^{3+} ion doped lead phosphate glasses, the interpretation in terms of fractals is not yet settled [67].

On the other hand it is difficult to unambiguously state that a structure is fractal based only on results compatible with Eq. (23). Yang et al. [68,69] have considered deterministic fractals and have shown that the direct transfer may lead to decays which, on a logarithmic scale, deviate from straight lines and exhibit wavy patterns. They point out that this waviness might render difficult the determination of a fractal exponent. Wavy patterns have also been observed for relaxation processes in hierarchical systems [70]. However, it may well happen that the waviness calculated for deterministic fractals is wiped out by the randomness which appears in statistical assemblies. Yang et al. [71] also simulated on a computer the ET among particles located on pores of spherical or cylindrical shapes with radii of the order of the typical transfer distance; the calculated kinetics shows fractal-like decays. Similar geometrical restrictions for ET have been studied by Baumann and Fayer [72] and by ourselves [73,74]. Levitz and Drake [75] have interpreted their measurements of ET processes for various porous silicas in terms of a cylindrical pore model.

ET processes have been studied for polymeric systems. Singlet ET mechanisms in anthracene-loaded copolymers of vinylnaphthalene show complicated temporal behaviors which are thought to be partly due to fractal-like distributions of naphthalene and anthracene molecules [76]. Triplet ET kinetics has been studied by Lin, Nelson and Hanson [77] for doped polymer films. The data fit their model better when fractal structures are taken into account. Further discussions of polymeric systems in terms of fractals are given in Refs. [78] and [79].

We close this section by mentioning other cases in which the ET model has initiated new ideas for the interpretation of related processes. Fractals have been considered in the discussion of the energy migration and trapping among cationic porphyrins adsorbed on anionic vesicle surfaces [80]. A scaling model has been developed for desorption processes from molecular-sieve materials, in which stretched exponential behavior is found [81]. Furthermore, a stretched exponential modulation model has been proposed for simulating vibrational dephasing in locally amorphous media [82]. Finally, hierarchical and fractal concepts have been used in the analysis of time resolved spectra taken from mixed crystals of dichlorobenzene and dibromobenzene [83].

4. CONCLUSIONS

In this paper we have pointed out the importance of fractals in chemistry. The discussion is of course incomplete since only a few of the many facets of fractals in chemistry could be considered. Thus we did not mention the role played by strange attractors in complicated chemical reaction systems and in coupled biochemical processes [84]; also we did not treat the fractal aspect in fully developed turbulence, an important modern problem [85,86]. For reaction dynamics on fractals which are fundamental for the understanding of the catalytic reaction processes in porous solids we refer to our companion article in this book [87].

Here we have centered on random patterns resulting from growth processes and have compared them to deterministic fractals. Furthermore, we have outlined several experimental techniques which are used to investigate the spatial mass distribution in disordered systems. In particular, we have derived the equations needed to analyze the fractal dimensions by means of time-resolved spectroscopy and we have presented several successful applications to glassy and polymeric systems. Also we have shown that the techniques have to be used carefully in order to obtain unambiguous results and insights.

There is no doubt that fractals have entered into almost all scientific disciplines. Fractal and scaling concepts have become standard for physicists and chemists. It has, however, to be emphasized that not all irregular structures are ultimately fractals or, switching to a time scale, not all non-exponential evolutions indicate scaling. Furthermore, selfsimilarity is generally bounded by lower and upper cut-offs, a fact that renders delicate the interpretation of experimental data. Therefore care is required in order to attain a sound physical understanding.

ACKNOWLEDGMENTS

The authors thank Professors K. Dressler, J. Friedrich and D. Haarer for many fruitful discussions. The support of the Deutsche Forschungs-gemeinschaft (SFB 213) and of the Fonds der Chemischen Industrie and grants of computer time from the ETH-Rechenzentrum are gratefully acknowledged.

REFERENCES

1. B.B. Mandelbrot, 'The Fractal Geometry of Nature', Freeman, San Francisco (1982).
2. S. Kirkpatrick, Rev. Modern Phys. 45, 574 (1973).
3. G. Deutscher, R. Zallen, and J. Adler (eds), 'Percolation Structure and Processes', Ann. Isr. Phys. Soc. Vol. 5, Hilger, Bristol (1983)
4. D. Stauffer, 'Introduction to Percolation Theory', Tanylor and Francis, London (1985).
5. S. Havlin and D. Ben-Avraham, Adv. Phys., in press.
6. P.-G. de Gennes, 'Scaling Concepts in Polymer Physics', Cornell University, Ithaca (1979).
7. L. Pietronero and E. Tosatti (eds), 'Fractals in Physics', North-Holland, Amsterdam (1986).
8. H.J. Herrmann, Phys. Rep. 136, 153 (1986).
9. F. Family and D.P.Landau (eds), 'Kinetics of Aggregation and Gelation', North-Holland, Amsterdam (1984).
10. J. Fricke (ed), 'Aerogels', Springer, Berlin (1986).
11. I. Zschokke (ed), 'Optical Spectroscopy of Glasses', D. Reidel, Dordrecht (1986).
12. E.W. Montroll and F.M. Shlesinger, in 'Nonequilibrium Phenomena II: From Stochastics to Hydrodynamics, (eds J.L. Lebowitz and E.W. Montroll), North Holland, Amsterdam (1984).

18

13. A. Blumen, J. Klafter, and G. Zumofen, p. 399 in Ref. 11.
14. H.E. Stanley and N. Ostrowsky (eds), 'On Growth and Form', Martinus Nijhoff, Kluwer Academic, Boston (1986).
15. R. Jullien and R. Botet, 'Aggregation and Fractal Aggregates", World Scientific, Singapore (1987).
16. W. Sierpinski, Compt. Rend. (Paris) **160**, 302 (1915); **162**, 629 (1916); 'Oeuvres Choisies' (S. Hartman et al.), Editions Scientifiques, Warsaw (1974).
17. R. Hilfer and A. Blumen, J. Phys. **A17**, L537, (1984); **A17**, L783 (1984).
18. R. Hilfer and A. Blumen, p. 33 in Ref. 7.
19. S. Alexander and R. Orbach, J. Phys. Lett. **43**, L625 (1982).
20. H.E. Stanley, J. Stat. Phys. **36**, 843 (1984).
21. Z. Alexandrowicz, Phys. Lett. **80A**, 284 (1980).
22. G. Mackay and N. Jan, J. Phys. **A17**, L757 (1984).
23. W. von Niessen and A. Blumen, J. Phys. **A19**, L289 (1986).
24. M. von Smoluchowski, Phys. Z. **17**, 585 (1916).
25. T. Witten and L. Sander, Phys. Rev. Lett. **47**, 1400 (1981).
26. P. Meakin, Phys. Rev. **A27**, 604 (1983); **A27**, 1495 (1983).
27. P. Meakin, Phys. Rev. **A33**, 3371 (1986); **A33**, 4199 (1986).
28. L. Turkevich and H. Scher, Phys. Rev. Lett. **55**, 1026 (1985).
29. M. Eden, Proc. Fourth Berkeley Symp. on Math. Stat. and Prob., (ed F. Neyman), Vol IV, University of Berkeley (1961), p. 223.
30. R. M. Brady and R.C. Ball, Nature **309**, 225 (1984).
31. M. Matasushita, M. Sano, Y. Hayakawa, H. Honjo, and Y. Sawada, Phys. Rev. Lett. **53**, 1033 (1984).
32. L. Niemeyer, L. Pietronero, and. H.J. Wiesmann, Phys. Rev. Lett. **52**, 1033 (1984).
33. H.J. Wiesmann and L. Pietronero, p. 151 in Ref. 7.
34. J. Nittman, G. Daccord, and H.E. Stanley, Nature **314**, 141 (1985).
35. R. Jullien, M. Kolb, and R. Botet, J. Physique Lett. **45**, L211 (1984).
36. D.A. Weitz and J.S. Huang, Phys. Rev. Lett. **52**, 1433 (1984).
37. M. Tence, J.P. Chevalier, and R. Jullien, J. Physique **47**, 1989 (1986).
38. H.J. Herrmann, D.P. Landau, and D. Stauffer, Phys. Rev. Lett. **49**,412 (1982).
39. D. Dhar, J. Math. Phys. **18**, 577 (1977).
40. A. Blumen, G. Zumofen, and J. Klafter, in 'Structure and Dynamics of Molecular Systems', (eds R. Daudel, J.-P. Korb, J.-P. Lemaistre and J. Maruani), Reidel, Dordrecht (1985) p. 71; G. Zumofen, A. Blumen, and J. Klafter, ibid, p. 87.
41. R. Rammal and G. Toulouse, J. Physique Lett. **44**, L13 (1983).
42. J. Klafter and A. Blumen, J. Chem. Phys. **80**, 875 (1984).
43. S.R. Forrest and T.A. Witten, J. Phys. **A17**, L109 (1979).
44. D.W. Schafer, J.E. Martin, P. Wiltzius, D.S. Cannell, Phys. Rev. Lett. **52**, 2371 (1984).
45. D.A. Weitz and M. Oliveria, Phys. Rev. Lett. **52**, 1433 (1984).
46. D.A. Weitz, J.S. Huang, M.Y. Lin, and J. Sung, Phys. Rev. Lett. **54**, 1416 (1985).
47. D.A. Weitz and M.Y. Lin, Phys. Rev. Lett. **57**, 2037 (1986).

48. J. Texeira, p. 145 in Ref. 14.
49. S.H. Sinha, T. Freltoft, and J. Kjems, p. 87 in Ref. 9.
50. D.W. Schaefer, J.E. Martin, and K.D. Keefer, J. Physique (Paris) Colloq. 46, C3-127 (1985).
51. J.E. Martin, J. Appl. Crystallogr. 19, 25 (1986).
52. H.D. Bale, and P.W. Schmidt, Phys. Rev. Lett. 53, 596 (1984).
53. P. Pfeifer and D. Avnir, J. Chem. Phys. 79, 3558 (1983).
54. D. Avnir and P. Pfeifer, J. Chem. Phys. 79, 3566 (1983).
55. D. Avnir, D. Farin, and P. Pfeifer, Nature 308, 261 (1984).
56. P. Pfeifer, p. 47 in Ref. 7
57. T. Förster, Z. Naturf. A4 321 (1949).
58. D.L. Dexter, J. Chem. Phys. 21 836 (1953).
59. U. Even, K. Rademan, J. Jortner, N. Manor, and R. Reisfeld, Phys. Rev. Lett. 52, 2164 (1984).
60. D.W. Schaefer, B.C. Bunker, and J.P. Wilcoxon, Phys. Rev. Lett. 58, 284 (1987).
61. U. Even, K. Rademan, J. Jortner, N. Manor, and R. Reisfeld, Phys. Rev. Lett. 58, 285 (1987).
62. P. Wiltzius, F.S. Bates, S.B. Dierker, and G.D. Wignall, Phys. Rev. A36, 2991 (1987).
63. Wei Shi, S. Wolfgang, T.C. Strekas, and H.D. Gafney, J. Phys. Chem. 89, 974 (1985).
64. D. Rojanski, D. Huppert, H.D. Bale, X. Dacai, P.W. Schmidt, D. Farin, A. Seri-Levy, and D. Avnir, Phys. REv. Lett. 56, 2505 (1986).
65. D. Pines-Rojanski, D. Huppert, and D. Avnir, Chem. Phys. Lett. 139, 109 (1987).
66. R.T. Brundage, and W.M. Yen, Phys. Rev. B34, 8810 (1986).
67. D.L. Smith, H.J. Stapleton, and M.B. Weissman, Phys. Rev. B33, 7417 (1986).
68. C.L. Yang, P. Evesque, and M.A. El-Sayed, J. Phys. Chem. 90, 1284 (1986).
69. C.L. Yang, M.A. El-Sayed, J. Phys. Chem. 90, 5720 (1986).
70. A. Blumen,, G. Zumofen, and J. Klafter, Ber. Bunsenges. Phys. Chem. 90, 1048 (1986).
71. C.L. Yang, P. Evesque, and M.A. El-Sayed, J. Phys. Chem. 89, 3442 (1985).
72. J. Baumann and M.D. Fayer, J. Chem. Phys. 85, 4087 (1986).
73. J. Klafter and A. Blumen, J. Lumin. 34, 77 (1985).
74. A. Blumen, J. Klafter, and G. Zumofen, J. Chem. Phys. 84, 1397 (1986).
75. P. Levitz and J.M Drake, Phys. Rev. Lett. 58, 686 (1987).
76. F. Bai, C.-H. Chang, and S.W. Weber, Macromolecules, 19, 2484 (1986).
77. Y. Lin, M.C. Nelson, and D. M. Hanson, J. Chem. Phys. 86, 1586 (1987).
78. D.R. Coulter, A. Gupta, A. Yavrouian, G.W. Scott, D. O'Connor, O. Vogl, and S.-C. Li, Macromolecules, 19, 1227 (1986).
79. H.F. Kauffmann, B. Mollay, W.-D. Weixelbaumer, J. Bürbaumer, M. Riegler, E. Meisterhofer, and F. R. Aussenegg, J. Chem. Phys. 85, 3566 (1986).
80. A. Takami and N. Mataga, J. Phys. Chem. 91, 619 (1987).

81. A.R. Kerstein, Phys. Rev. B31, 1612 (1985).
82. W.G. Rothschild, M. Perrot, F. Guillaume, Chem. Phys. Lett. 128, 591 (1986).
83. T. Kirkski, J. Grimm, and C. von Borczyskowski, J. Chem. Phys. 87, 2062 (1987), and preprints.
84. P. Bergé, Y. Pomeau, and C. Vidal, 'Order within Chaos: Towards a Deterministic Approach to Turbulence', Hermann and Wiley & Sons, New York, (1986).
85. J. Klafter, A. Blumen and M.F. Shlesinger, Phys. Rev. A35, 2081 (1987).
86. A. Blumen, G. Zumofen, and J. Klafter, to be published and J. Lumin. in press.
87. A. Blumen, G. Zumofen, and J. Klafter, this book.

FRACTAL CONCEPTS IN REACTION KINETICS

A. Blumen[&], G. Zumofen[$] and J. Klafter[*]

[&]Physics Institute and BIMF, University of Bayreuth,
D-8580 Bayreuth, West Germany

[$]Laboratorium für Physikalische Chemie, ETH-Zentrum,
CH-8092 Zürich, Switzerland

[*]School of Chemistry, Tel-Aviv University, Ramat-Aviv,
69978 Israel

ABSTRACT. This paper focusses on the kinetics of the A + A → 0 and the
A + B → 0 diffusion-limited reactions by modelling the dynamics through
random walks. The disorder aspect is introduced through hierarchical
structures: geometrical, temporal and energetical disorder are considered. While the geometrical disorder is analyzed using deterministic
fractals, the temporal disorder is accounted for through continuous-time
random walks (CTRW), whose waiting-time distribution displays long-time
tails. Energetic disorder is modelled using ultrametric spaces (UMS),
systems with hierarchically distributed energy barriers. Results for the
three models are presented and deviations from the Smoluchowski-type
behavior are explained. Combinations of two distinct disorder aspects
are also considered and discussed in terms of subordination.

1. INTRODUCTION

Reaction dynamics under diffusion-limited conditions are a classical
topic of physical chemistry [1,2] but only recently, due to the emer-
gence of new approaches for treating randomness analytically and due to
enhancements in computing capacity, extensive work on the role of
disorder and fluctuations on reaction kinetics has become feasible
[3-7].
　　These investigations have revealed many qualitative deviations from
the accepted Smoluchowski-type decay laws, when reactions in confined
geometries, such as obtain for limited dimensionalities or for porous
media, are studied.
　　Let us start by recalling several facets of randomness. Thus,
transport properties of spatially random systems (mixed crystals,
alloys) are triggered by a distribution of microscopic (site-to-site)
transfer rates (temporal disorder) and by different interactions with

21

A. Amann et al. (eds.),
Fractals, Quasicrystals, Chaos, Knots and Algebraic Quantum Mechanics, 21–52.
© *1988 by Kluwer Academic Publishers.*

the surroundings (energetic disorder). Treating the full microscopic problem is a hard task, which calls for large-scale numerical calculations.

Let us note from the beginning that there is little hope in describing such a complexity in terms of perturbations from ideal, non-random patterns. This point has been forcefully stressed by P.W. Anderson in the Les-Houches lectures on ill-condensed matter [8]. He remarked that multiple scattering theories (effective-medium or coherent-potential approximations) are paradigms [9] of an old attitude, which tries to model random systems through regular ones; the new attitude is exemplified by localization and percolation, which typify disordered situations as genuinely distinct problems [8].

In this contribution we adopt this line of thought and present several classes of models for disorder, which have come into close scrutiny only during the last decade. Each class may be viewed as arising due to a particular aspect of randomness. Thus fractals [10,11] exemplify the spatial disorder, continuous-time processes [12-16] the temporal and ultrametric structures [17-20] the energetic disorder.

Basic to our understanding of disorder is the temporal evolution of the systems under investigation. A comparison of relaxation patterns reveals that the decays which are observed in disordered materials seldom have kinetical chemical counterparts, the reason being the implicit underlying assumption in the kinetic scheme of a 'well-stirred' reactor. The lack of such a homogenization for disordered media leads then to interesting deviations from the kinetic picture.

The structure of this paper is as follows: we first introduce in the next sections the concepts of fractals and of ultrametric spaces and exemplify the dynamics using random walks; at this point the ideas behind the continuous-time random walks (CTRW) and their temporal-scaling behavior will also become clear. Then we center on bimolecular-kinetics of the $A + A \rightarrow 0$ and $A + B \rightarrow 0$ type, which we model through random walks. The disorder aspect is depicted through the hierarchical (self-similar) structures already discussed. First-of-all, however, we start by presenting the standard chemical-kinetics scheme.

2. THE BASIC KINETIC APPROACH

In this section we recall the basic kinetic scheme. As will turn out, the decay laws which follow from the kinetic formalism are not adequate for describing the intriguing relaxation forms found in disordered systems.

General irreversible reactions have the form:

$$A_1 + A_2 + \ldots + A_n = \sum_{i=1}^{n} A_i \xrightarrow{k} 0 \qquad (1)$$

to which in the kinetic scheme corresponds the system of (in general) nonlinear differential equations

$$\frac{dA_i(t)}{dt} = -k \prod_{i=1}^{n} A_i(t) \ . \tag{2}$$

Here the $A_i(t)$ are the concentrations of the i-th molecular species, and we will denote their initial values by A_{i0}.

The simplest case of (1) is the unimolecular reaction $A \xrightarrow{k} P$ (n=1), whose solution is exponential:

$$A(t) = A_0 \, e^{-kt} \ . \tag{3}$$

Bimolecular reactions $A + B \xrightarrow{k} 0$ (n=2) have, following Eq. (2), the kinetic set of equations:

$$\frac{dA(t)}{dt} = -k \, A(t)B(t) = \frac{dB(t)}{dt} \tag{4}$$

whose general solution, setting $C = B_0 - A_0$, is:

$$\frac{1 + C/A(t)}{1 + C/A_0} = e^{Ckt} \ . \tag{5}$$

From Eq. (5) we infer for $B_0 \gg A_0$, that $C \simeq B_0$ and thus $C/A(t) \gg 1$:

$$A(t) \simeq A_0 \, e^{-B_0 kt} \ . \tag{6}$$

Thus the decay of the minority species is quasiexponential.

On the other hand, if $A_0 = B_0$ then $C = 0$ in Eq. (5). An expansion in small C leads to the decay

$$A(t) = \frac{A_0}{1 + A_0 kt} \tag{7}$$

from which at longer times, $t \gg (A_0 k)^{-1}$, an algebraic time dependence emerges:

$$A(t) \sim \frac{1}{kt} \tag{8}$$

We pause to note that a very similar behavior also obtains for the $A + A \to 0$ reaction, with the kinetic equation:

$$\frac{1}{2} \frac{dA(t)}{dt} = -k[A(t)]^2 \ . \tag{9}$$

Separation of the variables and integration lead to:

$$A(t) = \frac{A_0}{1 + 2A_0kt} \tag{10}$$

a form very akin to Eq. (7). The long-time behavior obeys here

$$A(t) \sim \frac{1}{2kt} . \tag{11}$$

Thus, from unimolecular and from bimolecular reactions one has as long-time decays either exponential or 1/t algebraic dependencies.

What is the reason behind this simple finding? The basic assumption underlying the general kinetic scheme of Eq. (2) is the 'well-stirred reactor' model in which all spatial dependencies due to the positions of discrete particles are neglected. Thus the use of Eq. (2) implies a homogeneous spatial distribution of particles during the whole course of the reaction. That such an assumption is untenable in general was discussed by us in previous works [7,21-23]. There we pointed out that non-homogeneous conditions are widespread. Furthermore, even under homogeneous initial conditions, the microscopic reactions create by themselves non-homogeneities and enhance already existing density fluctuations. Diffusion can only partly wipe out such effects [3, 6,7,22,23] but only when the diffusion length is large compared to the mean interparticle distance. At low particle densities, diffusion (or stirring) cannot create a homogeneous background.

When analyzing random materials one is confronted with richer decay patterns. Thus, exemplarily, one finds:

I) The stretched-exponential, Kohlrausch-Williams-Watts [24-26] law

$$\Phi(t) = \exp[-(t/\tau)^\alpha] \tag{12}$$

This form was found in measurements by optical bleaching of the reversible transformation of spiropyran into merocyanine in polymers [27]. The same form is also present when monitoring abstraction reactions of hydrogen atoms from matrix molecules [28,29]. Reactions of trapped hydrogen atoms in γ-irradiated sulfuric acid glasses have been shown [30] to be describable via time dependent rate coefficients of the form $k(t) \sim t^{\alpha-1}$, which correspond to Eq. (12).

II) The exponential-logarithmic relaxation pattern

$$\Phi(t) = \exp[-C\ell n^\beta(t/\tau)] \tag{13}$$

which is of considerable use in describing electron scavenging and electron-hole recombination, and which also appears in the analysis of relaxation phenomena related to hole-burning in glasses [31,32].

III) General algebraic decays

$$\Phi(t) \sim (t/\tau)^{-\xi} \tag{14}$$

as for instance reported for the relaxation processes of photogenerated carriers, which occur after electron-hole-pair creation in amorphous Si:H [33]. A different application concerns the effect of chromatographic tailing, in which dissimilar chemical species display different trapping-release patterns during their hindered diffusion through chromatographic columns [34].

3. RANDOM WALKS AND FRACTALS

In order to show deviations from the chemical kinetic scheme we turn our attention to relaxation mechanisms, for which a series of steps (mostly randomly taken) is necessary for the completion of a reaction. Typical for such behavior are generalized diffusion models, under which random walks over discrete geometrical structures are very prominent [35-38]. We begin in the framework of the pseudo-unimolecular reactions, in which we have a minority and a majority species, and we monitor deviations of the relaxation pattern from exponentiality. In the simplest models one has one A and several B particles, and the A particle is annihilated at the encounter of a B particle. Depending on which of the species performs the motion one distinguishes between the trapping model [36-43] (only the A moves), the target (scavenging) model [7,21-23,44-47] (only the B moves) and the moving targets [48] (both species move). We monitor the survival probability of the A particle averaged over all possible realizations of particle distributions and motions. We consider first random walks on regular lattices and extend then our treatment to fractals.

For simplicity, we discuss here only the target model explicitly and we follow the development of Ref. [22]. Interestingly, this model can be solved explicitly, both for regular lattices [47] and also for ultrametric spaces [49]. We focus on the fate of an immobile A molecule which gets annihilated by the mobile B species. At start the non-interacting B molecules are distributed randomly, with probability p over the structure. For a finite, relatively large system of N_T sites one has $p \simeq B_0/N_T$. At each step we move the B-molecules to neighboring sites, the different directions occurring with equal probability. The reaction is assumed to happen instantaneously when the first B-molecule lands on the site occupied by the A-particle. The survival law obtains by averaging over all possible initial distributions and over all realizations of the random walks for the B-particles. Several possibilities may now be envisaged in creating the initial distribution of B particles on the remaining lattice sites. In [21] and [47] we have presented the survival probabilities for B particles which follow binomial and Poisson distributions. Here we will exemplify the procedure using the last one, for which the occupancy of a site is taken to be distributed as

$$g(j) = e^{-p} \frac{p^j}{j!} \, , \qquad\qquad (15)$$

where $g(j)$ is the normalized probability of having j B particles at one site and p is their average number density.

We now denote by $F_m(r)$ the probability that a random walker starting from r reaches the origin O (assumed to be the site at which the A-particle is located) for the first time in the mth step. For regular lattices, because of the symmetry of the walk $F_m(r)$ is also the first-passage time from O to r, as defined by Montroll and Weiss [50]. In general, we call $H_n(r)$ the probability that a first passage from r to O occurred in the first n steps and have:

$$H_n(r) = \sum_{m=1}^{n} F_m(r) \qquad\qquad (16)$$

The probability therefore that a walker from r did not reach O in the first n steps is thus

$$\Phi_n(r) = 1 - H_n(r) \qquad\qquad (17)$$

Using Eq. (17) we obtain the survival probability of the A molecule by appropriately weighting products of the $\Phi_n(r)$ functions:

$$\Phi_n = \prod_r{}' \left\{ \sum_j g(j) [\Phi_n(r)]^j \right\} \, . \qquad\qquad (18)$$

Here the product extends over all structure sites, with the exception of the origin. Inserting now Eq. (15) and (17) into Eq. (18) leads to

$$\Phi_n = \prod_r{}' \left\{ \sum_j e^{-p} \frac{[p\Phi_n(r)]^j}{j!} \right\}$$

$$= \prod_r{}' \exp\left[-p + p\Phi_n(r)\right] = \exp\left[-p \sum_r{}' H_n(r) \right] \qquad (19)$$

For regular lattices Eq. (19) may be further simplified since, according to Eq. (III.2) and (III.3) of Ref. [50], one has

$$\sum_r{}' F_m(r) = S_m - S_{m-1} \qquad (m \geqslant 1) \, . \qquad\qquad (20)$$

Here S_m denotes the mean number of distinct sites visited by a random walker in m steps, with $S_0 = 1$. Eq. (20) may also be derived directly, by noting that the increase of S_m is given by the total number of new sites visited in the mth step. Introducing now Eqs. (16) and (20) into (19), one has **exactly**

$$\Phi_n = \exp[-p(S_n-1)] . \tag{21}$$

Now, for regular lattices one has for not too small n [36,51]:

$$d = 1 : S_n = a_1\sqrt{n} + a_2/\sqrt{n} + \ldots \tag{22}$$

$$d = 2 : S_n = \frac{a_1 n}{\ln(a_2 n)} + \ldots \tag{23}$$

$$d = 3 : S_n = a_1 n + a_2\sqrt{n} + \ldots \tag{24}$$

where the a_i are constants which depend on the lattice structure.

We now note that the long-time behavior of Eq. (21) stays exponential in three dimensions, but becomes nonexponential for d = 1 and d = 2. It is interesting to see that for d = 1 a Williams-Watts relaxation pattern, Eq. (12) emerges, with $\alpha = \frac{1}{2}$. This result is identical to former findings for the target problem, which used the Glarum model of dipolar relaxation in glasses as experimental application [45,46,52-55]. As long as only random walks on regular lattices are envisaged, it is not obvious how to extend the target model to obtain non-trivial Williams-Watts forms with other α values ($\alpha \neq \frac{1}{2}$, $\alpha \neq 1$). In the following we will show that general α values are the rule when fractals, CTRW or ultrametric structures are considered.

Thus, let us now turn our attention to fractals. The term fractal was coined by Mandelbrot [56] to describe a wide class of objects which display scale-invariance: The patterns are self-similar, i.e. basically unchanged under dila(ta)tion operations. As topological objects, some fractals have come into mathematical scrutiny already early in this century [57,58]. It is, undoubtedly, the merit of Mandelbrot, to have pointed out the ubiquitous presence of fractal patterns in nature [10]. On physical grounds, the **dynamical** aspects of fractals have turned out to be very important, as stressed by Alexander and Orbach [59]. These aspects are connected to scaling and renormalization-group ideas. Examples for fractals are linear and branched polymers, amorphous and porous materials, epoxy resins, diffusion-limited aggregates (DLA), and percolation clusters at criticality. Since the role of fractals in chemistry is discussed by us in our companion article in this book, we refer the interested reader to it [60].

As we have seen, the random walk properties for regular lattices depend on their dimension d. For fractals there are several, mostly non-integer parameters which play roles similar to the dimension. For us here two major parameters are of concern: the fractal [10] (Hausdorff) dimension d, which is related to the density of sites, and the spectral (fracton [59]) dimension d̃, which appears in connection with dynamical properties (such as heat conduction, wave propagation and also reaction and diffusion) on fractals. Thus denoting by N the number of lattice

28

points inside a sphere of radius R, one defines $\bar{d}$ as being $\lim_{R \to \infty} (\ell n\ N/\ell n\ R)$, i.e. one requires

$$N \sim R^{\bar{d}} . \tag{25}$$

Hence for Sierpinski gaskets (for the explanation, see [60]) embedded in d-dimensional spaces [10] one has:

$$\bar{d} = \frac{\ell n(d + 1)}{\ell n\ 2} , \tag{26}$$

and furthermore, based on a scaling argument [59]:

$$\tilde{d} = \frac{2\ell n(d + 1)}{\ell n(d + 3)} . \tag{27}$$

Therefore, for the Sierpinski gasket embedded in d = 2, $\bar{d}$ = 1.584 and $\tilde{d}$ = 1.365. In general, the relation $\tilde{d} \leqslant \bar{d} \leqslant d$ holds: furthermore, for Sierpinski gaskets $1 < \tilde{d} < 2$, as may be seen from Eq. (27).

In [61] and [62] we have simulated a series of random walks on Sierpinski gaskets embedded in Euclidean spaces of dimensions d = 2, 3, 4 and 6, the spectral dimensions being thus from Eq. (27), d = 1.36, 1.55, 1.65 and 1.77, respectively. For the first two moments of the distribution of distinct sites visited, S_n and σ_n^2 , we found that the relations

$$S_n \simeq an^{\tilde{d}/2} \qquad (\tilde{d} < 2) \tag{29}$$

and

$$\sigma_n^2 \simeq bn^{\tilde{d}} \qquad (\tilde{d} < 2) \tag{30}$$

were well obeyed in the range investigated (see Figures 1 and 2 of reference [62]). The relations:

$$S_n \sim \begin{cases} n^{\tilde{d}/2} & (\tilde{d} < 2) \tag{31a} \\ \\ n & (\tilde{d} > 2) \tag{31b} \end{cases}$$

were previously inferred from arguments which used the probability of returning to the origin and the compact or non-compact nature of the underlying lattice [59].

We now turn to the target problem on fractals. One can show that the solution has again the structure of Eq. (21), but that, due to the fact that different sites on a fractal are not equivalent, one has also to average over all possible locations for the A-molecule. For Sierpinski gaskets, averaging over the positions is a minor problem, and one has, to a very good approximation,

$$\Phi_n \simeq \exp[-pn^{\tilde{d}/2}] \qquad (\tilde{d} < 2) \tag{32}$$

and

$$\Phi_n \simeq \exp[-pn] \qquad (\tilde{d} > 2) \tag{33}$$

One may note the Williams-Watts form of Eq. (32) and the fact that now α may be varied from $\frac{1}{2}$ to unity by a judicious choice of the Sierpinski (or Sierpinski-type) gasket [63] underlying a walk. Hence fractals admit a straightforward extension of the Glarum model [52], discussed in the previous subsection. For $\tilde{d} < 2$, Eq. (32) also provides a way of experimentally determining the spectral dimension of fractal objects, by monitoring the relaxation due to multiple-step mechanisms in such materials [47].

Summarizing this section, we have shown that random walks on regular lattices can lead to stretched-exponential decays. These findings are even more pronounced for random walks on fractals, which extend in a natural way former results to continuously varying dimensions.

4. CONTINUOUS-TIME RANDOM WALKS (CTRW)

In continuous-time processes one relaxes the condition that changes, such as the steps of a walker, may occur only at preassigned times. In the CTRW formalism one has thus to introduce a waiting-time distribution $\Psi(t)$, which gives the probability density that the time between steps equals t. This method has been implemented in a classic work by Montroll and Weiss [50], who used it to include $\Psi(t)$ in the generating-function formalisms for random walks.

Let us begin by listing a few $\Psi(t)$ forms of wide use. The simplest situation obtains for a memoryless process in which the probability of remaining at a given site during the time interval from 0 to t is exponential:

$$\Psi(t) = e^{-bt} . \tag{34}$$

If now $\Psi(t)$ is the probability density that an event occurs at time t after the previous event has taken place, then obviously

$$\Psi(t) = -\frac{d}{dt} \Psi(t) \tag{35}$$

and Eq. (5.1) leads to the Poisson process

$$\Psi^P(t) = q \exp(-qt). \tag{36}$$

Slightly more complex forms obtain by using for $\Psi(t)$ expressions due to multipolar or exchange interactions in the presence of substitutional disorder [64]. These forms are well-behaved in that, for the, the first moment τ_1 :

$$\tau_1 = \int_0^\infty t\Psi(t)dt \qquad (37)$$

is finite. Much more interesting are distributions for which the integral in Eq. (37) is divergent. Thus Scher and Montroll [14] have modeled transport in amorphous media through a $\Psi(t)$ which displays a long-time tail:

$$\Psi(t) \sim t^{-1-\gamma} \qquad (38)$$

Interestingly, such an expression is intimately related to a fractal set of event times [16,65,66]. From the Poisson process, Eq. (36), one constructs readily a dilatationally symmetric distribution, by taking into account events on all time scales, in the following way:

$$\Psi(t) = \frac{1-N}{N} \sum_{j=1}^{\infty} N^j q^j \exp(-tq^j) , \qquad (39)$$

where $N < 1$. As is evident, the distribution (39) is a normalized sum of Poisson terms and

$$\Psi(qt) = \frac{\Psi(t)}{(Nq)} - \frac{(1-N)}{N} \exp(-tq) \qquad (40)$$

For later applications we need $q < N$, so that $q < 1$ and thus at longer times $\Psi(qt) = \Psi(t)/Nq$. The last expression is equivalent to Eq. (38) when $\gamma = \ell n\, N/\ell n\, q$ is set. Eq. (38) shows directly the temporal <u>scaling</u> of $\Psi(t)$, i.e. its fractal nature in time.

We now remark that for algebraic $\Psi(t)$ forms, scaling carries over to other quantities related to $\Psi(t)$. Let $X_n(t)$ denote the probability that exactly n events occurred in time t. This basic quantity of the CTRW formalism is connected to $\Psi(t)$ via its Laplace transform:

$$\mathcal{L}[X_n(t)] \equiv X_n(u) = [\Psi(u)]^n \frac{[1-\Psi(u)]}{u} , \qquad (41)$$

where $\Psi(u) = \mathcal{L}[\Psi(t)]$.

As shown in Ref. [22], for $\Psi(t)$ whose first moment τ_1 is infinite, the $X_n(t)$ coalesce at long times. For a given time t the set of curves with n less than a certain parameter $n_{max}(t)$ scales, i.e. one has approximately

$$X_n(t) = \begin{cases} X_0(t) \sim t^{-\gamma} & \text{for } n < n_{max}(t) \\ 0 & \text{otherwise.} \end{cases} \qquad (42)$$

On the other hand, for distributions whose first moments exist, scaling does not hold.

We now evaluate $S(t)$, the mean number of sites visited by a walker in continuous time. Qualitatively one expects in situations in which $\tau_1 < \infty$ the pattern to follow Eq. (22) to (24), with n replaced by t/τ_1. This result is indeed well-fulfilled, as backed by extensive studies on (small) deviations due to higher order terms [64]. On the other hand, when τ_1 is infinite and $\psi(t)$ scales, t/τ_1 is meaningless. Then another argument may be used. From the additional temporal averaging required, $S(t)$ is nothing but

$$S(t) = \sum_{n=0}^{\infty} S_n X_n(t) . \qquad (43)$$

We know now that in general in the long-time domain S_n has a power-law dependence on n, both for regular lattices (d = 2 excluded), and for fractals.

Setting $S_n \sim n^{\mathcal{E}}$ and using Eq. (42) it follows that

$$S(t) \sim \sum_{n=0}^{n_{max}} n^{\mathcal{E}} X_0(t) \sim X_0(t) n_{max}^{1+\mathcal{E}} . \qquad (44)$$

Now the $X_n(t)$ are normalized, a fact which determines the time dependence of $n_{max}(t)$, [66]

$$1 = \sum_{n=0}^{\infty} X_n(t) \approx \sum_{n=0}^{n_{max}} X_0(t) = X_0(t) n_{max} . \qquad (45)$$

For $X_0(t) \sim t^{-\gamma}$ it follows that $n_{max} \sim t^{\gamma}$ and hence, from Eq. (44), $S(t) \sim t^{\gamma\mathcal{E}}$, or in fractal notation

$$S(t) \sim \begin{cases} t^{\gamma \tilde{d}/2} & \text{for } \tilde{d} < 2 \\ t^{\gamma} & \text{for } \tilde{d} > 2. \end{cases} \qquad (46)$$

In Eq. (46) the two fractal exponents (γ for the temporal and $\tilde{d}/2$ for the spatial aspect) combine multiplicatively, i.e. the two processes subordinate [10,62,67].

We now consider the relaxation pattern of pseudo-unimolecular $A + B \rightarrow 0$, $A_0 \ll B_0$ reactions under continuous-time conditions. Starting with the target problem, which again turns out to be simpler than trapping, we have to extend the formalism to the CTRW situation. From Eq. (16) we now obtain the probability $H(t;r)$ of a visit from r to 0 in time t:

$$H(t;\mathbf{r}) = \sum_{n=0}^{\infty} X_n(t)H_n(\mathbf{r}) , \tag{47}$$

where the $X_n(t)$ are defined as in Eq. (41). The average over times in Eq. (47) parallels, of course, that of Eq. (43). Since from Eq. (16) and (20) it follows that

$$\sum_{\mathbf{r}}' H_n(\mathbf{r}) = S_n - 1 , \tag{48}$$

one has, in fact, using Eq. (43)

$$\sum_{\mathbf{r}}' H(t;\mathbf{r}) = S(t) - 1 . \tag{49}$$

Under the assumed independence of motion of the walkers in the target problem, the CTRW transformation of Eq. (18) and (19) carries through, and one obtains for the relaxation of the target:

$$\Phi(t) \sim \exp[-pS(t)] , \tag{50}$$

where $S(t)$ is given by Eq. (46).

We stop to note the appearance of a Williams-Watts stretched-exponential form in Eq. (50). Now the parameter α in Eq. (12) may take any value between 0 and 1, if one chooses γ accordingly. This is true even for regular lattices in arbitrary dimensions, so that the CTRW provides an adequate extension for the Glarum model of spin relaxation in glasses [52], as pointed out by Shlesinger and Montroll [45,65].

Notice that due to the multiplicative connection of γ and $\tilde{d}/2$ in Eq. (46), which stems from subordination, a measured Williams-Watts parameter α may be explained by an infinity of $(\gamma, \tilde{d}/2)$ pairs. Monitoring the target annihilation one cannot distinguish the separate roles played by the spatial disorder exemplified by $\tilde{d}/2$, and by the temporal disorder given by γ. This distinction may be drawn in the trapping problem, since here the CTRW forms lead to quite different decay patterns [22,67].

We proceed to consider the energetic disorder and the ultrametric spaces (UMS).

5. RANDOM WALKS ON ULTRAMETRIC SPACES

In general, sites in a disordered material are separated by energy barriers, whose height is random [23]. Thus, a reacting particle positioned on a certain site needs thermal energy to surmount the surrounding barriers. A given activation energy lets the particle visit only a subset (cluster) of sites around the starting point. One may then classify the sites through the energy required to reach them [18]: To such a classification corresponds an ultrametric space (UMS)[17-19,23,68-70]. As examples, Fig. 1 shows the regularly multifurcating UMS Z_2 and Z_3 . Note that only the points on the baseline belong to the UMS and that the structure above the baseline documents barrier heights and intersite connections. The height $d(x,y)$ of the barrier between sites x and y may be used as a generalized "distance": It may be easily verified that

$$d(x,y) \leqslant \max(d(x,w), d(y,w)) \tag{51}$$

holds for all UMS sites. Eq. (51) is the "strong" triangle inequality which leads to the name "ultra" metric. Z_2 and Z_3 have branching ratios $b = 2$ resp. $b = 3$, and, for simplicity, we have taken the barrier heights to be hierarchically arranged, so that all consecutive energy levels differ by Δ, the energies E being $E = m\Delta$, $m \in \mathbb{N}$.

We now consider a simple qualitative argument for the temporal dependence of $S(t)$, the mean number of sites visited during t, when the particles are thermally activated, so that the intersite transition rates are proportional to $R = \exp(-E/kT)$.

Let us focus on the time interval

$$e^{m\Delta/kT} \leqslant wt_m \leqslant e^{(m+1)\Delta/kT} . \tag{52}$$

During this time interval b^m points of the UMS Z_b are accessible to the walker, and one has

$$b^m \sim b^{(kT/\Delta)\ell n(wt_m)} = e^{\delta \ell n(wt_m)} = (wt_m)^\delta , \tag{53}$$

where we set

$$\delta = \frac{kT}{\Delta} \ell nb . \tag{54}$$

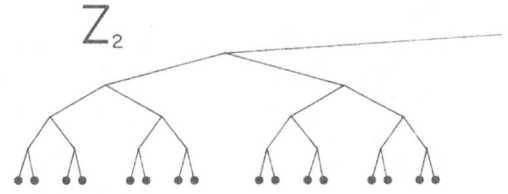

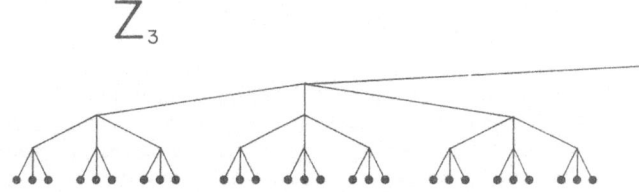

Figure 1. The ultrametric spaces (UMS) Z_2 and Z_3.

For $\delta < 1$, b^m increases more slowly than t_m, and the walker explores practically all accessible points. We are in the case of compact exploration. Therefore,

$$S(t) \simeq (wt)^\delta \sim t^\delta \quad (\delta < 1) . \tag{55}$$

On the other hand, for $\delta > 1$, b^m increases more rapidly than t_m, and the mean number of distinct sites visited stays proportional to t_m; hence

$$S(t) \sim t \quad (\delta > 1) . \tag{56}$$

From this simple qualitative argument we have obtained a result which parallels our findings for fractals and for CTRW, Eq. (31) and (46). Eq. (55) and (56) can also be obtained rigorously, as we proceed to show following the lines developed in Ref. [49].

The basic idea of Ref. [49] is that many results valid for regular lattices carry over exactly to regularly multifurcating UMS. Thus, let $P_{0,n}$ be the probability of being at the origin of the walk at the n-th step and $P(z)$ the corresponding generating function.

$$P(z) = \sum_{n=0}^{\infty} P_{0,n} z^n \tag{57}$$

Similarly, let $S(z)$ be the generating function of the mean number of distinct sites visited, S_n :

$$S(z) = \sum_{n=0}^{\infty} S_n z^n \tag{58}$$

For regular lattices, and also for regularly multifurcating UMS [49] (but not for fractals!) the following relation holds:

$$S(z) = \frac{1}{(1-z)^2 \, P(z)} \tag{59}$$

The proof proceeds as in Ref. [49]. Starting point is the relation:

$$P_{i,n} = \sum_{m=1}^{n} P_{ii,n-m} F_{i,m} + \delta_{i0} \, \delta_{n0} \tag{60}$$

where $F_{i,m}$ is the probability to reach site i for the first time in the m-th step. Eq. (60) states (beside the obvious initial condition) that in order to be in the n-th step at i one has to arrive there either at the n-th step or earlier, at m, followed by a return to i in n-m steps (whose probability is $P_{ii,n-m}$). Eq. (60) holds since the random walk is a homogeneous Markov process, invariant with respect to time trans-lation, and since event spaces corresponding to different first-time arrivals are disjoint sets. Due to translational invariance, for Bravais lattices one has $P_{ii,n-m} = P_{i-i,n-m} = P_{0,n-m}$. The same relation holds also for regularly multifurcating UMS, since all sites are equivalent. (On the other hand, in general for fractals the relation $P_{ii,n-m} = P_{0,n-m}$ is only approximate.) Hence also for UMS:

$$P_{i,n} = \sum_{m=1}^{n} P_{0,n-m} F_{i,m} + \delta_{i0} \delta_{n0} \qquad (61)$$

The course to Eq. (59) is now straightforward [50]. The increment Δ_m in newly visited sites at the m-th step is

$$\Delta_m = \sum_{i \neq 0} F_{i,m} \qquad (62)$$

with $\Delta_0 = 1$. Then S_n, the mean number of distinct sites visited in n steps is:

$$S_n = \sum_{m=0}^{n} \Delta_m \qquad (63)$$

Eqs. (61) and (62) give by summing over $i \neq 0$:

$$\sum_{m=0}^{n} P_{0,n-m} \Delta_m = 1 \qquad (64)$$

where the requirement of conservation of probability was used. Switching over to generating functions one has

$$P(z) \ \Delta(z) = (1-z)^{-1} \qquad (65)$$

Furthermore, from Eq. (63):

$$(1-z) \ S(z) = \Delta(z) = [(1-z)P(z)]^{-1} \qquad (66)$$

which is Eq. (59).

The second ingredient of Ref. [49] is the interplay between random walks and CTRW. Let us exemplify the basic idea using the $P_{0,n}$. Recalling that $X_n(t)$ is the probability of having performed exactly n steps during t, Eq. (41), one has in CTRW:

$$P_0(t) = \sum_{n=0}^{\infty} X_n(t) \ P_{0,n} \qquad (67)$$

which in Laplace-transformed form is

$$P_0(u) = \frac{1-\Psi(u)}{u} \sum_{n=0}^{\infty} [\Psi(u)]^n P_{0,n} \qquad (68)$$

This, apart from the factor $[1-\Psi(u)]/u$ is nothing else but the generating function $P(z)$, evaluated at $z = \Psi(n)$.

Hence, one may switch from $P_{0,n}$ to $P(z)$ to $P_0(t)$, if only one of them is known. Furthermore, due to Eq. (59), then also S_n , $S(z)$ and $S(t)$ follow.

Now, following former works on UMS [71-73], Bachas and Huberman [74] have succeeded in exactly solving the master equation for regularly multifurcating trees. They obtain:

$$P_0(t) = b^{-n} + (b-1) \sum_{m=1}^{n} b^{-m} \exp(-\lambda_m\, t/\tau) \qquad (69)$$

where $P_i(0) = \delta_{i0}$ and

$$\lambda_m = (1-R/b) \sum_{p=m}^{n-1} R^p + R^n \qquad (70)$$

with $R = \exp(-\Delta/kT)$. For $b=2$ and $\tau = 1/b$ the solution is that of Ref. [72]. For an infinite tree, $n \to \infty$, Eq. (69) takes the form of a Weierstrass-series [16, 22,65,66]:

$$P_0(t) = (b-1) \sum_{m=1}^{\infty} b^{-m} \exp[-CR^m(t/\tau)] \qquad (71)$$

see Eq. (39), with q replaced by R and N by $1/b$. In Eq. (71) $C = (b-R)/(b-bR)$. Note that $P_0(Rt) \sim bP_0(t)$, from which $P_0(t) \sim t^{-\delta}$ with $\delta = (\ln b)kT/\Delta$, i.e. Eq. (54) follows.

As discussed in Ref. [49], the solution (69) corresponds to the CTRW-process with a Poisson-distributed $\Psi(t)$, Eq. (36). It then follows exactly:

$$P_{0,n} = (b-1) \sum_{m=1}^{\infty} b^{-m} \exp(-nC'R^m) \qquad (72)$$

with $C' = (b-R)/(bR-R)$. By Laplace-transformation of Eq. (71) one has

$$P_0(u) \sim \begin{cases} u^{\delta-1} & \text{for } \delta < 1 \\ \text{const} & \text{for } \delta > 1 \end{cases} \qquad (73)$$

and then for the Poisson process:

$$S(u) \sim \begin{cases} u^{-\delta-1} & \text{for } \delta < 1 \\ u^{-2} & \text{for } \delta > 1 \end{cases} \tag{74}$$

from which in the time domain follows:

$$S(t) \sim \begin{cases} t^{\delta} & \text{for } \delta < 1 \\ t & \text{for } \delta > 1 \end{cases} \tag{75}$$

Since the random walk has a Poisson distribution of waiting times, $n \sim \lambda t$, and hence $S_n \sim n^{\delta}$ for $\delta < 1$ and $S_n \sim n$ for $\delta > 1$. This agrees with our previous analysis for random walks on UMS. One may note that for $\delta < 1$ also the relation

$$S(t) \sim 1/P_0(t) \tag{76}$$

holds, which is the hallmark of compact exploration [59,67,69,70].

Furthermore, now we are in the position of considering CTRW with broad waiting time distributions, Eq. (38). Starting point, as in Ref. [49] is the Laplace-transform of Eq. (71), together with Eqs. (59) and (68). One obtains

$$P_0(t) \sim \begin{cases} t^{-\gamma\delta} & \text{for } \delta < 1 \\ t^{-\gamma} & \text{for } \delta > 1 \end{cases} \tag{77}$$

For $\gamma < 1$ and $\delta < 1$ the two coefficients combine multiplicatively in $P_0(t)$, i.e. the two processes subordinate [67]. Remarkable is also the fact that for $\delta > 1$ the long-time behavior is dictated by the temporal disorder and that the energetic disorder is no longer important. This is similar to the previous finding for CTRW on fractals, with $\bar{d} > 2$, where for $\gamma < 1$ the spectral dimension becomes an irrelevant parameter.

Let us now turn to the mean number of distinct sites visited. From Eq. (43) together with Eq. (41) one has:

$$S(u) = \frac{1-\psi(u)}{u} \sum_{n=0}^{\infty} [\psi(u)]^n S_n \tag{78}$$

The S_n are as given after Eq. (44), $S_n \sim n^{\varepsilon}$, with $\varepsilon = \min(1,\delta)$. Now [67]

$$\sum_{n=0}^{\infty} n^{\varepsilon} z^n \approx \int_0^{\infty} x^{\varepsilon} e^{x \ln z} dx = \Gamma(\varepsilon+1) (-\ln z)^{-\varepsilon-1} \tag{79}$$

so that, for $1-\psi(u) \sim u^{\gamma}$

$$S(u) \sim u^{\gamma-1} u^{-\gamma(\varepsilon+1)} = u^{-\gamma\varepsilon-1} \tag{80}$$

i.e. $S(u) \sim u^{-\gamma\delta-1}$ for $\delta < 1$ and $S(u) \sim u^{-\gamma-1}$ for $\delta > 1$. Hence

$$S(t) \sim \begin{cases} t^{\gamma\delta} & \text{for } \delta < 1 \\ t^{\gamma} & \text{for } \delta > 1 \end{cases} \tag{81}$$

Thus emerges the central result of this section, that CTRW on UMS subordinate. Note furthermore that relation (76) $S(t) \sim 1/P_0(t)$ is now obeyed for <u>all</u> δ (i.e. for all temperatures).

From these results we proceed to show the findings for relaxation phenomena on UMS. As in the previous sections we restrict ourselves to the target problem, and refer the reader interested in the trapping problem to our former publications [22,23,68-70].

Let us start by again noting that all sites of our regularly multifurcating UMS (fixed branching ratio b and equidistant barriers) are equivalent. We again denote the site on which the target sits as the origin, r=0, and assign integer number $r \in N$, by counting, to the other sites. Eqs. (16) to (19) hold also for UMS, when the position of the site (r in Section 3) is reinterpreted to be the ordinal integer r. Furthermore, Eq. (20) holds on UMS since $\sum_r F_m(r) \equiv \sum_r F_{r,m}$ gives the

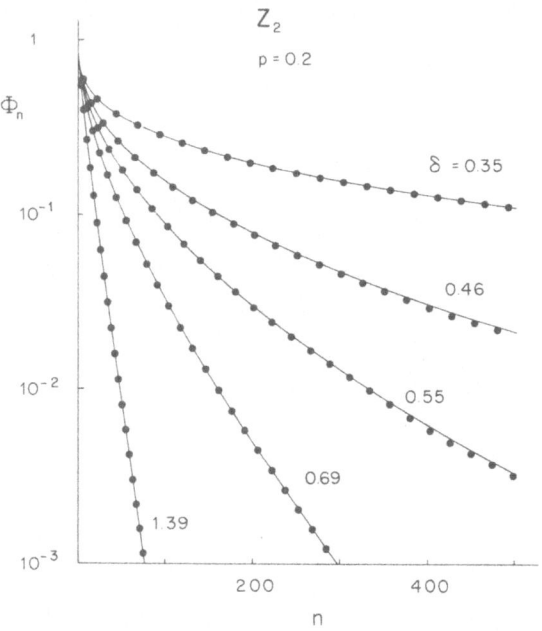

Figure 2. Decay law due to the annihilation of targets by random walkers on the UMS Z_2 . The concentration of walkers in p = 0.2 and the decays are monitored as a function of the temperature $\delta = (\ell n\ 2)kT/\Delta$, for the values Δ/kT = 0.5, 1, 1.25, 1.5 and 2. The full lines are the exact solution, Eq. (82), whereas the dots are the results of simulation calculations.

increase in the total number of visited sites in the mth step, see Eqs. (62) and (63), and thus equals $S_m - S_{m-1}$. Consequently, we rederive for UMS, as in Eq. (21), the exact decay law of A particles annihilated by moving B species:

$$\Phi_n = \exp[-p(S_n-1)] \tag{82}$$

where the B molecules were initially Poisson distributed, Eq. (15).
    From Eqs. (75) and (81) one has:

$$\Phi_n \simeq \exp(-Cpt^\eta) \tag{83}$$

which for $\eta < 1$ shows a stretched-exponential form, Eq. (12).
    In Figure 2 we present the decay in the target problem on the UMS Z_2 . The full lines give the decay obtained from S_n in conjunction with Eq. (82) whereas the dots indicate the direct simulation of the target annihilation. We have taken the density of the walkers to be $p = 0.2$, and the dynamics take place over the UMS Z_2 . As expected, in every case the agreement between the two forms is excellent. Note that depending on the temperature one has a crossover from exponential decays for $\eta \simeq 1$, i.e. at higher temperatures ($kT \ell n\ b > \Delta$), to stretched-exponential decays at lower temperatures ($kT \ell n\ b < \Delta$). These aspects stress the point that, in the range of parameters of the figure, random walks on UMS parallel findings for lattices with a spectral dimension $\tilde{d} = 2\delta$ [71]. Interestingly, one may therefore switch through the marginal behavior at $\delta = 1$ ($\tilde{d} = 2$) through a simple temperature change. Such a phase transition should be experimentally observable through the qualitative pattern (exponential vs. stretched exponential) of the corresponding relaxation behavior.
    To summarize this section, we have analyzed dynamical relaxation behaviors on UMS, and have pointed out the analogies to previous results for random walks and CTRW on regular lattices and on fractals. This section concludes our exposition on pseudo-unimolecular decays which obtain in disordered systems, when fractals, CTRW or ultrametric spaces are used as models. In the next section we present an overview of bimolecular relaxation patterns.

6. THE BIMOLECULAR REACTIONS A + A → 0 AND A + B → 0 ($A_0 = B_0$)

As stressed in Section 2, in which we investigated the chemical kinetic scheme for the A + A → 0 and for the strictly bimolecular A + B → 0 ($A_0 = B_0$) reactions, in both cases the decay follows the 1/t dependence under 'well-stirred' conditions. Our findings, Eqs. (8) and (11), were the consequence of a spatially homogeneous situation. In this section we determine the decay laws which apply in the presence of disorder, which will again be modeled through fractals, CTRW and UMS. The main type of relaxation pattern which will emerge from our studies is algebraic, $\Phi(t) \sim t^{-\xi}$, Eq. (14), where ξ may take any value between zero and one.

Here we start with the A + A → O reaction, since it will turn out to be
less influenced by fluctuations than the strictly bimolecular A + B → O,
$A_0 = B_0$, reaction. This reaction is of importance in energy transfer
problems, where it describes exciton up-conversion and annihilation
processes [22,75-78]. From the previous study of pseudo-unimolecular
reactions, we found that the kinetic exponential was modified by the
appearance of S(t). Thus one may expect that the A + A → O decay will
follow a 1/S(t) law at longer times [21,75] and thus, one may find as an
approximation to the decay,

$$\phi_n^{AA} \simeq (1 + 2pS_n)^{-1} . \tag{84}$$

In [21] we have established through numerical simulations that Eq. (84)
correctly describes the decay behavior both on regular lattices and on
fractals (Sierpinski gaskets). For the dynamical processes periodic
boundary conditions were used, so that the systems appeared infinite.
The particles were placed randomly on the lattice, following a binomial
(yes-no) distribution. At each step the walkers moved one by one to
neighboring sites. Any two particles that during this process happened
to occupy the same site were immediately removed.

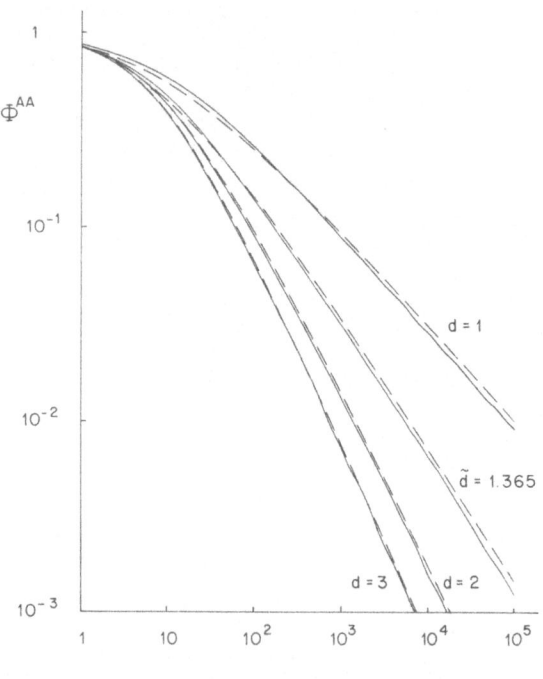

Figure 3. Decays due to the bimolecular reaction A + A → O. The full
lines are the simulation results for the linear chain (d = 1), for the
square (d =2), for the simple cubic lattice (d = 3) and for a two-
dimensional Sierpinski gasket ($\tilde{d}$ = 1.365). The dashed lines are Smoluchowski-type approximations, Eq. (84).

In Fig. 3 we display the results for the decay following the
A + A → 0 reaction on the linear chain, the square, the simple cubic
lattice and for the two-dimensional Sierpinski gasket. The full lines
give the simulation results obtained by averaging over 100 distinct
initial conditions and walks each, whereas the dashed lines are the
approximation, Eq. (84). We choose to plot $\ln \Phi_n^{AA}$ vs $\ln n$. In these
scales, at long times the decays should turn into straight lines. From
the figure this behavior is clearly apparent. Moreover, it turns out
that Eq. (84) is almost quantitative in the range investigated [21,22].

The same behavior obtains also for fractals, as we have exemplified
in [21], by analyzing the deviations of the Smoluchowski-type approximation, Eq. (84) from the simulated decays. An exact solution to the
many-body problem involved in the A + A → 0 reaction was found by Torney
and McConnell for d = 1 in the continuum, diffusion-equation limit [78].
In the range of Fig. 3 their expression is hardly distinguishable from
the discrete, random-walk result. Hence, in all cases investigated, the
long-time decay Φ_n^{AA} follows an algebraic form $\Phi_n^{AA} \sim n^{-\xi}$ (with $\xi = \tilde{d}/2$
for $\tilde{d} < 2$ and $\xi = 1$ for $\tilde{d} > 2$). Heuristically one may view Eq. (84), the

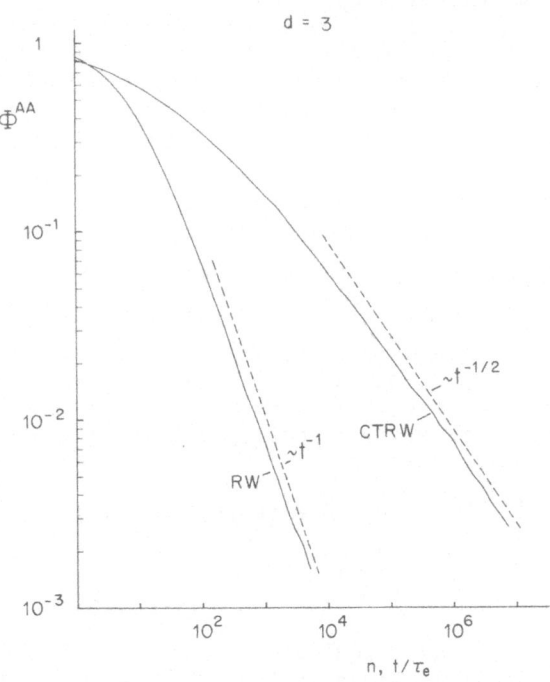

Figure 4. Decay laws Φ^{AA} for the A + A → 0 reaction on a simple cubic
lattice, both for simple RW and for CTRW with $\gamma = 1/2$. The initial
particle density is p = 0.1. The full lines are the simulation results
averaged over 100 runs. The theoretical long-time slopes are indicated
by dashed lines.

solution of a many-body problem, as being related to the probability of encounter of two particles, which itself is expressible, via S_n , by the volume visited by each.

We now turn to the $A + A \rightarrow 0$ reaction under CTRW-conditions. The previous discussion for regular lattices and for fractals has lead to

$$\phi_n^{AA}(t) \sim \begin{cases} n^{-\tilde{d}/2} & \text{for } \tilde{d} < 2 \\ n^{-1} & \text{for } \tilde{d} > 2. \end{cases} \tag{85}$$

From our viewing S_n as a measure of explored volume it is therefore tempting to envisage that in the CTRW scheme S_n gets replaced in Eq. (84) by $S(t)$, and therefore, for CTRW one should find:

$$\phi^{AA}(t) \sim \begin{cases} t^{-\gamma\tilde{d}/2} & \text{for } \tilde{d} < 2, \ \gamma < 1 \\ t^{-\gamma} & \text{for } \tilde{d} > 2, \ \gamma < 1. \end{cases} \tag{86}$$

Eq. (86) is then another example of subordination [10,62,67].

Our numerical simulations support Eq. (86) well [66]. In Fig. 4 we present the decay of the $A + A \rightarrow 0$ reaction under CTRW conditions. We start from walkers on a simple cubic lattice, d = 3, and use for $\psi(t)$ a form with $\gamma = 1/2$. In the figure we present the corresponding decay law and contrast it with the simple RW results. Whereas at longer times the random walk decay follows $1/t$, for the CTRW we find at longer times a $t^{-1/2}$ dependence, as may be verified by inspection of Fig. 4, in which these asymptotic slopes are also indicated. As a further example we have performed simulations on several Sierpinski gaskets and on the linear chain [66]. The long-time decay behavior indeed follows the form $t^{-\tilde{d}/4}$ for CTRW with $\gamma = 1/2$ instead of $t^{-\tilde{d}/2}$ for the simple RW decay. All findings are consistent with $\phi^{AA}(t) \sim [S(t)]^{-1}$, i.e. with Eq. (86).

To conclude our study of the $A + A \rightarrow 0$ reaction we also consider the influence of energetic randomness, as displayed by ultrametric spaces (UMS). At longer times, we expect, paralleling the previous discussion, the relaxation pattern

$$\phi^{AA}(t) \sim \begin{cases} t^{-\delta} & \text{for } \delta = (kT/\Delta) \ \ell n \ b < 1 \\ t^{-1} & \text{else.} \end{cases} \tag{87}$$

The results of a typical calculation are presented in Fig. 5. The reaction depicted again takes place on the UMS Z_2 at a somewhat low temperature, so that $\delta = 0.347$. The initial density of particles is $p_A = 0.2$. The overall decay of ϕ_n^{AA} at longer times (larger n) is quite well described by $1/n^{0.35}$, as indicated by the slope parallel to the decay. Thus, on UMS the trend found for regular lattices and for fractals continues: ϕ_n^{AA} may be well approximated through $\phi_n^{AA} \sim S_n^{-1}$.

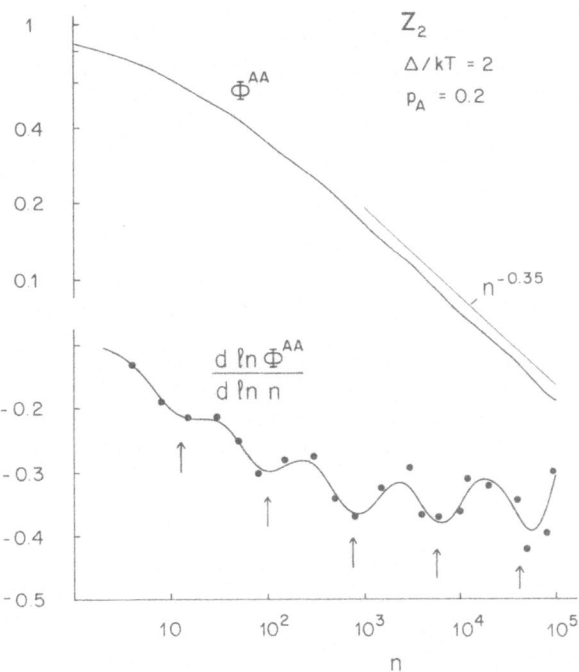

Figure 5. Decay pattern for the A + A → 0 reaction on the UMS Z_2 . The upper section shows the decay for Δ/kT = 2 and for an initial particle density p_A = 0.2. Also indicated is the slope of the decay for n large. In the lower section the variations of the slopes are given by dots, the full line is the result of a smoothing procedure. The arrows mark theoretically predicted minima.

Superimposed on this algebraic decay one finds, however, slight fluctuations. To present them in a magnified form we have plotted in the lower part of Fig. 5 the derivative $d(\ln \Phi^{AA})/d(\ln n)$. The dots are the numerically evaluated slopes taken from the simulation, and the full line is the result of a smoothing procedure. The arrows indicate the estimates for the locations of the minima determined by a detailed analysis; in fact the minima may be again related to the cluster structure of Z_2 .

Increasing the temperature so as to have $\delta > 1$ again leads to a 1/t pattern, as also expected from Eq. (87), together with Eq. (75). As before, an increase in temperature leads one to the kinetic result.

We now turn our attention to the strictly bimolecular reaction A + B → 0, $A_0 = B_0$. We note from the start that the simplicity of the A + A → 0 reaction does not carry over; the long-time regime is different from that of Eq. (84) to (87). The difference results from spatial fluctuations, which get <u>enhanced</u> by the chemical A + B → 0 reaction. The reason for this effect is that at longer times, due to the progress of the reaction, large regions containing only A or only B molecules appear. Then the diffusion no longer provides an efficient stirring, and the reactions proceed more slowly since mainly only molecules at the boundaries of the A and B regions are prone to react. The expected decay at longer times then follows [4,6,7,21-23]:

$$\phi_n^{AB} \sim \begin{cases} n^{-\tilde{d}/4} & \text{for } \tilde{d} < 4 \\ n^{-1} & \text{for } \tilde{d} > 4. \end{cases} \tag{88}$$

The marginal dimension for the $A + B \to 0$, $A_0 = B_0$ reaction is thus four [4,6].

In Fig. 6 we present a snapshot of a simulation calculation on a Sierpinski gasket ($\tilde{d} = 1.36$), after 3×10^5 steps have elapsed. In the calculation we started with some 10^5 particles, placed such that the initial concentrations $p_A = p_B$ were 0.05. The separation of the A and B

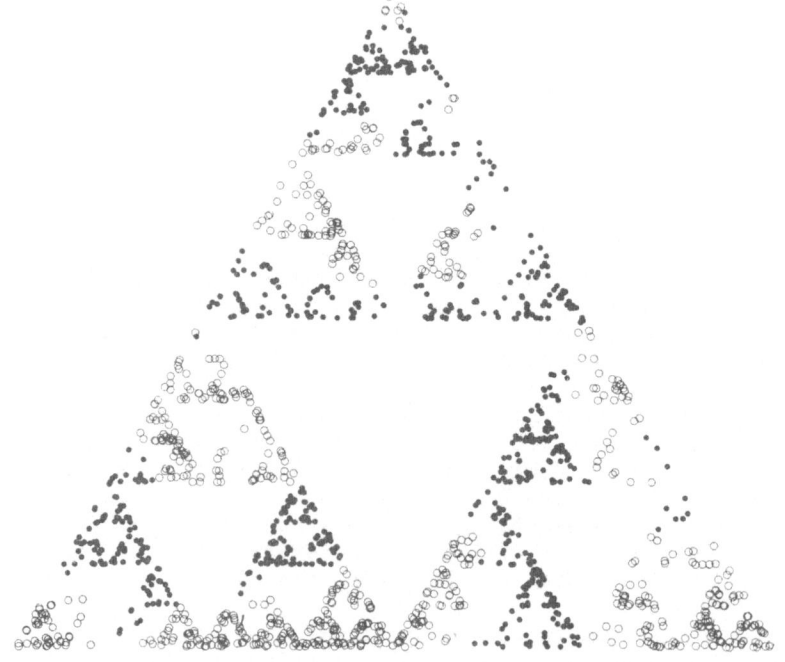

Figure 6. Snapshot of the strictly bimolecular reaction $A + B \to 0$, $A_0 = B_0$ on a two-dimensional Sierpinski gasket ($\tilde{d} = 1.365$) at the 12th iteration stage (some 10^6 lattice sites). The initial concentrations are $A_0 = B_0 = 0.05$ and correspond to some 10^5 particles. The picture shows the situation after 3×10^5 steps. The remaining A and B particles (around 700 each) are indicated by circles and dots. (Reproduced from Ref. [48]).

species into distinct regions is clearly evident. This picture of particle segregation, taken from Ref. [48], (the work reporting our previous results being [79]) finds its parallel also under steady-state conditions, where one has a different marginal dimension, see Ref. [80].

In this time regime the decay law ϕ_n^{AB} has crossed over from the Smoluchowski-type pattern, Eq. (84) and (85), valid for smaller n, to the decay behavior of Eq. (88). As we have demonstrated in [7] and [21], see Fig. 7, a similar behavior obtains also for other Sierpinski gaskets, imbedded in spaces of higher Euclidean dimensions; we have considered Sierpinski gaskets with spectral dimension $\tilde{d}$ = 1.55, 1.65 and 1.77. Similar findings were also reported for strictly bimolecular reactions on stochastic fractals [81]. We remark that with increasing $\tilde{d}$ the Smoluchowski-type region of Eq. (84) increases. Hence, an increase in the spectral dimension pushes to later times the crossover to the form given by Eq. (88).

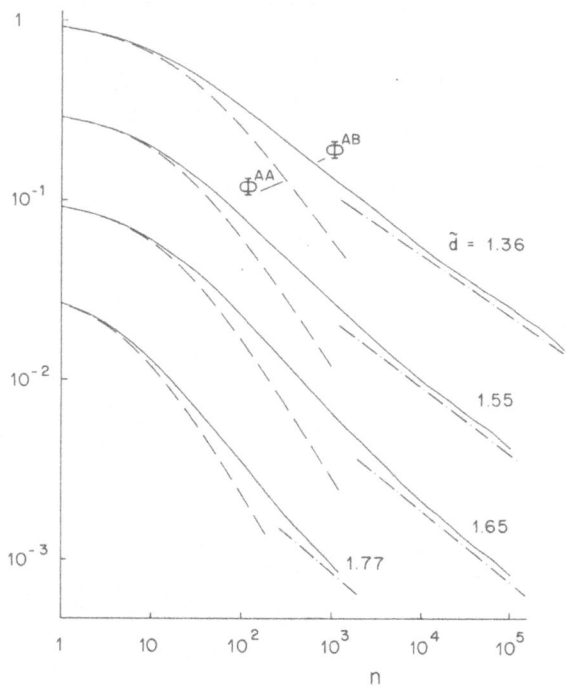

Figure 7. Decay laws due to reactions on fractals (Sierpinski gaskets) of different spectral dimensions $\tilde{d}$. Full lines for the reaction A + B → 0, $A_0 = B_0$ with expected slopes at longer times denoted by dash-dotted lines. Dashed lines for the reaction A + A → 0. For both types of reactions the initial concentrations were p = 0.1.

To conclude let us summarize our comparison of the bimolecular reactions A + A → 0, A + B → 0 (A_0 = B_0) on regular lattices and on fractals. For the first we find agreement with the Smoluchowski-type form, Eq. (84), for the whole decay range studied, whereas the second obeys such a form only in the initial time domain and crosses over to a lower decay $t^{-\tilde{d}/4}$ ($\tilde{d}$ < 4) at longer times. Thus, bimolecular reactions display a richer behavior than that predicted by the standard kinetic approach.

Now the role of the CTRW remains to be assessed. As in previous cases, we expect the decay form to subordinate through the time variable to $\Psi(t)$, and expect hence:

$$\Phi^{AB}(t) \sim t^{-\gamma\tilde{d}/4} \qquad \text{(for } \tilde{d} < 4, \gamma < 1). \qquad (89)$$

Simulation calculations [66] for particles moving on a linear chain under the influence of a waiting-time distribution with γ = 1/2 support the conjectured subordination displayed by Eq. (89) as we show in Fig. 8.

We now address the question of the A + B → 0, A_0 = B_0 reaction on UMS. One expects here, based on our previous knowledge of this reaction

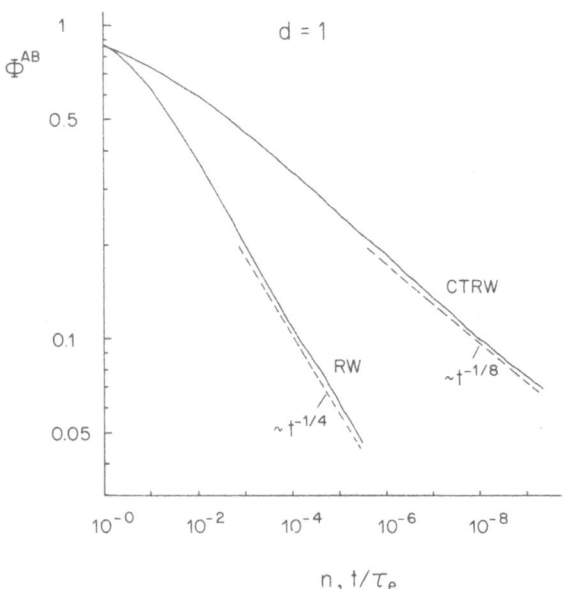

Figure 8. Decay law $\Phi^{AB}(t)$ for the strictly bimolecular A + B → 0, A(0) = B(0), with particles moving according to CTRW for a $\Psi(t)$ with γ = 1/2 on the linear chain. The long-time slopes are also indicated.

and on the relation $\tilde{d} = 2\delta$,

$$\phi_n^{AB} \sim \begin{cases} t^{-\delta/2} & \text{for } \delta < 1 \\ t^{-1} & \text{for } \delta > 2 \end{cases} \qquad (90)$$

with a crossover region for δ between 1 and 2. Simulation calculations [23] show, indeed, that Eq. (90) offers a very reasonable description of the long-time behavior for $\delta < 1$ and for $\delta > 2$.

In Fig. 9 we display the decay pattern ϕ_n^{AB} which obtains for Z_3 and $p_A = p_B = 0.1$ [23]. Here we averaged over five different realizations of initial conditions and walks. We take Δ/kT to be 0.5; 1; 1.5 and 2 so that δ varies between 0.55 and 2.2.

To monitor the algebraic forms we plot in logarithmic scales $\ell n \, \phi^{AB}$ vs. $\ell n \, n$. In Fig. 9 we have indicated through dashed lines the best fitted slopes to the decays as dashed lines. It is now evident by

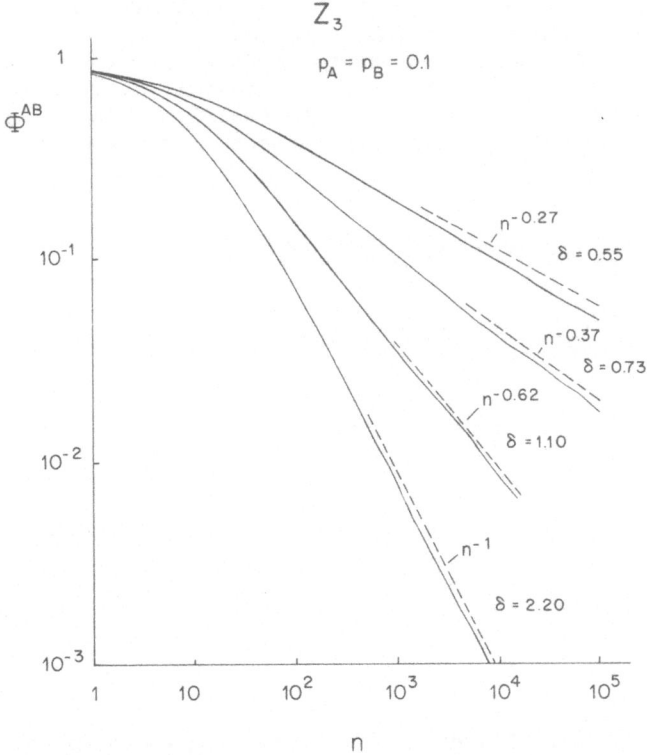

Figure 9. Decay pattern of the $A + B \rightarrow 0$, $A_0 = B_0$ reaction on the UMS Z_3. The initial number densities are $p_A = p_B = 0.1$. The temperature parameter Δ/kT varies from 0.55 to 2.2. The slopes for large n are indicated by dash-dashed lines. (Reproduced from Ref. [23])

inspection that for $\delta = 0.55$ and for $\delta = 0.73$ the fitted exponent to the algebraic decay is around $\delta/2$, i.e. δ is only one <u>half</u> of its value for the A + A → 0 reaction. On the other hand, for quite large δ, such as $\delta = 2.2$, we again recover the $1/t$ kinetic form.

The findings of Fig. 9 can again be understood from the fact that in the strictly bimolecular scheme in the course of the reaction large regions containing only A- or only B-molecules appear. At long times the decay proceeds more slowly, since only molecules at the boundary between the A- and the B-regions can react. We document this for UMS in Fig. 10. In this Figure we present the course of a A + B → 0, $A_0 = B_0$ reaction on Z_3 , where the five patterns are snapshots after $n = 1, 50, 9 \cdot 10^2, 3 \cdot 10^4$ and $2 \cdot 10^5$ steps. The UMS sites are arranged horizontally; shown is a fifth of the Z_3 structure at the 12th hierarchical stage, as in Fig. 9. Initially, the A and B particles are randomly distributed with $p_A = p_B = 0.1$. The course of the reaction magnifies the fluctuations; from a rather homogeneous initial situation one has at later stages well-separated A- and B-domains.

To conclude this section devoted to bimolecular reactions on regular lattices, fractals and UMS, also under continuous-time conditions, we note that several microscopic models of bimolecular type lead to power-law forms, Eq. (84) to (90). In general, such decays may be well distinguished from other relaxation behaviors, such as stretched exponentials, by a sufficiently large dynamical range of measurements.

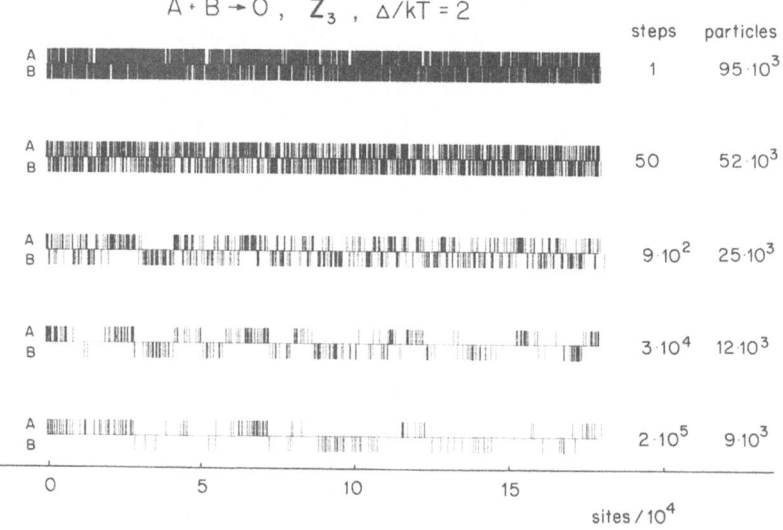

Figure 10. Evolution of the A + B → 0, $A_0 = B_0$ reaction on the UMS Z_3. The snapshots are taken after the number of steps n indicated. The UMS is arranged horizontally and the A and B positions are indicated as up and down dashes, respectively. Displayed is one-fifth of the Z_3 structure at the twelfth stage. (Reproduced from Ref. [23])

However, distinguishing between different power-law decays may not be easy. Thus, additional experimental information such as concentration and temperature dependence may be necessary in order to pinpoint the microscopic relaxation behavior of a specific material which displays a power-law form.

7. CONCLUSIONS

In this work we have emphasized the fractal description of disorder. Randomness as found in glasses, displays not only spatial but also temporal and energetic facets. We have found it advisable to use models tailored to disorder. Thus spatial randomness may be modeled through fractal structures, temporal randomness through waiting-time distributions in the framework of continuous-time random walks (CTRW) and energetic randomness through ultrametric spaces (UMS). The advantage of these models is that they are very flexible and mathematically tractable.

The dynamics were exemplified through pseudo-unimolecular and bimolecular chemical reactions. The ordered state in chemical kinetic schemes is the one given by the 'well-stirred reactor' in which the spatial distribution of reactants is completely homogeneous. To such a distribution correspond clear-cut, simple relaxation patterns: the exponential decay for pseudo-unimolecular, the $1/t$ form for bimolecular reactions.

As we have shown, deviations from such relaxation behaviors are widespread, and extend from stretched-exponential to algebraic decays. Moreover, one may readily obtain such decay forms from theoretical approaches which include randomness. In all cases considered we were able to show the decay laws for the pseudo-unimolecular and for the bimolecular reactions investigated, either in analytically closed form or as a result of computer simulations. Disorder helps to differentiate between several reactions schemes, whose decays are similar under 'well-stirred' conditions. Thus the target problem, in which the minority species is stationary, has a more rapid decay than the trapping problem, in which only the minority species moves. The $A + A \rightarrow 0$ reaction has a marginal dimension of 2, whereas the marginal dimension of the $A + B \rightarrow 0$, $A_0 = B_0$ strictly bimolecular reaction is 4.

Underlying the basic models for disorder are scaling (fractal) symmetries, which lead to unexpected connections. Thus, the δ parameter which relates the temperature, the activation energies and the branching ratio of a UMS has connotations of an effective spectral dimension, $\delta = \tilde{d}/2$ for a related fractal space. The same parameter δ is immediately reinterpretable as determining a waiting-time distribution in the CTRW picture. Furthermore, CTRW processes may be applied to fractals and UMS; generally the CTRW parameter γ and the spectral dimension $\tilde{d}$ (or, equivalently, the UMS parameter δ) combine multiplicatively in the relaxation patterns: the processes subordinate.

In our opinion, the main challenge, the open frontier, is the judicious application of these findings to experiments. For this a close cooperation between theory and experiment is absolutely mandatory.

50

ACKNOWLEDGMENTS

The authors are indebted to Professors K. Dressler, J. Friedrich and D. Haarer for many stimulating discussions. A.B. acknowledges the cooperation of Dipl. Phys. G. Köhler in establishing the recent findings on UMS. The support of the Deutsche Forschungsgemeinschaft (SFB 213) and of the Fonds der Chemischen Industrie and grants of computer time from the ETH-Rechenzentrum are gratefully mentioned. Permission to reproduce the figures as indicated is kindly acknowledged.

REFERENCES

1. S. Chandrasekhar, Rev. Mod. Phys. 15, 1 (1943) and ref. mentioned, such as M. von Smoluchowski, Phys. Z. 17, 557, 585 (1916); Z. Phys. Chem. (Leipzig) 92, 129 (1917).
2. D.F. Calef and J.M. Deutch, Ann. Rev. Phys. Chem. 34, 493 (1983).
3. B. Ya Balagurov and V.G. Vaks, Zh. Exp. Theor. Fiz. 65, 1939 (1973). [English translation Sov. Phys. JETP. 38, 968 (1974)].
4. A.A. Ovchinnikov and Ya. B. Zeldovich, Chem. Phys. 28, 215 (1978).
5. M.D. Donsker and S.R.S. Varadhan, Commun. Pure Appl. Math. 28, 525 (1975); 32, 721 (1979).
6. D. Toussaint and F. Wilczek, J. Chem. Phys. 78, 2642 (1983).
7. G. Zumofen, A. Blumen, and J. Klafter, J. Chem. Phys. 82, 3198 (1985).
8. P.W. Anderson in 'Ill-Condensed Matter' (eds R. Balian, R. Maynard and G. Toulouse), North Holland, Amsterdam (1979), p. 162.
9. T.S. Kuhn, 'The Structure of Scientific Revolutions' (2nd edn), Univ. of Chicago Press, Chicago (1970).
10. B.B. Mandelbrot, 'The Fractal Geometry of Nature', Freeman, San Francisco (1982).
11. K.J. Falconer, 'The Geometry of Fractal Sets', Cambridge Univ. Press (1985).
12. H. Scher and M. Lax, Phys. Rev. B7, 4491 (1973).
13. H. Scher and M. Lax, Phys. Rev. B7, 4502 (1973).
14. H. Scher and E.W. Montroll, Phys. Rev. B12, 2455 (1975).
15. G. Pfister and H. Scher, Adv. Phys. 27, 747 (1978).
16. E.W. Montroll and M.F. Shlesinger in 'Nonequilibrium Phenomena II: From Stochastics to Hydrodynamics' (eds J.L. Lebowitz and E.W. Montroll), North Holland, Amsterdam (1984).
17. N. Bourbaki, 'Eléments de mathématique, Topologie générale', Chap. IX, CCLS, Paris (1974).
18. A.D. Gordon, 'Classification', Chapman and Hall, London (1981).
19. W.H. Schikhof, 'Ultrametric Calculus', Cambridge Univ. Press (1984).
20. R. Rammal, G. Toulouse, and M.A. Virasoro, Rev. Modern Phys. 58, 765 (1986)
21. A. Blumen, G. Zumofen, and J. Klafter in 'Structure and Dynamics of Molecular Systems' (eds R. Daudel et al.), Reidel, Dordrecht (1985), p. 71.
22. A. Blumen, J. Klafter, and G. Zumofen in 'Optical Spectroscopy of Glasses', (ed. I. Zschokke), Reidel, Dordrecht (1986), p. 199.

23. A. Blumen, G. Zumofen, and J. Klafter, Ber. Bunsenges. Phys. Chem. 90, 1048 (1986).
24. R. Kohlrausch, Ann. Phys. (Leipzig) 12, 393 (1847).
25. G. Williams and D.C. Watts, Trans. Faraday Soc. 66, 80 (1970).
26. G. Williams, Adv. Polymer Sci. 33, 59 (1979).
27. R. Richert and H. Bässler, Chem. Phys. Lett. 118, 235 (1985).
28. V.L. Vyazovkin, B.V. Bol'shakov, and V.A. Tolkatchev, Chem. Phys. 75, 11 (1983).
29. T. Doba, K.U. Ingold, and W. Siebrand, Chem. Phys. Lett. 103, 339 (1984).
30. A. Plonka, J. Kroh, W. Lefik, and W. Bogus, J. Phys. Chem. 83, 1807 (1979).
31. J. Friedrich and D. Haarer, Angew. Chem. Int. Ed. Engl. 23, 113 (1984).
32. J. Friedrich and A. Blumen, Phys. Rev. B32, 1434 (1985).
33. J. Tauc, Semicond. Semimet. 21B, 299 (1984).
34. G.H. Weiss, Sep. Sci. Techn. 17, 1609 (1982-83).
35. M.N. Barber and B.W. Ninham, 'Random and Restricted Walks', Gordon and Breach, New York (1970).
36. G.H. Weiss and R.J. Rubin, 'Random Walks: Theory and Selected Applications', Adv. Chem. Phys., 52, 363 (1983).
37. Papers presented at the Symposium on Random Walks, J. Stat. Phys. 30, No. 2 (1983).
38. M.F. Shlesinger and B.J. West (eds), 'Random Walks and their Applications in the Physical and Biological Sciences', Amer. Inst. Phys., New York (1984).
39. G. Zumofen and A. Blumen, Chem. Phys. Lett. 88, 63 (1982).
40. H.E. Stanley, K. Kang, S. Redner, and R.L. Blumberg, Phys. Rev. Lett. 51, 1223 (1983).
41. K. Kang and S. Redner, Phys. Rev. Lett. 52, 955 (1984).
42. R. Kopelman, P.W. Klymko, J.S. Newhouse, and L.W. Anacker, Phys. Rev. B29, 3747 (1984).
43. J. Klafter, A. Blumen, and G. Zumofen, J. Phys. Chem. 87, 191 (1983).
44. E.W. Montroll and J.T. Bendler, J. Stat. Phys. 34, 129 (1984).
45. M.F. Shlesinger and E.W. Montroll, Proc. Natl. Acad. Sci. USA 81, 1280 (1984).
46. S. Redner and K. Kang, J. Phys. A17, L451 (1984).
47. A. Blumen, G. Zumofen, and J. Klafter, Phys. Rev. B30, 5379 (1984).
48. A. Blumen, G. Zumofen, and J. Klafter, J. Physique Colloque, Tome 46, C7-3 (1985).
49. G. Köhler and A. Blumen, J. Phys. A20, 5627 (1987).
50. E.W. Montroll and G.H. Weiss, J. Math. Phys. 6, 167 (1965).
51. G. Zumofen and A. Blumen, J. Chem. Phys. 76, 3713 (1982).
52. S.H. Glarum, J. Chem. Phys. 33, 639 (1960).
53. P. Bordewijk, Chem. Phys. Lett. 32, 592 (1975).
54. J.E. Shore and R. Zwanzig, J. Chem. Phys. 63, 5445 (1975).
55. J.L. Skinner, J. Chem. Phys. 79, 1955 (1983).
56. B.B. Mandelbrot, 'Les objets fractals: forme, hasard et dimension', Flammarion, Paris (1975).
57. W. Sierpinski, Compt. Rend. (Paris) 160, 302 (1915); 162, 629

52

(1916); 'Oeuvres Choisies' (S. Hartman et al.), Editions
Scientifiques, Warsaw (1974).
58. P. Urysohn, Verh. Koning. Akad. Wetensch. (Amsterdam) 1. Sect. 13, 4
(1927).
59. S. Alexander and R. Orbach, J. Phys. Lett. 43, L625 (1982).
60. G. Zumofen, A. Blumen, and J. Klafter, this book.
61. A. Blumen, J. Klafter, and G. Zumofen, Phys. Rev. B28, 6112 (1983).
62. J. Klafter, A. Blumen, and G. Zumofen, J. Stat. Phys. 36, 561
(1984).
63. R. Hilfer and A. Blumen in 'Fractals in Physics' (eds L. Pietronero
and E. Tossatti), North Holland, Amsterdam (1986), p.33.
64. A. Blumen and G. Zumofen, J. Chem. Phys. 77, 5127 (1982).
65. M.F. Shlesinger, J. Stat. Phys. 36, 639 (1984).
66. A. Blumen, J. Klafter, and G. Zumofen in 'Fractals in Physics' (eds
L. Pietronero and E. Tossatti), North Holland, Amsterdam (1986), p.
399.
67. A. Blumen, J. Klafter, B.S. White, and G. Zumofen, Phys. Rev. Lett.
53, 1301 (1984).
68. G. Zumofen, A. Blumen, and J. Klafter, J. Chem. Phys. 84, 6679
(1986).
69. A. Blumen, J. Klafter, and G. Zumofen, J. Phys. A19, L77 (1986).
70. A. Blumen, G. Zumofen, and J. Klafter, J. Phys. A19, L861 (1986).
71. S. Grossmann, F. Wegner, and K.H. Hoffmann, J. Physique Lett. 46,
L575 (1985).
72. A.T. Ogielski and D.L. Stein, Phys. Rev. Lett. 55, 1634 (1985).
73. B.A. Huberman and M. Kerszberg, J. Phys. A18, L331 (1985).
74. C.P. Bachas and B.A. Huberman, Phys. Rev. Lett. 57, 1965 (1986).
75. P.W. Klymko and R. Kopelman, J. Phys. Chem. 87, 4565 (1983).
76. P. Argyrakis and R. Kopelman, J. Chem. Phys. 83, 3099 (1985) and
references therein.
77. D.C. Torney and H.M. McConnell, Proc. Roy. Soc. London A387, 147
(1983).
78. D.C. Torney and H.M. McConnell, J. Phys. Chem. 87, 1941 (1983).
79. A. Blumen, G. Zumofen, and J. Klafter in 'Photoreaktive Festkörper'
(eds H. Sixl et al.), Wahl Verlag, Karlsruhe, 1984.
80. L.W. Anacker and R. Kopelman, Phys. Rev. Lett. 58, 289 (1987).
81. P. Meakin and H.E. Stanley, J. Phys. A17, L173 (1984).

FRACTAL CHARACTER OF CHEMICAL REACTIONS IN DISORDERED MEDIA

Panos Argyrakis
Department of Physics 313-1
University of Thessaloniki
GR-54006 Thessaloniki, Greece

ABSTRACT. The concept of fractals, as it is applied to problems in physical chemistry, is presented. A short discussion is given to identify these structures, then a short review of the work pertaining to fractal transport mechanisms of single particle diffusion. The more recent problem of chemical reactions on fractals is dealt more at depth. We discuss bimolecular reactions of the A+A and A+B types. We show how the diffusion-limited mechanism affects normal reactions on regular and on fractal lattices. Stirring of such reactions seems to provide a quantitative way for controlling their rate, and in a sense provide a new approach to catalysis. Finally, some related applications to fractal chaotic motion in polyatomic molecules, and entropy functions as a measure of disorder are also briefly discussed.

1. INTRODUCTION

Fractals have provided practically all natural sciences with insight for solutions to problems that up to recently were considered to be too complex to be solved. Problems in physical chemistry certainly have not lagged behind in this curious endeavor. Some areas of interest that are discussed here include: Energy transfer in mixed crystals, intramolecular energy redistribution in polyatomic molecules, exciton-exciton annihilation events, and more recently, mechanisms of chemical reactions. This list is by no means exhaustive, but makes only a small fraction of the reported applications.

1.1. Fractals

What is a fractal? A fractal is a geometrical structure that at first look seems to be too complicated, irregular, and

53

A. Amann et al. (eds.),
Fractals, Quasicrystals, Chaos, Knots and Algebraic Quantum Mechanics, 53–64.
© 1988 by Kluwer Academic Publishers.

random. When carefully viewed one begins to realize the presence of tractable properties that are inherent in it, and help us to systematically study them. There have been several introductions and/or reviews on this subject [1-4], where one could find a plethora of fractal figures, shapes, clusters, aggregations, etc. There are two categories of fractals, both well studied: The deterministic fractals, which are exact and repeatable structures, such as the Sierpinski gasket [for a figure see Ref. 2] and the random fractals, such as a percolation cluster (made of a lattice with randomly placed open and closed sites) or an aggregated structure (made by repeated addition of identical units to a core), etc. The common feature that all these structures possess is that they do not occupy the entire underbedding space (such as, for example, a molecular crystal), but leave a large number of blanks. However, the amount and arrangement of occupied and blank space obey some relations that make fractals useful. The most important is that of self-similarity, i.e. a fractal "looks alike" under any scale of magnification, the only limit being the size of the unit cell that makes this structure. For random fractals this is true only in the statistical sense (average of many realizations). The result is that a fractal structure that "sits" on a 2-dim space will have a dimension D that is less than 2. The Sierpinski gasket (a structure made of three equilateral triangles stacked together to make a larger equilateral triangle) has a fractal dimension of $D=\ln3/\ln2 = 1.58$. For a percolation cluster at the critical point the fractal dimension is found by considering several sections of the lattice with a different linear size λ each time, and then by calculating the number of sites M that belong to this cluster in each section. The fractal (Hausdorff) dimension D_f comes in the relation:

$$M \sim \lambda^{D_f} \tag{1}$$

and thus:

$$D_f \sim \ln(M(\lambda))/\ln(\lambda) \tag{2}$$

$D_f=1.89$ for a 2-dim lattice, while $D_f=2.5$ for a 3-dim simple cubic lattice [5,7]. We see that the dimensionality now has a non-integer value. This does not agree with our intuitive notion of dimensions, but we become accustomed to it when we see their physical significance and implications in the following sections.

1.2. Diffusion via random walks

The diffusion of a single particle in space is one of the most studied, interesting and intriguing problems in physics. A pertinent model that has been heavily used is that of the random walk, sometimes called the ant-in-the-labyrinth, or the drunk man's walk. One uses mathematical properties, such as the number of distinct sites visited at least once in a t-step walk, S_t, the mean-square displacement, R_t^2, to monitor such a process. Up to 1982 these models provided satisfactory answers [6], but only for homogeneous space, as for example is a perfect crystal lattice. It was then recognized [7] that even in an inhomogeneous space a relation may be found, if this space has a fractal dimension. The conjecture originally proposed [7] was that S_t scales as:

$$S_t \sim t^{D_s/2} \tag{3}$$

while R_t^2 scales as:

$$R_t^2 \sim t^{2/(2+\theta)} \tag{4}$$

D_s is a new dimension called the spectral dimension, while $(2+\theta)$ is the diffusion constant dimension. These laws were studied in detail [8,9 and references therein], and showed that $D_s=1.30\pm0.02$ (d=2) and $D_s=1.33\pm0.02$ (d≥3). The θ exponent was shown to be $\theta=0.89\pm0.05$. All details of this work have already been reported in the past, and will not be presented here. The important point ot remember is that diffusion properties behave predictably on fractals, and the large volume of work reported provides very accurate answers to all these questions.

1.3. Fractal Entropy

The quantities S_t or R_t^2 described above give the overall range of the random walk, while it makes no difference how many times has a particle visited the same site. An additional piece of information can come, however, from the occupational frequency for each site. We introduce the quantity i_k, which is the number of times that each site k has been visited in the random walk. The probability P_k of visiting the kth site is: $P_k=i_k/t$. Then:

$$I_t = -\sum_{k=1}^{S_t} P_k \ln P_k \tag{5}$$

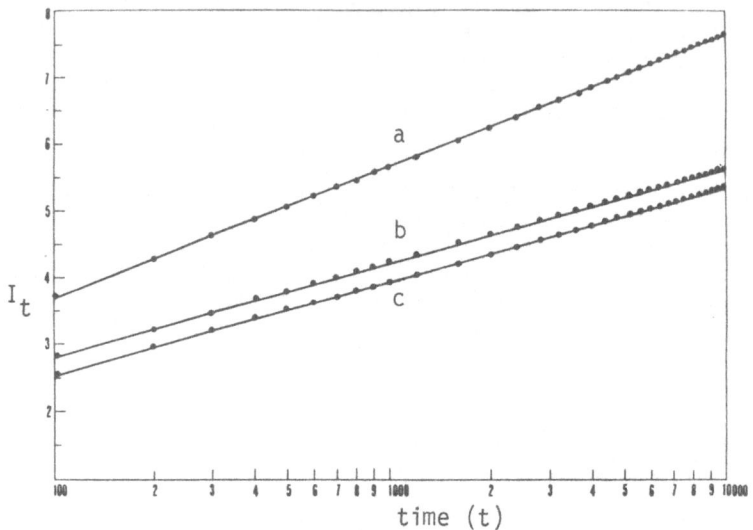

time (t)

Figure 1. Plot of the I(t) as a function of time for (a)square lattice, (b)percolation cluster (myopic ant model), (c)percolation cluster (blind ant model). Lattice size: 300x300, average after 1000 realizations.

is the function containing this new information, which due to the form PlnP bears out an entropy-like character [10]. It is conjectured that for a random walk this function shows scaling properties in the same spirit as the S_t scaling law. We now define D_I:

$$D_I = I_t/\ln t \qquad\qquad (6)$$

as the information dimension, in parallel with the fractal and spectral dimensions discussed above. The value of this new exponent is derived from a plot of I_t as a function of time t, given in Figure 1. In this plot I_t is first calculated from the i_k function. Then it is plotted as a function of time. For comparison purposes the perfect lattice case is also included (line a). For the fractal lattice the calculation is done using the myopic ant [11] model (line b), and the blind ant model (line c). The least-squares values of the slopes of the straight lines give: $D_I=0.89\pm0.02$ (perfect lattice) and $D_I=0.62\pm0.02$ (fractal lattice). For the perfect lattice the dimension D=0.89, instead of the expected D=1.00, due to the well known logarithmic and other correction terms [12]. For the fractal lattice, as given above, the exponent is $D_s/2$, i.e. ≃0.65, which is in close agreement with this new result. Thus, the information dimension presents an alternate approach to the well-known sets of fractal exponents for random walk problems.

1.4. Fractals and chaos

The scaling-law-type relation that gives the dimension of a process, as discussed above, may be applied to the complex problem of intramolecular vibrational energy redistribution in a polyatomic molecule. It is well known [10] that given certain initial conditions the associated atomic motion may well lead (using classical mechanics methods) to periodic or chaotic behavior, or anywhere in-between these two extremes. Preliminary work [13] shows that a fractal exponent resulting from such a scaling law is quite appropriate in quantitatively describing the onset of chaos. As with previous examples this exponent has non-integer values in the crossover region, while it is an integer in the completely periodic and chaotic regimes. However, this is not the focus of this paper, and this subject will not be further discussed here. Several other papers in this volume deal primarily with this area.

1.5. Chemical Kinetics

The usual laws in chemical kinetics equations make the implied assumption that the reactants are free to move in a homogeneous 3-dim space. The main two questions are now addressed in the following way: Suppose that in a reaction the motion of the reacting species is constrained in a space that has lower dimensionality than usual, as is the case with fractal structures. (1)How are, then, these laws modified to properly describe the new reaction rate? (2)What, if anything, can one do to influence this new behavior? Since the reaction in these cases is severely limited by the diffusion process we call such reactions Diffusion-Limited Reactions (DLR).

This effect has easily been observed in the following equations. For the simple A+A reaction the rate equation is: $-d\rho/dt=k\rho^2$, and its solution is: $\rho^{-1} - \rho_0^{-1}$ =kt, where ρ_0 is the initial density at t=0. For this same reaction on a percolating cluster at the critical threshold point calculations have shown [14] that the solution is: $\rho^{-1} -\rho_0^{-1}=(k/f)t^f$, where f≃0.65, i.e. f≈$D_s$/2. This result immediately hints that one-particle motion shows the same space exploration characteristics as multiple reacting particles. This is shown in Figure 2 where time is eliminated, and one plots the single particle property S_t

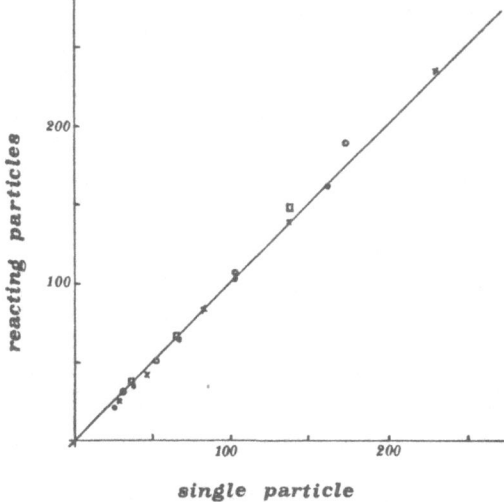

reacting particles

single particle

Figure 2. Single particle S(t) vs. reacting particles $(1/\rho - 1/\rho_o)$ for 2-dim lattices. Filled circles: C=0.60, x: C=0.65, cicles: C=0.70, squares: C=0.80. Only the larest cluster is used. Several time intervals are considered up to t=2000 steps.

vs. the reacting particle density function $(\rho^{-1} - \rho_o^{-1})$ for a range of cluster concentrations. This plot is after 2000 steps, but the same characteristics are seen for any t we tried. The straight line confirms the above assumption [15].

2. METHOD OF CALCULATION

The techniques for computer simulations for a single particle diffusion have been previously discussed in detail [8,9]. The most complicated part is that of the generation of the percolating cluster, which has been well described in the past. The random walk is a rather easy process to simulate, even with a personal microcomputer. In the reaction kinetics the same original method is utilized as with a single particle. The difference is that at time t=0 a certain initial density of particles is placed on the lattice and all particles move (one at a time) one step before an overall time-step is consumed. Only one particle is allowed to occupy each site. Upon encounter, if two particles occupy the same coordinates both are immediately removed from the lattice. Cyclic boundary conditions are employed. Usually the initial density is 0.05 particles/ site. The results reported here pertain to transient kinetics [14], since the only participating reactants are gererated at t=0. Also of interest is the case of steady-state kinetics [16], where a steady source of particles keeps replenishing the reactants so that the rate of reaction is constant, and the reaction is at equilibrium. It seems that the transient and steady-state cases have certain differences that are quite intriguing [16].

3. RESULTS

3.1. Kinetics via Diffusion

The effect of dimensionality on the reaction rate is shown in a plot of $(1/\rho - 1/\rho_o)$ as a function of time, for several dimensionalities, as shown in Figure 3 (a and b). Here, in a log-log plot we derive the f values from the slopes of the straight lines as:

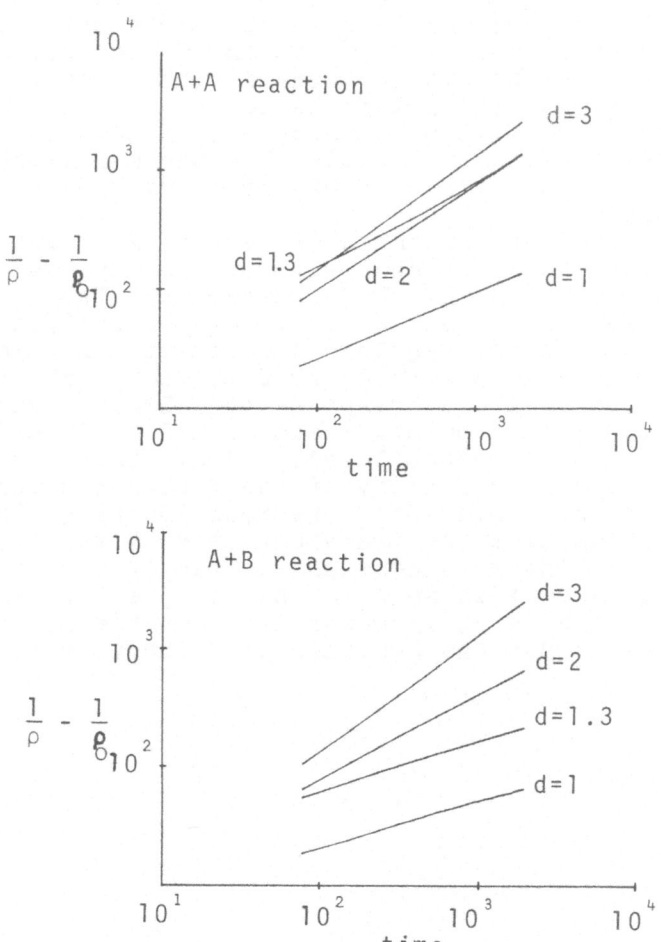

Figure 3. (a)Top part (b)Bottom part.

Table 1
f values

            A + A   A + B
----
d=3   0.98   1.00
d=2   0.89   0.72
d=1.3   0.74   0.42
d=1   0.57   0.38

These slopes are for early time behavior (t goes to 2000 steps here). It is expected that in the asymptotic limit of long times these f values may be somewhat different. One, of course, runs here to the finite-size-lattice problem, i.e. there are too few particles left in the long time limit for good statistics. To improve this one needs to go to extremely large lattices, but again this becomes impractical for calculations, and thus there is a compromise between these two trends.

We notice here that in these plots the $1/\rho$ factor is included, unlike some recent work where the simple ρ-function was plotted [17,18]. We observe in Table 1 the dramatic drop from the classical value of $f \approx 1$ (d=3) as we go down to d=1, as both regular and fractal lattices are examined. The d=1,2,3 correspond to dimensions of regular lattices, while the d=1.3 corresponds to a 2-dim percolation cluster at the critical point, which is C=0.60 (C being the open site occupational probability). From this trend we conclude that the dimensionality of the reactant space plays the most important role for the reaction rate. The same trend is seen if we plot the "rate constant" K as a function of time, Figure 4. The rate constant is simply the function $K=(1/\rho - 1/\rho_0)/t$, and it is seen in this figure that only for the cubic lattice K is faily constant. For the percolating lattices we clearly see the decrease as a function of time.

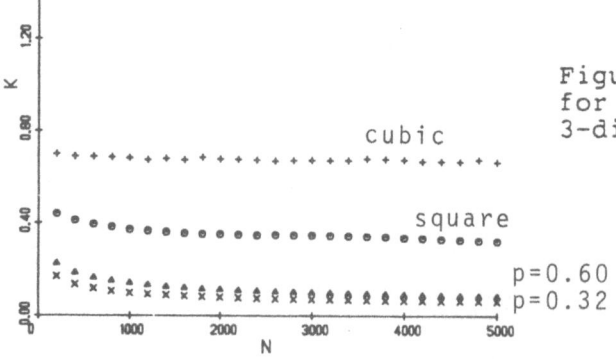

Figure 4. Rate constant for DLR A+A for 2-dim and 3-dim lattices as shown.

Thus, the conclusion is that for fractal lattices there does not exist any diffusion constant for a reaction.

Also of interest for the A+B reaction is the question of segregation of the A from the B species during the course of the reaction, something that does not show in the f exponent. There are some very good pictorials showing this trend [16,17]. Even though this segregation is quite natural, it has not been explained yet quantitatively. It is nevertheless important in our understanding of the reaction mechanism, and some effort is placed in this direction. Establishing certain universal quantitative criteria for segregation is not an obvious process.

3.2. Kinetics with stirring

Mixing of reactants during the course of a reaction is not a new idea. It has espesially been useful in continuous flow systems [19]. In most reactions mixing occurs internally with the course of time, and this constitutes a homogeneous reaction. This is not the case, however, for the reactions described in the previous section (DLR), which due to the effective lower dimensionality show more pronounced clustering properties, segregation of reactants, and reduced reaction rates. One realizes that these effects are due to the spatial and temporal correlations that are continuously been built in the reaction system. Such correlations are a product of -the constraints imposed by the space heterogeneity. It is natural, therefore, to attempt to break up such effects, and hopefully be able to control a reaction and make it more homogeneous. Such a break-up is attained by fully stirring (mixing) all reactants during the course of reaction. All particles rerandomize their positions on the lattice without any knowledge of their previous position. We call this the well stirred case. We observe as a result in Figure 5 that the "rate constants" of Figure 4 are indeed constant now that full stirring takes place, even for lattices of lower dimensionality. The f values of a well-stirred reaction is f≈1, regardless of dimensionality, for normal and fractal lattices [20].

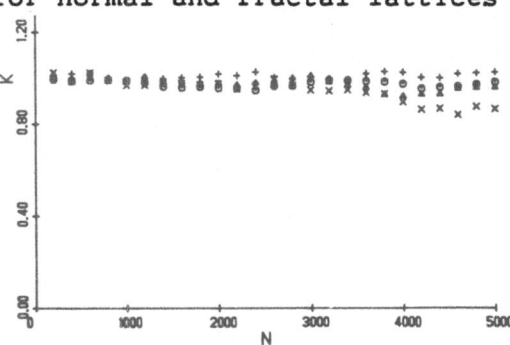

Figure 5. Rate constant for well-stirred reaction A+A for 2-dim and 3-dim lattices as shown.

More interesting is the case of partial stirring that bridges the two extremes of the diffusion-limited and the well-stirred reactions. Figure 6 shows the effect of partial stirring for a 1-dim A+A reaction. Here only a certain small percentage of all remaining particles is stirred continously. When this percentage is small (1%) the case is similar to the diffusion-limited reaction. As it goes up we approach the case of a well-stirred reaction. This crossover can be followed by the change in the slopes of these lines. In Figure 7 we present the case of local stirring. Here all particles are stirred continously, but only close to their previous positions, say ±5 sites

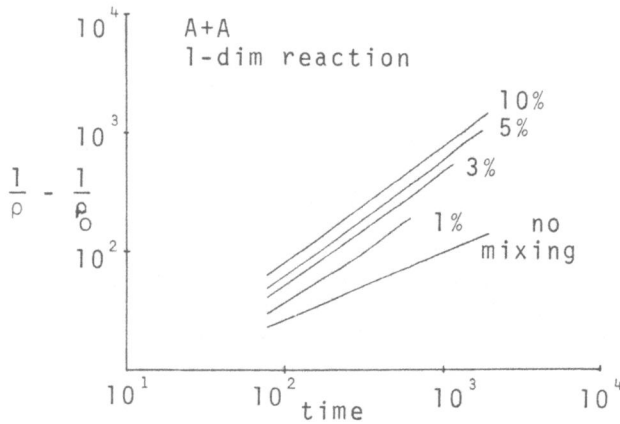

Figure 6

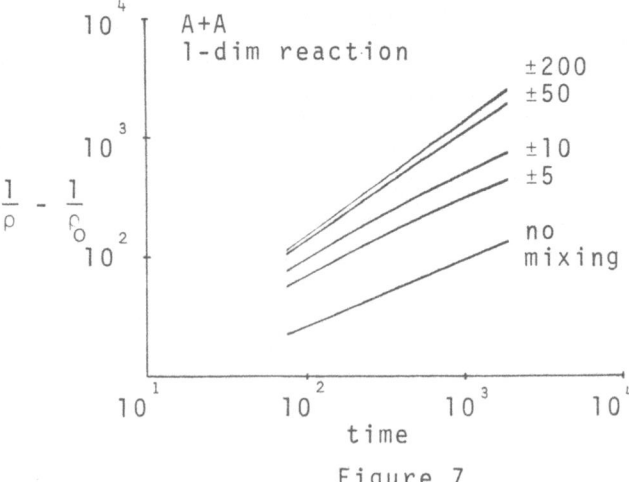

Figure 7

away. As this number increases to ±200 sites we approach the case of well-stirred reactions.

4. CONCLUSIONS

Fractals have provided a new tool in studies of disordered chemical systems. They re-define the concept of dimensionality of a system by making it more general. They bring out an "inherent symmetry" that random inhomogeneous systems possess, albeit not apparent, and in a way they bridge the gap between crystalline and amorphous materials. Chemical reactions on fractals depend on the diffusion-limited process of particle motion, which is a dimensionality-dependent process. The new equations that pertain now have a fractal dimension dependence. The diffusion-limited process can, however, be controlled with external mixing. By carefuly choosing a mixing mechanism one can determine at will the rate of reaction (between the two limits of DLR anf total mixing). Thus, this may offer a new way of catalyzing chemical reactions in constrained spaces.

REFERENCES

(1) B.B.Mandelbrot, The fractal geometry of nature, Freeman, San Francisco, 1982.

(2) R.Pynn and A.Skeltorp, Scaling phenomena in disordered systems, Plenum, N.Y., 1985

(3) L.Pietronero and E.Tosatti, Fractals in Physics, Springer, Berlin, 1986.

(4) S.H.Liu, Sol. St. Phys., $\underline{39}$,207(1986).

(5) R.Orbach, Nature, $\underline{231}$,814(1986).

(6) E.W.Montroll and G.H.Weiss, J. Math. Phys., $\underline{6}$,167(1965).

(7) S.Alexander and R.Orbach, J. Phys. (Paris), $\underline{43}$,L625(1982);R.Rammal and G.Toulouse, J. Phys. (Paris), $\underline{44}$,L13(1983).

(8) P.Argyrakis and R.Kopelman, J. Chem. Phys., $\underline{81}$,1015(1984); J. Chem. Phys., $\underline{83}$,3099(1985).

(9) P.Argyrakis, in Structure and Dynamics of Molecular Systems, R.Daudel, J.P.Korb, J.P.Lemaistre and J.Maruani, eds, D.Reidel Publ. Co., p. 209(1986).

(10) B-L. Hao, Chaos, World Scientific Publ. Co.,

64

Singapore, 1984.

(11) C.D.Mitescu and J.Roussenq, Ann. Israel Phys. Soc., 5,81(1983). These two models are briefly explained here: In the blind ant model when the particle is on the perimeter of the percolation cluster an attempt to move outside this cluster consumes one time-step. In the myopic ant model such an attempt consumes no time at all, a new attempt is made until the particle moves to a regular cluster site.

(12) G.Zumofen and A.Blumen, J. Chem. Phys., 76,3713(1982).

(13) P.Argyrakis and S.Farantos, To be published.

(14) J.S.Newhouse, P.Argyrakis, and R.Kopelamn, Chem. Phys. Lett., 107,48(1984).

(15) P.Argyrakis and R.Kopelman, in Advances in Chemical Reaction Dynamics, P.Rentzepis and C.Capellos, eds, D.Reidel, p. 339 (1986).

(16) L.W.Anacker and R.Kopelman, Phys. Rev. Lett., 58,289(1987).

(17) D.Toussaint and F.Wilczek, J. Chem. Phys., 78,2642(1983).

(18) K.Kang and S.Redner, Phys. Rev. A 32,435(1985).

(19) E.B.Nauman and B.A.Buffham, Mixing in continous flow systems, J.Wiley, 1983.

(20) P.Argyrakis and R.Kopelman, J. Phys. Chem., 91,2699(1987).

REACTION KINETICS FOR DIFFUSION CONTROLLED AGGREGATION

F. Leyvraz
Instituto de Fisica-UNAM
Apdo. postal 20-364
01000 Mexico D.F.
Mexico

ABSTRACT. Irreversible coagulation (or aggregation) processes are described first by rate equations, for which a scaling theory is described. It is then argued that the range of validity of this description does not necessarily include the case where diffusion is the rate-limiting step. A simplified model to simulate this latter case is then described and shown to deviate from the predictions of the rate equations if the space dimension d (or the spectral dimension d, for fractal substrates) is less than or equal to two. Finally, some open problems in the treatment of more realistic models are discussed.

1. INTRODUCTION

Irreversible growth phenomena have attracted considerable interest in recent years[1], both due to their quite common occurrence in the most varied fields of physics and to the wealth of remarkable and often unexpected behavior they display. Among these, irreversible coagulation, i.e. growth of clusters by bonding of two smaller clusters, has been thoroughly investigated. The reaction scheme can be given as follows:

$$A_i + A_j \xrightarrow[K(i,j)]{} A_{i+j} \qquad (1)$$

where the A_i denote clusters of mass i, or i-mers, and the $K(i,j)$ denote size-dependent reaction rates. This type of reaction is of central importance in various fields of physics:
-cloud physics[2], where coalescence of droplets plays an important role.
-colloid and aerosol science[3], since it is well-known that such particles, when coming in very close contact, usually form a bond due to van der Waals interaction, that is frequently strong enough to be considered irreversible.
-chemistry, in the study of polymerization[4], in particular gelation.
Furthermore, it has come up in such fields as astrophysics[5] (coalescence of gas clouds), dairy research (milk curdling), coagulation of blood[6] as well as numerous other applications.
 The study of coagulation is itself divided in two rather distinct parts: the study of the cluster size distribution and the study of the morphology of the individual clusters. In the following, we shall be primarily concerned with the former. It should

A. Amann et al. (eds.),
Fractals, Quasicrystals, Chaos, Knots and Algebraic Quantum Mechanics, 65–72.
© 1988 by Kluwer Academic Publishers.

be pointed out, however, that there are fascinating problems connected with the latter: in particular, if the shape of the cluster is not allowed to relax after the reaction, it has been found by computer simulations that the clusters thus generated are fractals[7], and that their fractal dimension differs depending on the nature of the cluster size distribution[8]. Such findings have also been reported experimentally[9].

In the following, we will study the time-dependent cluster size distribution, in particular for large times, where the behavior is independent of the initial conditions. In Sec. 2, we will review what is known about the rate equations describing the coagulation process. In Sec. 3, we introduce a model due to Kang and Redner[10] for diffusion-limited coagulation that lends itself well to computer simulations and can also be analyzed rather thoroughly in one dimension. This will show deviations from the predictions of the rate equations. Finally, in Sec. 5, we analyze some open problems associated with more realistic models.

2. THE RATE EQUATIONS AND SCALING

In order to describe the complex phenomena mentioned above, the simplest approach is to use rate equations, that is, we assume that there are no spatial correlations in the system and that the reactions occur with a frequency determined by the concentration of the reactants only. Hence we have, if $c_j(t)$ is the concentration of A_j at time t;

$$\dot{c}_j = \frac{1}{2} \sum_{k=1}^{j-1} K(k, j-k) c_k c_{j-k} - c_j \sum_{k=1}^{\infty} K(j, k) c_k \qquad (2)$$

Clearly, an exact solution of these equations for general $K(i, j)$ is beyond our reach. However, it has turned out to be possible to analyze the behavior of eq. (2) in the limit of both large cluster sizes and large times. In this case, it is presumed that the system has forgotten the initial conditions and that its behavior is therefore only dictated by the large-scale behavior of the $K(i, j)$. More precisely, it has been assumed[12,13] that in this limit the cluster size distribution $c_j(t)$ reduces to a function of a single variable $j/s^*(t)$, where $s^*(t)$ is some measure of the typical cluster size, that is:

$$c_j(t) \simeq s^*(t)^{-2} \Phi \left(j/s^*(t) \right) \qquad (3)$$

where the prefactor $s^*(t)^{-2}$ is necessary to ensure that the total mass in the system does indeed remain of order one as $t \to \infty$.

This is known as the scaling Ansatz and has been extremely fruitful in discussing the large-time behavior of eq. (2). Put in eq. (2), the Ansatz yields an integro-differential equation for $\Phi(x)$[11,12], which can then, in principle, be solved numerically. The crucial advantage of this formulation, however, lies in the fact that the small- and large-x behavior of the function $\Phi(x)$ can be described in considerable detail under very general assumptions on the rate constants $K(i, j)$. This has been done by van Dongen and Ernst[11], who show that two exponents describing the $K(i, j)$ are sufficient for this purpose:

$$\begin{aligned} K(i, j) &\simeq j^{\lambda} &\quad \text{for } i \simeq j \to \infty \\ K(i, j) &\simeq j^{\nu} &\quad \text{for } i \ll j \to \infty \end{aligned} \qquad (4)$$

Summarizing their results, they obtain for λ, $\nu \leq 1$-which is necessary to exclude gelation singularities at finite times[14-16], which do not concern us here-the following growth law for the typical size:

$$s^*(t) \simeq t^z$$
$$z = \frac{1}{1 - \lambda} \tag{5}$$

Further, if $\lambda > \nu$, $\Phi(x)$ has a power-law behavior for small x, indicating a very polydisperse cluster size distribution. One has in this case:

$$\Phi(x) \simeq x^{-\tau} \qquad (x \to 0)$$
$$\tau = 1 + \lambda \tag{6}$$

On the other hand, if $\lambda < \nu$, $\Phi(x)$ has a quasi-exponential behavior for small x, i.e. small clusters are strongly suppressed and the cluster size distribution is clearly peaked around a well-defined typical size. More precisely:

$$\Phi(x) \simeq \exp(-x^{\lambda-\nu}) \qquad (x \to 0) \tag{7}$$

The case $\lambda = \nu$ is far more intricate, and for the details we refer to van Dongen and Ernst[11]. One case, however, is simple enough and should be mentioned. If $K(i,j) = K$ is independent of i and j, the model has been solved exactly by Smoluchowski[17] for monodisperse initial conditions, yielding:

$$c_j(t) = \frac{4}{(Kt+2)^2} \left(\frac{Kt}{Kt+2} \right)^{j-1} \tag{8}$$

which can be rewritten as:

$$c_j(t) \simeq (s^*(t))^{-2} \, \Phi(j/s^*(t))$$
$$\Phi(x) = e^{-x} \qquad s^*(t) = \frac{Kt+2}{2} \tag{9}$$

thus implying $z = 1$ and $\tau = 0$. This also is found to extend to arbitrary initial conditions, if they decay sufficiently fast for large j.

Two other exponents are also frequently studied, as they are easily defined and observed:

$$c_1(t) \simeq t^{-w} \qquad (t \to \infty)$$
$$\sum_{k=1}^{\infty} c_k(t) \simeq t^{-\alpha} \qquad (t \to \infty) \tag{10}$$

Thus, the exponent w describes the long-time decay of very small clusters (compared to the typical size) and α describes the behavior of the total number of clusters. These are related to τ and z by the following relations[13]

$$(2 - \tau)z = w$$
$$\alpha = z \qquad \text{if } \tau \leq 1$$
$$\quad = w \qquad \text{if } \tau \geq 1 \tag{11}$$

as is readily seen from the scaling Ansatz.

Qualitatively, this amounts to saying that a power-law polydisperse size distribution is generated by any aggregation mechanism where coalescences between large clusters occur at a faster rate than between large and small clusters. If the opposite is the case, a well-peaked size-distribution ensues. This observation has been qualitatively confirmed by computer simulations[18,19], as well as experimentally[9] on various systems. Thus fast (i.e. diffusion-limited) aggregation in colloids is usually described in terms of the expression for the rates originally derived by Smoluchowski[17]

$$K(i,j) = (D(i) + D(j))(R(i) + R(j))^{d-2} \qquad (d > 2) \tag{12}$$

where $D(i)$ is the diffusion constant of an i-mer, $R(i)$ is its radius and d the spatial dimension. For this expression, if $D(i) \sim R(i)^{-1}$, one has $\lambda < \nu$ and hence a well-peaked distribution. On the other hand, if coalescence is the rate-limiting step, it is found that large clusters are about as likely to react among each other as with small clusters[20] and a power-law polydisperse cluster size distribution is indeed observed[19]

3. DIFFUSION CONTROLLED REACTIONS: A SIMPLE MODEL

Despite the impressive successes of rate equations in predicting the qualitative features of aggregating systems, the reason for their applicability is often obscure. Indeed, the crucial assumption involved is that no spatial correlations exist in the system. If it is thoroughly stirred on a time scale far less than the typical time needed for two clusters to aggregate, this assumption is quite sensible and the validity of the rate equations can in fact be derived by standard methods. In the case where diffusion is the rate-limiting step, however, this is clearly untenable. Indeed, the reaction generates a void around each growing cluster, which itself will eventually affect the growth. To take these effects into account, Kang and Redner[10] developed the following model, which they called the particle coalescence model (PCM): point particles are put at random on a lattice and perform independent random walks. Each particle originally has mass one. Whenever two particles meet at one lattice point, they combine with probability p to form a particle (still a point) with a mass equal to the sum of the masses of the reacting particles. Clearly, in this model, the rates $K(i,j)$ should be chosen to be size-independent ($K(i,j) = p$). Thus, the exponents predicted by the rate equations are:

$$\tau = 0 \qquad z = 1 \qquad w = 2 \qquad \alpha = 1 \tag{13}$$

This model has been extensively studied both numerically and analytically[10,21,22]. The result is that, in three dimensions, the exponents predicted by the rate equations are indeed correct, but the exact solution given by eq. (8) does not apply. In particular the rate constant K cannot be determined unambiguously from p alone. Rather, it also depends on the structure of the underlying lattice.

In two dimensions, the exponents are again as predicted by the rate equations, but with logarithmic corrections strongly suggesting that $d = 2$ is in fact the upper critical dimension for this model, a fact later proved by Peliti[21]. For $d = 1$, on the other hand, many surprising things occur: the exponents are given by

$$\tau = -1 \qquad z = \frac{1}{2} \qquad w = \frac{3}{2} \qquad \alpha = \frac{1}{2} \tag{14}$$

Further, it is seen that the asymptotic behavior of the various quantities involved does not depend on the probability of reaction p. Thus, in particular, for the total number of clusters, one observes

$$\sum_{k=1}^{\infty} c_k(t) \propto (Dt)^{-1/2} \tag{15}$$

where D is the diffusion constant of the clusters (assumed in this model to be size independent). This behavior can be understood intuitively by noticing that, in time t, a one dimensional random walk visits on the order of $\sqrt{t}$ distinct sites. It therefore presumably visits each of them on the order of $\sqrt{t}$ times. Thus, if two particles come in close contact, they will do so an infinite number of times as time goes on, so that they eventually coalesce irrespective of the value of p.

Analytically, the model has also been studied, in particular in the limiting case where p is equal to one. The main result proved therein concerns the quantity $\sum_{k=1}^{\infty} c_k(t)$, for which it is shown that, for arbitrary d

$$\sum_{k=1}^{\infty} c_k(t) \propto \frac{1}{S(t)} \tag{16}$$

where $S(t)$ is the average number of distinct sites visited by a random walk in d dimensions. This implies, if $\tau \leq 1$, which in this case applies:

$$s^*(t) \propto \left(\sum_{k=1}^{\infty} c_k(t) \right)^{-1} \propto S(t) \tag{17}$$

which, on an intuitive level, is quite satisfactory: it means that a particle, having visited $S(t)$ sites, has grown as much as if it had been the only growing particle, all the others remaining stationary. These rigorous results are also in agreement with the value of the exponent z obtained by a combination of numerical work and plausible arguments.

A generalization of the above model to arbitrary mass dependent reaction rates was also performed[22]. If the particles are to remain pointlike during the growth process, there are essentially two ways of effecting such a mass-dependence of the reaction rates:

i) the probability p for the two reactants to coalesce is made to depend in the required way on the masses i and j of the reactants. This, however, encounters a slight problem: since p clearly must remain less than one, and since the $K(i,j)$ usually grow to infinity as $i, j \to \infty$, it is necessary to multiply the $K(i,j)$ by an appropriate (system size dependent) constant, leading to an increase in computation time and a corresponding difficulty in attaining the asymptotic long-time regime.

ii) for $K(i,j) = Di) + D(j)$, a much simpler method consists in making i-mers diffuse with a size-dependent diffusion constant $D(i)$.

Using the first technique led to the following results: in three dimensions, the exponents predicted by the rate equations were indeed recovered: the rates used were $K(i,j) = (ij)^{\lambda/2}$, so that these exponents were

$$\tau = 1 + \lambda \qquad z = \frac{1}{1 - \lambda}$$
$$\alpha = w = 1 \tag{18}$$

since, in this case $\nu = \lambda/2$. On the other hand, in one dimension, the result appeared not to differ from the exponents for *size-independent* reactions, given in eq. (14). This presumably reflects the p-independence of the one dimensional aggregation process, as was mentioned above. Since the basic time scale is not influenced at all by the value of p, it should perhaps not come as a surprise that mass-dependent reaction rates fail to modify the exponents, which are primarily determined by the properties of one-dimensional diffusion.

If the diffusion properties are of such overrriding importance in the one-dimensional problem, we would expect the second method described above (changing the diffusion coefficients) to give different results. Numerically, the following value for z was obtained for $K(i,j) = i^\lambda + j^\lambda$

$$z = \frac{1}{2 - \lambda} \tag{19}$$

which is indeed a new result, different from both the rate equation prediction as well as from the constant case. Eq. (19) has the following plausible derivation: let us assume that a generalization of equation (17) holds also in this case. In one dimension, one has

$$S(t) = \sqrt{Dt} \tag{20}$$

But, in our case, D is mass-dependent, and hence also time-dependent. We have, for a typical particle, i.e. one which is always the typical size:

$$D(t) = (s^*(t))^\lambda \propto t^{z\lambda} \tag{21}$$

and hence

$$s^*(t) \propto \sqrt{D(t)t} \propto t^{(z\lambda+1)/2} = t^z \tag{22}$$

from which eq. (19) follows. This shows clearly that in one dimension the two apparently equivalent ways of generating size dependent rates are, in fact, profoundly different: an ad hoc modification of reaction probabilities does not affect the exponents, whereas a modification in the mechanism of diffusion affects them strikingly. This is in marked contrast to the three dimensional case, where the exponents predicted by the rate equations are routinely recovered and no difference between the two methods of inducing mass-dependent rates is noticed.

4. OPEN PROBLEMS

From the foregoing, it would at first appear that in three dimensions-the physically usual case-the rate equations describe everything correctly except for a rescaling of the reaction rates. While such a result would be very pleasing, it is subject to several qualifications:

i) the reactions need not always occur in three dimensional space: they may occur, for example, in a disordered medium[23] which may be well modelled by a fractal. Under such circumstances it can be shown (numerically and by plausible arguments[23,24]) that if d_s is the spectral dimension of the fractal one has

$$\sum_{k=1}^{\infty} c_k(t) \propto t^{-d_s/2} \qquad (t \to \infty) \tag{23}$$

if $d_s < 2$. The behavior of the other exponents has not yet been studied to my knowledge.
ii) the PCM does not take into account the fact that clusters grow in size as they
coalesce. Thus, it is applicable as long as the cluster radius is much less than the
typical distance between clusters. As soon as this limitation is dropped, however,
things become considerably more complex. First, it is not allowable to neglect
many-body collisions at large times: this comes from the fact that the probability of
(say) a three-body collision is of the order of ϕ^2, where ϕ is the volume fraction of
aggregates, if those are compact. But the volume fraction remains constant in time, so
that neglecting many-body effects is never justified. If, on the other hand, the
clusters are fractal, ϕ now means an effective volume fraction, which actually grows
with time. But it is well-known that at this stage, the clusters stop growing
fractally[7].

Second, it is not at all clear that in this case the limiting dimension will be
two. The following argument may show the problem: in the PCM, as time goes on, the
distance r between nearest neighbors increases with time. But the probability for two
nearest neighbors to coalesce then goes as $r^{-(d-2)}$, i.e. it goes to zero in the case
where rate equations apply and not otherwise. This is indeed the physical origin of the
validity of the rate equations, since systematic coalescence with nearest neighbors
clearly induces strong correlations. But, in the case we are considering now, both the
distance between nearest neighbors and the diameters of the reacting particles
increase. If the growth is compact, and if the diffusion constant is assumed to be
proportional to the inverse radius, one can calculate that the probability of reacting
with a nearest neighbor remains bounded for all times in any dimension.

Thus, it does not appear warranted to believe that rate equations will still be
valid in three dimensions if the cluster sizes are assumed to grow. Again, if the
growth is fractal, the situation is even worse.
iii) a further limitation of the above remarks is given by the fact that we have only
considered diffusive dynamics. In many real systems convective effects must be taken
into account, introducing many new effects. For the question of the validity of the
rate equations, however, these mechanisms are generally more likely to be well
described by them, as they mix clusters more efficiently than diffusion.

Summarizing, we have described the rate equations approach to aggregation
processes and seen that its qualitative predictions are in good agreement with
observed facts (much as the Curie-Weiss theory of a ferromagnet correctly predicts the
existence of exponents, scaling, etc...). The question then arises as to whether it was
quantitatively reliable (i.e. if the exponents had the right values) particularly in
the diffusion-limited case, where the basic assumptions of the model do not seem to
hold. A simple model (PCM) showed that we could expect rate equations to be correct
above two dimensions, but not below. The mechanism leading to the discrepancies in one
dimension was discussed and the question, as to whether rate equations may rightly be
applied to more complex models remains open.

5. REFERENCES

1. For a recent presentation of the subject, see e.g. *On Growth and Form*, H. E.
Stanley and N. Ostrowski eds., M. Nijhoff, Dordrecht (1986)
2. H. R. Pruppacher and J. D. Klett, *Microphysics of Clouds and Precipitation*, Reidel,
Dordrecht (1978)
3. S. K. Friedlander, *Smoke, Dust and Haze*, Wiley, New-York (1977)

4. P. J. Flory, *Principles of Polymer Chemistry*, Cornell University, Ithaca, New-York (1953); W. H. Stockmayer, J. Chem. Phys. **11**, 45 (1943)

5. G. B. Field and W. C. Saslow, Astrophys. J. **142**, 568 (1965); J. Silk and S. D. White, Astrophys. J. **223**, L59 (1978)

6. F. W. Wiegel and A. S. Perelson, J. Stat. Phys. **29**, 813 (1982)

7. P. Meakin, Phys. Rev. Lett. **51**, 1119 (1983); M. Kolb, R. Botet and R. Jullien, Phys. Rev. Lett. **51**, 1123 (1983)

8. R. Jullien and M. Kolb, J. Phys. A: Math. Gen. **17**, L639 (1984)

9. D. Weitz and M. Oliveira, Phys. Rev. Lett. **52**, 1433 (1984); D. W. Schaefer, J. Martin, P. Wiltzius and D. S. Cannell, Phys. Rev. Lett. **52**, 2371 (1984); D. Weitz, J. S. Huang, M. Y. Lin and J. Sung, Phys. Rev. Lett. **54**, 1416 (1985)

10. K. Kang and S. Redner, Phys. Rev. A30, 2833 (1984); Phys. Rev. A32, 435 (1985)

11. P. G. J. van Dongen and M. H. Ernst, Phys. Rev. Lett. **54**, 1396 (1985);

12. S. K. Friedlander, J. Meteor. **17**, 373 (1960); J. Meteor. **17**, 479 (1960)

13. T. Vicsek and F. Family, Phys. Rev. Lett. **52**, 1669 (1984)

14. F. Leyvraz and H. R. Tschudi, J. Phys. A: Math. Gen. **14**, 3389 (1981); J. Phys. A: Math. Gen. **15**, 1951 (1982); F. Leyvraz, J. Phys. A: Math. Gen. **16**, 2661 (1983)

15. E. M. Hendriks, M. H. Ernst and R. M. Ziff, J. Stat. Phys. **31**, 519 (1983)

16. P. G. J. van Dongen, J. Stat. Phys. **44**, 785 (1986)

17. M. v. Smoluchowski, Phys. Z. **17**, 593 (1916)

18. R. Jullien, M. Kolb and R. Botet in *Kinetics of Aggregation and Gelation*, F. Family and D. P. Landau eds., Elsevier, North Holland, Amsterdam (1984)

19. W. D. Brown and R. C. Ball, J. Phys. A: Math. Gen. **18**, L517 (1985)

20. R. C. Ball, D. Weitz, T. Witten and F. Leyvraz, Phys. Rev. Lett. **58**, 274 (1987)

21. L. Peliti, J. Phys. A: Math. Gen. **19**, 973 (1986)

22. K. Kang, S. Redner, P. Meakin and F. Leyvraz, Phys. Rev. A31, 1171 (1986)

23. R. Kopelman, P. W. Klymko, J. S. Newhouse and L. W. Anacker, Phys. Rev. B 29, 3747 (1984)

24. P. Meakin and H. E. Stanley, J. Phys. A: Math. Gen. **17**, L173 (1984)

QUASICRYSTALS: FROM PERIODIC TO NON-PERIODIC ORDER IN SOLIDS

Peter Kramer*
Institut für Theoretische Physik
Auf der Morgenstelle 14
7400 Tübingen
West Germany

ABSTRACT. Experiments on metal alloys as $Al_4 Mn$ since 1984 enforce the concept of non-periodic long-range order in condensed matter. Theoretical principles for this order are formulated. Orientational order with non-crystallographic point symmetry can be obtained by lifting the symmetry group from E^3 to E^n, $n > 3$. The cells of a lattice Y in E^n form a Euclidean cell complex with a metrical dual Y^*. A rearrangement of Y into klötze yields on a subspace E^m, $m < n$ an in general non-periodic quasilattice and tiling. In a quasicrystal model, these tiles become the cells of the long-range order. The Fourier theory for the quasicrystal model allows one to express the transform through integrals over the new cells.

INTRODUCTION

In the present contribution we describe new theoretical tools for the physics of non-periodic long-range order. In the ideal non-periodic order to be described, condensed matter is well organized by the broken periodic symmetry of a lattice in E^n, $n > 3$. In section 1 it is explained how orientational order forces us to embed E^m, $m = 2,3$, into E^n. Given a lattice Y in E^n, one now needs a geometric prescription for the embedding of E^3 with broken periodic symmetry. The clue to this problem is formulated in section 3 with the topological concepts of Euclidean cell complexes in E^n, their skeletons of dimension m, $n - m$, metrical duality and klotz constructions. A survey of applications and additional results is given in section 4.

*Work supported by the Deutsche Forschungsgemeinschaft

A. Amann et al. (eds.),
Fractals, Quasicrystals, Chaos, Knots and Algebraic Quantum Mechanics, 73–91.

1. GROUPS, REPRESENTATIONS, AND ORIENTATIONAL ORDER

The notions of group action and group representation are useful for the description of orientational order in condensed matter.

1.1 Group action

Let G be a group and X a set, called the space X, with elements $x \in X$ called points. The set of all bijections $X \to X$ forms a group G_X. The group G is called a transformation group on X if it is (isomorphic to) a subgroup of G_X, $G < G_X$. We call the map

$$G \times X \to X$$

the group action and consider it in more detail. Given a point $x \in X$, the orbit at x under G is the subset

$$Gx = \{x' \mid x' = gx \quad \text{for some } g \in G\}$$

G is easily seen to link any pair of points from Gx. We say that G acts transitively on Gx, and that Gx is a homogeneous space under G. The property of points to be on the same orbit is an equivalence relation on X, and so X is partitioned into disjoint orbits. The inner properties of an orbit are characterized by its stability group. The stability group G_0 of a point $x_0 \in X$ on the orbit Gx_0 is the subgroup $G_0 < G$,

$$G_0 = \{h \mid hx_0 = x_0 \, , \quad h \in G\}$$

The stability groups of different points on the same orbit turn out to be conjugate subgroups. Consider the set G/H of all left cosets of a subgroup $H < G$,

$$G/H = \{\underset{\alpha}{\cup} \, g_\alpha H \mid g_\alpha H \neq g_{\alpha'} H\}$$

The map $G \times (G/H) \to G/H$ given by

$$(g, \, g'H) \to g \, g'H$$

is an action of G on the coset space G/H, and, since all cosets can be reached by this action, it is a transitive action.

<u>1.1 Prop.</u> Any homogeneous space X under a transformation group G is in one-to-one correspondence to a coset space G/H. The map $X \to G/H$ is compatible with the group action. Proof: Let H be the stability group of the point $x_0 \in X$ and construct the function

$$f: X \to G/H \, , \quad f(x) = g'H \quad \text{if} \quad x = g'x_0$$

Then f is bijective and we get the commutative diagram

$$
\begin{array}{ccc}
x & \xrightarrow{\ f\ } & g'H \\[2pt]
{\scriptstyle g}\big\downarrow & & \big\downarrow{\scriptstyle g} \\[2pt]
gx & \xrightarrow[\ f\ ]{} & g\,g'H
\end{array}
$$

So, any concrete orbit Gx on X has an abstract counterpart of the form
G/H, and, more important, all homogeneous spaces under G can be classified by group/subgroup pairs G/H.

Return to the full space X. Orbits on X with conjugate stability
group have the same abstract form G/H and form strata. The reconstruc-
tion of X from the orbits can be done with the help of a transversal or
fundamental domain FD,

$$
FD = \{\, x_\alpha \mid Gx_\alpha \cap Gx_{\alpha'} = \emptyset, \ \bigcup_\alpha Gx_\alpha = X \,\}.
$$

So the transversal contains exactly one point from each orbit, and X is
reconstructed by letting G act on all points of FD.

1.2 Group representations

A linear representation of a group G is a homomorphism of G into the
group of all linear invertible operators on a linear space E. The representation is irreducible if it leaves no proper subspace of E invariant.
We shall deal exclusively with finite or with infinite discrete groups.
Representations we denote by

$$
D: g \to D(g) \, ,
$$

and their character by

$$
\chi(g) = \text{trace } (D(g)), \qquad \chi(e) = \dim(D)
$$

We turn to the group/subgroup analysis of representations. The restriction of a representation D of G to a subgroup H < G is called the sub-
duced representation $D \downarrow H$. Let D^a and D^α denote irreducible representations of G and H. Then

$$
D^a \downarrow H = M(\underset{\oplus\alpha}{\textstyle\sum} D^\alpha) \, M^{-1}
$$

and

$$
\chi \downarrow H = \underset{\alpha}{\textstyle\sum} \ m(a \downarrow \alpha) \, \chi^\alpha
$$

where $m(a \downarrow \alpha)$ is the multiplicity of D^α in $D^a \downarrow H$. The converse construction of representations of G from those of H is called induction.

<u>1.2 Def.</u> Let $c_1 = e$, c_2, ..., c_r denote generators of the cosets of G/H,
and let D be a representation of H. The matrix elements of the induced
representation $D \uparrow G$ are

$$D \uparrow (g)_{ij} = D(h) \; \delta(c_i^{-1} g c_j, h \in H) \quad , \quad i,j = 1,\ldots,r$$

To understand the construction note that, for given $g \in G$ and c_j, the equation $g c_j = c_j h$ has a unique solution c_j, h. The homomorphism property is easily verified, and we have in terms of the orders of G and H

$$\dim (D \uparrow G) = \dim (D) \; |G|/|H|.$$

Subduction and induction are linked by Frobenius reciprocity. For D^a and D^α as given above, consider the multiplicity $m(\alpha \uparrow a)$ of the irreducible representation D^a of G in the induced representation $D^\alpha \uparrow G$. Reciprocity states that

$$m(\alpha \uparrow a) = m(a \downarrow \alpha)$$

and moreover allows one to construct most of the properties of induction from those of subduction.

1.3 Orientational order

We turn to the action of translation and point groups on a Euclidean space $\mathbb{E}^n$. We shall work with the cyclic group of order n, the dihedral group D(n) of order 2n, the symmetric group S(n) of order n!, and with the hyperoctahedral group $\Omega(n)$:
1.3 Def. For the elements $f \in \Omega(n)$, assign a sign $\varepsilon_i(f) = \pm 1$ and a permutation $i \to f(i)$, $i = 1,2,\ldots,n$. Then the defining representation of $\Omega(n)$ is

$$D_{ij}(f) = \varepsilon_i(f) \quad \delta_{i,f(j)} \, .$$

The hyperoctahedral group $\Omega(n)$ is the holohedry of the hypercubic lattice spanned by n orthonormal vectors. We denote the hypercubic space group by $(T, \Omega(n))$ where T is the translation group.
    We consider now the non-periodic familiar Penrose pattern in $\mathbb{E}^2$. Its orientational order is characterized by the action of the cyclic group C(5) given by

$$D(g_5) = \begin{bmatrix} c & -s \\ s & c \end{bmatrix} \quad , \quad \begin{aligned} c &= \cos (2\pi/5) \\ s &= \sin (2\pi/5) \end{aligned}$$

This action on $\mathbb{E}^2$ is non-crystallographic, i.e. it cannot be an element of a space group in $\mathbb{E}^2$. Now clearly C(5) can act crystallographically on $\mathbb{E}^5$. To see this consider the representation of g_5 of the form

$$\tilde{D}(g_5) = \begin{bmatrix} 0 & 0 & 0 & 0 & 1 \\ 1 & 0 & 0 & 0 & 0 \\ 0 & 1 & 0 & 0 & 0 \\ 0 & 0 & 1 & 0 & 0 \\ 0 & 0 & 0 & 1 & 0 \end{bmatrix}$$

Clearly this is a special element of the group $\Omega(5)$ and hence we have the subgroup relation $C(5) < \Omega(5)$. Moreover, it is easy to construct a matrix m with the property

$$
m \, \tilde{D}(g_5) \, m^{-1} = \begin{bmatrix} c & -s & 0 & 0 & 0 \\ s & c & 0 & 0 & 0 \\ 0 & 0 & c' & -s' & 0 \\ 0 & 0 & s' & c' & 0 \\ 0 & 0 & 0 & 0 & 1 \end{bmatrix} \, ,
\qquad
\begin{aligned}
c &= \cos(2\pi/5) \\
s &= \sin(2\pi/5) \\[6pt]
c' &= \cos(4\pi/5) \\
s' &= \sin(4\pi/5)
\end{aligned}
$$

so that $\tilde{D}$ is reduced to real irreducible form. This similarity transformation applied ot $\mathbb{E}^5$ provides a subspace $\mathbb{E}^2$ where $C(5)$ acts through its non-crystallographic matrix representation given above. So this action of $C(5)$ can, through the embedding

$$C(5) < \Omega(5) < (T, \Omega(5)),$$

be interpreted as the action of a subgroup of the point group on an irreducible subspace $\mathbb{E}^2$ in $\mathbb{E}^5$. Through this excursion, we have found a hypercubic lattice Y in $\mathbb{E}^5$. No translation vector of Y can be in the subspace $\mathbb{E}^2$, but we shall see that the lattice Y produces a quasilattice in $\mathbb{E}^2$ which organizes the non-periodic long-range order seen in the Penrose pattern.

We can now formulate a way to associate non-crystallographic orientational order with a lattice Y and a space group (T,G) in $\mathbb{E}^n$ as proposed by Kramer and Neri [1]:
(I) Consider for a lattice Y in $\mathbb{E}^n$ the holohedry group (T,G) where T is the translation group and G the point group. For the hypercubic lattice in $\mathbb{E}^n$, the point group is the hyperoctahedral group $\Omega(n)$ described in section 4, and the space group is the semidirect product

$$T \wedge \Omega(n)$$

where T is the translation group. The point group transforms $\mathbb{E}^n$ into $\mathbb{E}^n$ by a linear representation D^n of dimension n.
(II) Choose a subgroup $H < G$ such that
(a) the subduction from G to H of D^n has the real orthogonal form

$$D^n \downarrow H = D^m + D^{n-m} \quad , \quad 1 \leqslant m < n \, ,$$

and decompose $\mathbb{E}^n$ in correspondence to this subduction as

$$\mathbb{E}^n \rightarrow \mathbb{E}^m + \mathbb{E}^{n-m}$$

so that $\mathbb{E}^m$ and $\mathbb{E}^{n-m}$ are invariant under the action of G.
(b) No translation vector of T leaves $\mathbb{E}^m$ invariant. So what we require is a subgroup chain

$$H < G < (T,G)$$

such that the representation D^m of H does not admit the embedding into a space group of $\mathbb{E}^m$, i.e. H acting in $\mathbb{E}^m$ is non-crystallographic. The technical problem in implementing this program is the insufficient information on crystallography in $\mathbf{E}^n$. This problem can be partially solved by induced representation theory, compare Haase, Kramer, Kramer and Lalvani [2]:

Consider a non-crystallographic group H of $\mathbf{E}^m$ (m = 2,3) with a representation D^β of dimension m, and a subgroup K < H. If $D^\beta \downarrow K$ contains the representation D^α of K, we can induce D^β from D^α. For simplicity assume that D^α be real, and one-dimensional. Then the induced representation $D^\alpha \uparrow H$ has dimension $r = |H|/|K|$. Looking at the explicit form, it is easy to see that we get for any $h \in H$ a signed permutation matrix. Comparison with the hyperoctahedral group yields the embedding

$$K < H < \Omega(r) \quad , \qquad\qquad r = |H|/|K|$$

which can be extended to the hypercubic space group in $\mathbf{E}^r$, so that now

$$K < H < \Omega(r) < T \wedge \Omega(r)$$

So we have found, starting from H, the construction required.

As the first example we choose the cyclic group H = C(5), generated by an element g_5 of order 5. The one-dimensional complex irreducible representations are

$$D^k(g_5) = \zeta^k \ , \qquad k = 0,\pm1,\pm2 \qquad\qquad \zeta = \exp(2\pi i/5)$$

We need the real orthogonal representations which for the element g_5 are

$$
\begin{aligned}
&D^0(g_5) \quad 1 \\
&(D^1 + D^{-1}) \rightarrow \begin{bmatrix} c & s \\ s & c \end{bmatrix}, \quad c = \cos(2\pi/5), \quad s = \sin(2\pi/5) \\
&(D^2 + D^{-2}) \rightarrow \begin{bmatrix} c' & s' \\ s' & c' \end{bmatrix}, \quad c' = \cos(4\pi/5), \quad s' = \sin(4\pi/5)
\end{aligned}
$$

The only subgroup of C(5) is the trivial group K = I. Induction from I yields the regular representation of C(5),

$$(I \uparrow C(5))_{ij} \ (g_5) = \delta(c_i^{-1} \ g_5 \ c_j, \ e)$$

Choosing $c_j = (g_5)^j$ one finds the condition

$$(g_5)^{-i+1+j} = e \ , \qquad j - i \equiv 1 \bmod 5$$

so that g_5 is represented by a cyclic matrix of dimension 5. The chain now reads

$$I < C \ (5) < \Omega(5) < (T,\Omega(5))$$

Now we subduce the defining representation of $\Omega(5)$ of dimension 5 denoted by $\tilde{D}^5$ to C(5). But $\tilde{D}^5$ is just the induced representation, and so by use of reciprocity for the regular representation one finds

$$\tilde{D}^5 \downarrow C(5) = m^{-1}(D^0 + (D^1 + D^{-1}) + (D^2 + D^{-2}))m$$

where m is the reducing matrix of dimension 5 and the brackets refer to the orthogonal representations. In the corresponding decomposition of E^5 we have

$$E^5 \rightarrow E^1 + E^2 + E^2 = E^2 + (E^1 + E^2)$$

and we choose the subspace E^2 which carries the natural non-crystallographic representation $(D^1 + D^{-1})$ of $C(5)$.

1.4 Non-periodic quasilattices associated with the icosahedral group A(5)

The icosahedral group as an abstract group is isomorphic to the subgroup A(5) of even permutations of the symmetric group S(5). The irreducible representation of S(5) are characterized by the partitions [5], [41], [32], [31], [221], [21³], [1⁵]. Under restriction to A(5), the representations for associate partitions become equivalent, [1⁵] ~ [5], [21³] ~ [41], [221] ~ [32], but for the self-associate partition [31²] we get two irreducible representations of dimension 3 which we denote by [31²₊] and [31²₋]. The representation [31²₊] is the one which contains the symmetry operations of the regular dodecahedron and icosahedron, it contains 6 5-fold, 10 3-fold and 15 2-fold rotation axes and is non-crystallographic.

As subgroups K of the icosahedral group H = A(5) consider the dihedral subgroups [2] $\mathcal{D}(1)$, $1 = 5,3,2$ obtained by combining a 2-fold rotation axis with C(1). Analysis of the subduction from the non-crystallographic representations [31²₊] of A(5) shows that [31²₊] $\downarrow$ D(1) contains the non-trivial one-dimensional representations of $\mathcal{D}(1)$. The representations $D^\alpha \uparrow$ A(5) induced from these representation of the dihedral groups are of dimension $r = |A(5)|/|\mathcal{D}(1)| = 60/(21)$, and they also have the form of permutation matrices with signs $D^\alpha(k) = \pm 1$. This allows for the embedding into hyperoctahedral groups and corresponding hypercubic space groups according to

$$D(1) < A(5) < \Omega(60/(21)) < (T,\Omega(60/(21)))$$

in a hypercubic lattice of E^n, $n = 60/(21) = 6,10,15$. If now we subduce the representation D^n of $\Omega(n)$ to A(5), we obtain always by reciprocity

$$D^n \downarrow A(5) = m^{-1}(D^{[31²₊]} + D^{n-3})m$$

where D^{n-3} always contains the irreducible representation $D^{[31²₊]}$, along with other irreducible representation for $1 = 10,15$. The corresponding decomposition of E^n is

$$E^n \rightarrow E^3 + E^{n-3}$$

where E^3 carries the representation [31²₊] of A(5).

2. LATTICES, DUAL CELL COMPLEXES, AND NON-PERIODIC LONG-RANGE ORDER

Starting from the action of a translation group T on $\mathbf{E}^n$, we shall use the concept of a Euclidean cell complex and its metrical dual to re-arrnage the lattice Y in $\mathbf{E}^n$ in a way which displays non-periodic long-range order in a subspace $\mathbf{E}^m$, $m < n$.

2.1 Translations, Lattices, Euclidean cell complex

Let the translation group T act on $\mathbf{E}^n$. This action is not transitive and we shall consider the orbits. The transversal or fundamental domain FD is the crystallographic unit cell. For the hypercubic translation group the unit cell is the hypercube. For n orthogonal unit vectors $b_1,\ldots,b_n$, it is given by

$$h(n) = \{y \mid -\frac{1}{2} \leqslant y \cdot b_i \leqslant \frac{1}{2} \quad , i = 1,\ldots,n\}$$

The p-boundaries of h(n) are defined as the p-dimensional polytopes

$$h(p; a(1)\ldots a(p)) = \{y \mid -\frac{1}{2} \leqslant y \cdot b_{a(i)} \leqslant \frac{1}{2} \, , \quad i = 1,\ldots,p$$
$$y \cdot b_{a(i)} = \pm\frac{1}{2} \, , \quad j \neq i \quad \}$$

where $k \to a(k)$ is a permutation of S(n). Now we introduce the notion of an Euclidean cell complex taken from algebraic topology, compare Munkres [3].

<u>2.1 Def.</u> An n-dimensional Eucliden cell complex is a collection K of convex polytopes in $\mathbf{E}^n$ such that
(1) Every face of a polytope from K is in K.
(2) The intersection of any two polytopes from K is a face of each of them.

Now consider the translated copies of the hypercube along with its translated p-boundaries.

<u>2.2 Prop.</u> The action of the hypercubic translation group T on the hypercubic and its p-boundaries, $p = 0,\ldots,n-1$ yields in $\mathbf{E}^n$ a Euclidean complex which we denote by Y.

We use one more concept:

<u>2.3 Def.</u> The p-skeleton $K^{(p)}$ of a Euclidean cell complex is the subcomplex of all faces with dimension $r \leqslant p < n$.

So we can speak of the p-skeleton $Y^{(p)}$ of the hypercubic cell complex Y.

The subdivision of $\mathbf{E}^n$ into hypercubes and boundaries is significant with respect to the action of T: It is easy to see that the interior points of h(n) all form part of a fundamental domain. Points on boundaries are always translated into points on boundaries, but one must introduce additional restrictions to select a fundamental domain for points on boundaries.

2.2 Metrical Duality

There are several notions of duality in algebraic topology. We shall
need a specialized concept of duality which we call metrical duality.
2.4 Def. Two Euclidean cell complexes Y, Y* in E^n are called the metrical dual to one another if for $p = 0,\ldots,n$ the cells of the p-skeleton
of Y and of the (n-p)-skeleton of Y* always appear as local pairs of
polytopes h(p;), h*(n-p;) with the properties
(1) h(p;) and h*(n-p;) intersect in a single point q,
(2) h(p;) and h*(n-p;) are spanned, with respect to q, by mutually orthogonal sets of vectors in E^n.
It is not hard to construct the metrical dual Y* for the hypercubic cell
complex Y: Construct a second hypercubic lattice whose origin is shifted,
with respect ot Y, to a vertex of the hypercube h(n). This new lattice
with its p-boundaries h*(n-p;) provides the metrical dual Y* to Y. In
Fig.1 we show for n = 2 the square lattice Y and its metrical dual Y*. In
Fig.2 we show for n = 2 a second lattice Y with cells and boundaries. Its
metrical dual Y* is different from Y.

2.3 Klotz Constructions

We shall consider in particular the hypercubic lattices Y and their metrical duals Y*. Consider a hyperplane in E^n perpendicular to a unit
vector e. Then E^n may be decomposed into orthogonal subspaces

$$E^n \to E_1^{n-1} + E_2^1 \quad , \quad E_1^{n-1} \perp E_2^1$$

parallel or perpendicular to the hyperplane. We use the subscripts 1,2
for linear spaces and for the orthogonal projection of vectors and polytopes. Now we state
2.5 Def. For any pair of matrical dual boundaries h(n-1;i) and h*(1;i)
from Y and Y*, define the klotz

$$kl((n-1) + 1;i) \quad \text{as the n-dimensional polytope}$$

$$kl((n-1) + 1;i) = \{y \mid y = y_1 + y_2 , \quad y_1 \in h_1(n-1;i) , \quad y_2 \in h_2^*(1;i)\}$$

Since the center $q_i = \frac{1}{2} k_i$ of a klotz is on the boundary, this polytope
in general has points inside and outside of h(n). We call $h_1(n-1;i)$ and
$h_2^*(1;i)$ the 1- and 2- chart of the klotz.
2.6 Prop. Select n klötze from h(n) such that their centers are not related by translation vectors from T. Then these klötze intersect at most
in boundaries of dimension n-1 and form a new fundamental domain FD
under the translation group T.
    Examples of the klotz construction are given in Figs. 3,4. Note
that the klötze are never in face-to-face position, a condition required
for Euclidean cell complexes. The key properties of the klötze are that
their boundaries are parallel or perpendicular to the intersecting hyperplane. For the proof of Prop. 2.6 compare [4,5,6]. Since the klötze
form a fundamental domain, their translated copies provide a periodic

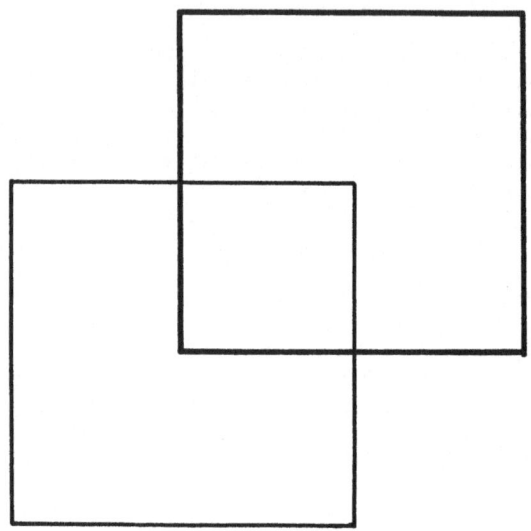

Fig.1   The cell   of the square lattice Y is indicated with
heavy lines. The second square is the cell of the metrical
dual lattice Y*.

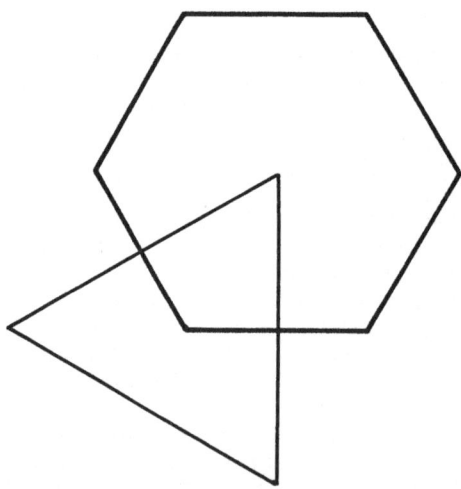

Fig.2   The   cell   of the hexagonal lattice Y is indicated
with heavy lines. The triangle is a cell of the metrical
dual lattice Y*.

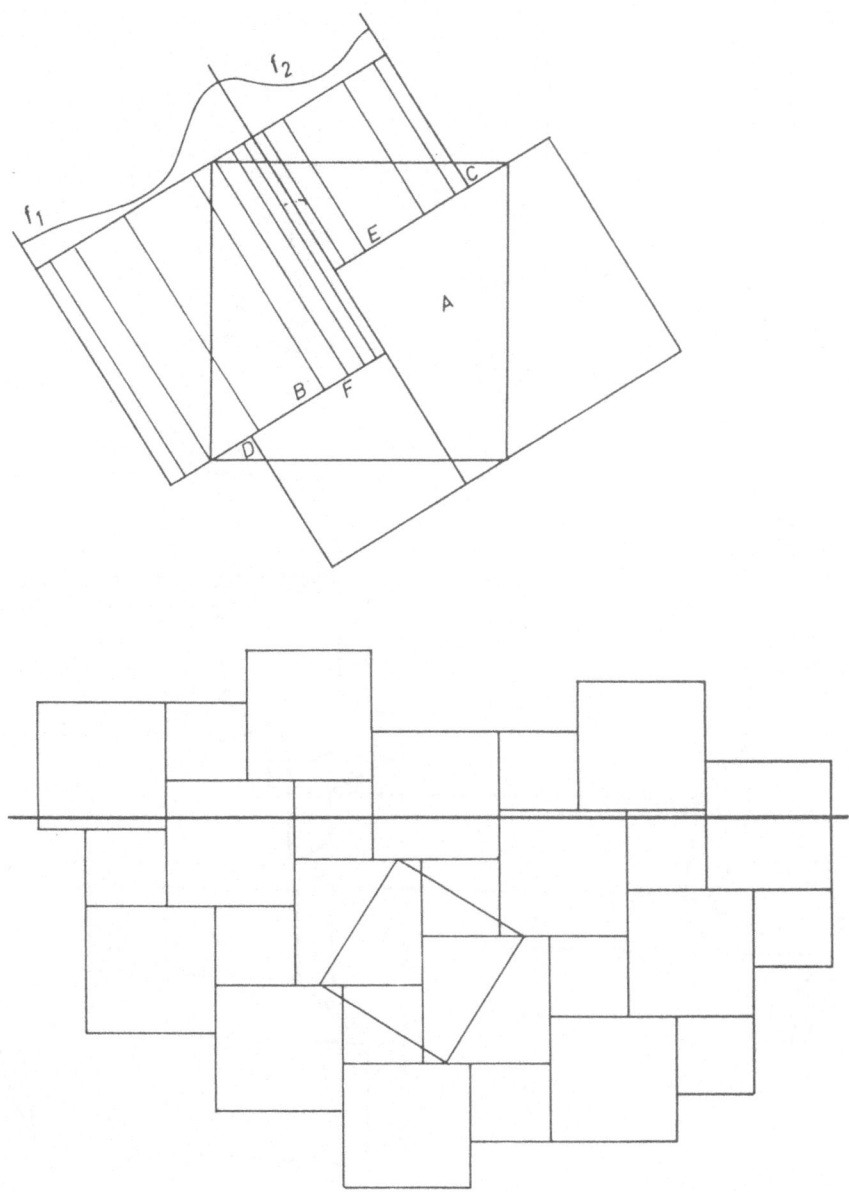

Fig.3   The square lattice has two klötze which in the quasicrystal model provide two tiles or cells. The klötze support   the densities f_1, f_2. The lower part of the figure
shows the tiling by the klötze and an intersecting line.

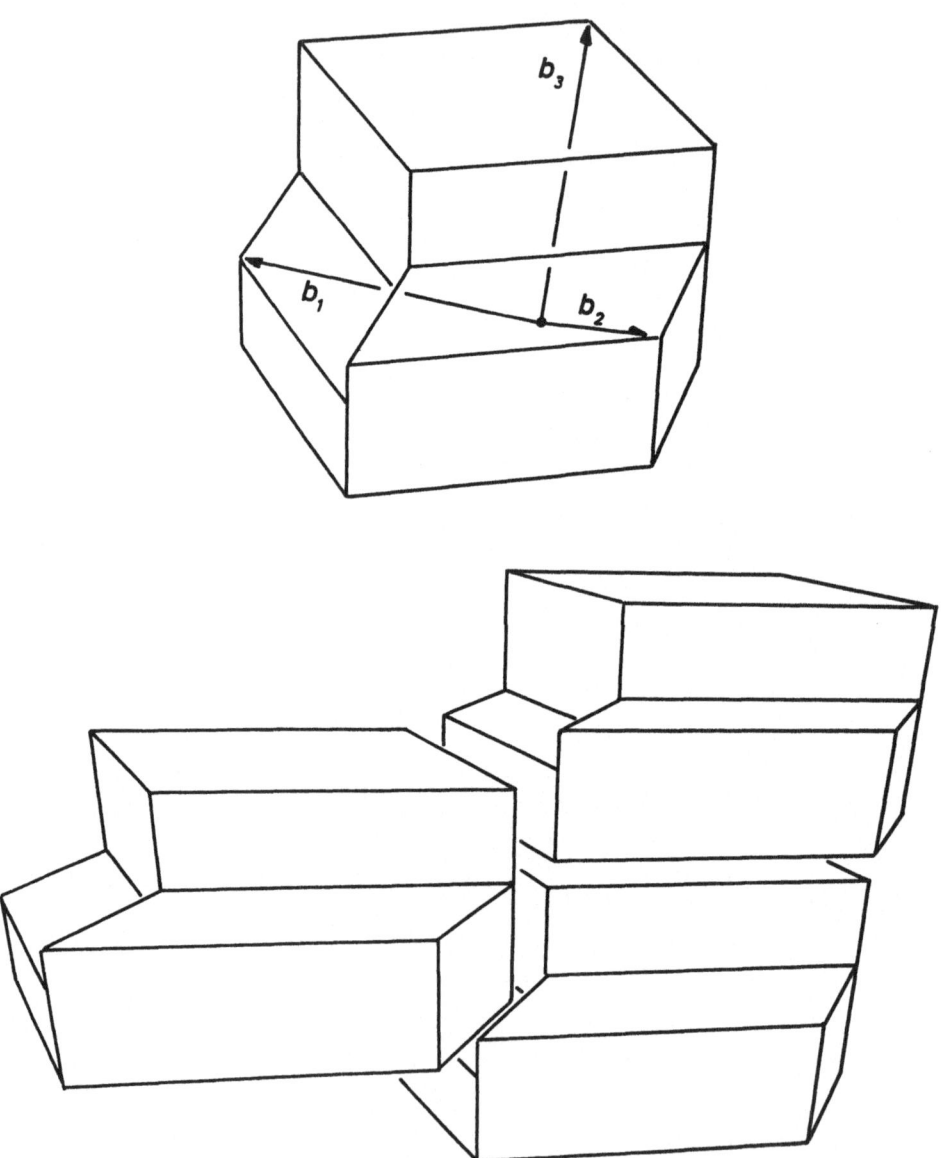

Fig. 4   The cubic lattice has three klötze which in the
plane provide three quasicrystal cells. Each klotz has a
parallelogram base. The three klötze tile 3-space periodically as indicated in the lower part of the figure.
The tiling in a plane is in general non-periodic.

space filling of $\mathbb{E}^n$. Consider now the intersection of a plane for $n = 3$
or a line for $n = 2$ with the periodic klotz tiling. If the intersection
hits a klotz, it has the form of the 1- chart $h_1(n-1;i)$, and one finds
in general

2.7 Prop. The intersection of a hyperplane with the periodic klotz
tiling of $\mathbb{E}^n$ provides a tiling of the hyperplane $\mathbb{E}_1^{n-1}$. The n tiles
have the form of the 1-charts $h_1(n-1;i)$. The tiling of $\mathbb{E}_1^{n-1}$ is non-periodic if no translation vector of T is parallel to the hyperplane.

For the Penrose quasilattice in $\mathbb{E}^2$ and for the icosahedral quasilattice in $\mathbb{E}^3$ one needs the following generalization of Def. 2.5 for a
decomposition

$$\mathbb{E}^n \rightarrow \mathbb{E}_1^m + \mathbb{E}_2^{n-m}$$

2.8 Def. For any pair of metrical dual boundaries $h(m;)$ and $h^*(n-m;)$
from Y and Y*, define the klotz $kl(m + (n-m);)$ as the n-dimensional polytope

$$kl(m + (n-m);) = \{y \mid y = y_1 + y_2 \ , \ y_1 \ \epsilon \ h_1(m;) \ , \ y_2 \ \epsilon \ h_2^*(n-m;) \}$$

In the Penrose case one needs $n = 5$, $p = 2$ and in the icosahedral case
$n = 6$, $p = 3$. In the latter case it is shown in [7] that 20 klötze
$kl(3 + 3;)$ form a new fundamental domain for the hypercubic lattice in
$\mathbb{E}^6$. The corresponding proof for the Penrose case is much simpler.

2.4 Non-periodic Long-range Order

So far, the construction described in the last subsection allows one to
construct, at least for hypercubic lattices in $\mathbb{E}^n$, a klotz tiling of
$\mathbb{E}^n$ adapted to the splitting $\mathbb{E}^n \rightarrow \mathbb{E}_1^m + \mathbb{E}_2^{n-m}$. In $\mathbb{E}^m$, this klotz tiling
provides an in general non-periodic tiling which reflects the broken
periodic symmetry of $\mathbb{E}^n$. The number of tiles equals the number of
klötze which form the new fundamental domain in $\mathbb{E}^n$. We inquire now
about the significance of the tiles in $\mathbb{E}^m$ with respect to atomic positions. If in $\mathbb{E}^n$ we introduce a periodic function f^P, its domain can be
chosen as the collection of klötze. The atomic density supported on the
intersection of a klotz with $\mathbb{E}^m$ would not be stable since the intersection with translated copies of a given klotz occur at different values of the coordinates perpendicular to $\mathbb{E}^m$. We restrict the periodic
function f^P to obtain the quasicrystal model:

2.9 Def. A quasicrystal model in $\mathbb{E}^m$ is defined by restricting the periodic function f_i^P on the klotz i by

$$f_i^P \ (x_1, x_2) = f_i^P \ (x_1) \ .$$

This definition has the consequence that the density f^P supported on the
intersection of $\mathbb{E}^m$ with the klötze becomes stable on each type of tile.
So the result of the construction is a non-periodic long-range order
with several cells. Each cell or tile supports a fixed atomic density.
The Fourier transform is discussed in the next section.

3. FOURIER THEORY, ALMOST PERIODICITY, AND QUASICRYSTAL MODELS

The Fourier transform of a periodic in $\mathbb{E}^n$ reduces to the Fourier series. On an intersection with the periodic lattice one finds the conditions of almost periodicity, of which the quasicrystal model is a special case.

3.1 The Fourier Transform and Series

We give here a general derivation of the Fourier transform for an m-dimensional linear cut through a lattice in $\mathbb{E}^n$ which supports a periodic function.

The Fourier transform of a function f supported on $\mathbb{E}^n$ we define by

$$
\begin{aligned}
F^n(f)(k): &= \tilde{f}(k) \\
&= (2\pi)^{-n} \int d^n y f(y) \exp(-ik \cdot y),
\end{aligned}
$$

and its inverse by

$$
\begin{aligned}
f(y) &= (F^n)^{-1}(\tilde{f})(y) \\
&= \int d^n k \tilde{f}(k) \exp(ik \cdot y) \ .
\end{aligned}
$$

Let

$$
\mathbb{E}_n \to \mathbb{E}_1^m + \mathbb{E}_2^{n-m}
$$

be an orthogonal decomposition and denote the orthogonal projections of all vectors or polytopes by the subscripts 1,2 respectively.

To describe the intersection of objects in $\mathbb{E}^n$ with $\mathbb{E}^m$ introduces for a fixed vector c_2 the *cut function*

$$
v: \ v(y) = \delta^{n-m}(y_2 - c_2)
$$

with the Fourier transform

$$
\tilde{v}(k_1, k_2) = (2\pi)^{-n+m} \delta(k_1) \exp(-ik_2 \cdot c_2) \ .
$$

For the Fourier transform of a function f on the intersection one gets by convolution

$$
F^n(fv)(k_1, k_2) = (2\pi)^{-n+m} \int d^{n-m} l_2 \tilde{f}(k_1, k_2 - l_2) \exp(-il_2 \cdot c_2).
$$

Now one can express the Fourier transform of the function f with respect to the subspace $\mathbb{E}^m$ in the form

$$
F^m(f(,c_2)(k_1)
$$

$$
= (2\pi)^{n-m} F^n (fv)(k_1, 0)
$$

$$
= \int d^{m-m} l_2 \ \tilde{f}(k_1, -l_2) \ \exp (-il_2 \cdot c_2).
$$

Consider a translation group T acting on $\mathbf{E}^n$, the lattice Y generated by T, a function f^P periodic with respect to T supported on the fundamental domain FD, and denote by T^R the translation group of the reciprocal lattice. Then the Fourier integral of f^P collapses to the Fourier
series which can be expressed in the form

$$\tilde{f}^P(k) = (\text{vol}(FD))^{-1} \sum_{k \in T^R} \delta^n(k - k^R) a(k^R)$$

$$a(k^R) = \int_{FD} d^n x \, f^P(x) \exp(-ik^R \cdot x) \quad .$$

This result can also be obtained from the orthogonality relation for the
irreducible representations of the translation group.

Let k be a vector form the first Brillouin zone BZ of the reciprocal lattice Y^R. The volumes of the Brillouin zone and of the unit cell
FD of Y are related by

$$(\text{vol}(BZ))^{-1} = (2\pi)^{-n} \text{vol}(FD)$$

The irreducible representations of the translation group T are given by

$$D^k(b) = \exp(-i\,k \cdot b) \quad , \quad k \in BZ$$

3.1 Prop. The orthogonality and completeness relations of these representations are

$$(2\pi)^{-n} \text{vol}(FD) \sum_{b \in T} \overline{D^{k'}(b)} \, D^k(b) = \delta^n(k'-k),$$

$$(2\pi)^{-n} \text{vol}(FD) \int_{BZ} d^n k \, \overline{D^k(b')} \, D^k(b) = \delta(b',b)$$

3.2 Prop. Let k be a general vector in Fourier space, and T^R the translation group of Y^R, then

$$\sum_{b \in T} \exp(-i\,\tilde{k} \cdot b)$$
$$= (2\pi)^n (\text{vol}(FD))^{-1} \sum_{k^R \in T^R} \delta^n(\tilde{k} - k^R)$$

Proof: Suppose that $\tilde{k}$ has the decomposition $\tilde{k} = \overset{o}{k}{}^R + k$, $\overset{o}{k}{}^R \in T^R$, $k \in BZ$.
Now the orthogonality relation yields

$$\sum_{b \in T} \exp(-i\,\tilde{k} \cdot b) = \sum_{b \in T} \exp(-i\,k \cdot b)$$

$$= (2\pi)^n (\text{vol}(FD))^{-1} \delta^n(\tilde{k} - \overset{o}{k}{}^R)$$

which agrees with the term $k^R = \overset{o}{k}{}^R$ of the sum given above. The result
must hold for any reciprocal lattice vector, and so we get the result.

If a general point y in $\mathbb{E}^n$ is decomposed as

$$y = b + x, \qquad b \in T, \qquad x \in FD,$$

then Prop. 3.2 yields the collapse from the Fourier integral to the Fourier series.

3.2 Almost Periodicity

Now we combine the results on a cut in $\mathbb{E}^n$ of dimension m with the collapse for a periodic function f^P to obtain

3.3 Prop. The Fourier transform of a periodic function f^P on the subspace $\mathbb{E}^m$ is given by

$$F^m(f^P(,c_2))$$

$$= (vol(FD))^{-1} \sum_{(k_1^R, k_2^R) \in T^R} \delta^m(k_1 - k_1^R) \exp(i\, k_2^R \cdot c_2) a(k_1^R, k_2^R)$$

This Fourier transform has its amplitudes on the discrete set of points $k_1 = k_1^R$. The class of functions with Fourier amplitudes $f(k)$ supported on a discret set of points was studied by H. Bohr [8]. Bohr showed that, in position space, the corresponding functions have a relatively dense set of translation vectors and so are almost periodic.

Consider now the application of this Fourier theory to the models developed in section 2.3. Upon introduction of the klötze $kl(m+(\acute{n}-m);i)$ as the fundamental domain, the periodic function f^P can be specified in parts f_i^P defined on a klotz. But, since the intersection of $\mathbb{E}^m$ varies with the position of the klotz in $\mathbb{E}^n$, a repeated tile in $\mathbb{E}^m$ does not support a stable set of values f_i^P. Only if the intersection of a klotz occurs at almost the same relative position x_2 there is an almost repetition of the density. We restrict the density to the quasicrystal model by requiring for all i inside a klotz

$$f_i^P(x_1, x_2) = f_i^P(x_1)$$

Now of course we get stable values of the density on each tile, and we obtain

3.3 Prop. For a quasicrystal with 1 klötze, the Fourier transform is given according to Prop. 3.4 with the Fourier series coefficients

$$a(k_1^R, k_2^R)$$

$$= \sum_{i=1}^{1} \left(\int_{h_1(m;i)} d^m x_1\, f_i^P(x_1)\, \exp(-i\, k_1^R \cdot x_1) \right.$$

$$\times \left. \int_{h_2^*(n-m;i)} d^{n-m} x_2\, \exp(-i\, k_2^R\, x_2) \right)$$

The Fourier coefficients for each klotz factorize into an integral over
the density on the tile and a kinematical factor which is the characteristic function of a polytope in $\mathbb{E}^{n-m}$. The result can be extended to
the boundary points of the klötze. Among these boundaries there are the
vertices which provide the model of a single point atom per quasicrysal
cell.

4. SURVEY OF APPLICATIONS

We mention here some applications and recent progress made in the field
of quasicrystals

4.1 From Non-periodic to Periodic Order

If the intersecting plane $\mathbb{E}^m$ in $\mathbb{E}^n$ contains translation vectors from
T, a periodic structure is expected in $\mathbb{E}^m$. For the cubic primitive lattice in $\mathbb{E}^3 \rightarrow \mathbb{E}^2$ the quasicrystal model in a plane perpendicular to a
3-fold axis reduces to the periodic standard model [5]. For $\mathbb{E}^6 \rightarrow \mathbb{E}^3$,
there is a choice of $\mathbb{E}^3$ which yields the face-centered cubic periodic
lattice as a cut through the primitive hypercubic lattice in $\mathbb{E}^6$ [9]. By
a continuous rotation, the subspace $\mathbb{E}^3$ can be transformed into the representation space of the icosahedral non-crystallographie quasilattice.
This rotation preserves the full tetrahedral subgroup of both the cubic
and icosahedral groups.

4.2 Thermodynamics, Landau Theory

The phenomenological Landau theory of phase transitions in periodically
ordered matter employs group/subgroup relations, compare Birman [10].
With the embedding of non-periodic lattices into periodic lattices in
$\mathbb{E}^n$, one can try to implement the group/subgroup scheme for non-periodic
quasicrystals.

4.3 Electron States in Quasicrystals

In periodic order, the Bloch states of electrons are characterized by
the unitary irreducible representations of the translation group. An analysis of electron propagation in a discretized non-periodic potential
is given by Kohmoto [11].

4.4 Icosahedral Quasilattices

A variety of icosahedral quasilattices can be constructed by induction.
The quasilattices in $\mathbb{E}^3$ obtained from hypercubic lattices in $\mathbb{E}^{10}$ and
$\mathbb{E}^{15}$ have a rather complex cell structure [2]. The full diffraction
theory of the icosahedral quasilattice obtained from $\mathbb{E}^6 \rightarrow \mathbb{E}^3$ is derived
in [6].
For a review from the crystallographic point of view compare Mackay
[12]. Atomic models for the icosahedral phase were proposed by Guyot and
Audier [24,25]. Large icosahedral quasicrystals have been prepared by

Dubost, Lang, Tanaka, Sainfort and Audier [26]. Some additional papers on theory and experiment are included in the list of references.

ACKNOWLEDGMENTS

The author is indebted to R.W. Haase, U. Thomas and D. Zeidler for discussions and to L. Kramer for technical assistance at Tübingen. The work was stimulated by conversations with A. Janner and T. Janssen, Nijmegen and with A.L. Mackay, London

REFERENCES

a)   Related to text

[1] Kramer, P. and Neri, R., Acta Cryst. A40(1984)580
[2] Haase, R.W., Kramer, L. Kramer, P. and Lalvani, H., Acta Cryst.
     A43(1987)574-587
[3] Munkres, J.R., Elements of Algebraic Topology, Addison-Wesley,
     Menlo Park, 1984
[4] Kramer, P., Mod.Phys.Lett.B1(1987)7-18
[5] Kramer, P., Int.J.Mod.Phys.B1(1987)145-165
[6] Kramer, P., submitted for publication
[7] Kramer, P., submitted for publication
[8] Bohr, H., Fastperiodische Funktionen, Berlin 1932
[9] Kramer, P., Acta Cryst. A43(1987)486-489
[10] Birman, J.L., Physica 114A(1982)564-571
[11] Kohmoto, M., Kadanoff, L.P. and Tang, Ch., Phys.Rev.Lett.50(1983)
     1870-1872
[12] Mackay, A.L., J.Microsc. 146(1987)233-243

b)   Historical and parallel sources of the theory

[13] Kepler, J.,   Harmonices Mundi Libri V, 1619
[14] Kowalewski, G.,   Der Keplersche Körper und andere Bauspiele,
     Leipzig 1938
[15] Penrose, R., Bull.Inst.Math.Appl.10(1974)266
[16] Mackay, A.L., Sov.Phys.Cryst.26(1981)257
[17] Kramer, P., Acta Cryst.A38(1982)257-264
[18] Levine, D. and Steinhardt, P.J., Phys.Rev.Lett. 53(1984)2447-2480
[19] Kramer, P., Z.Naturf.40a(1985)775-788
[20] Kramer, P., Z.Naturf.41a(1986)879-911
[21] Elser, V., Acta Cryst. A42(1986)36-43
[22] Gähler, F. and Rhyner, J., J.Phys.A Gen. A19(1986)267-277
[23] Janner, A. and Janssen, T., Physica 99A(1979)27-76

c)   Atomic order and morphology of quasicrystals

[24] Guyot, P. and Audier, M.   Phil.Mag. B 52(1985)L15-L19
[25] Audier, M. and Guyot, P.   Phil.Mag. B 53(1986)L43-L51
[26] Dubost, B., Lang, J.M., Tananka, M., Sainfort, P. and Audier, M.
     Nature 324(1986)48-50

CRYSTALLOGRAPHY OF QUASICRYSTALS

A. Janner
Institute for Theoretical Physics, University of Nijmegen
Toernooiveld
6525 ED Nijmegen, The Netherlands

ABSTRACT.   A working definition is adopted which makes quasicrystals a
special case of incommensurate crystal structures and includes the case
of Penrose-like tilings. Embedding of quasicrystals in a higher-
dimensional space allows to recover lattice periodicity and thus a
crystallographic symmetry group. Natural is to attach to the space an
Euclidean metrics. One then gets space groups of corresponding
dimension. It is also possible to embed the same quasicrystal in a
space having indefinite metrics. The symmetry is then given by what one
could call Minkowskian crystallographic groups, allowing lattices with
a point group symmetry of infinite order. The corresponding generators
induce self-similarity transformations on the quasicrystal lying in the
pseudo light cone.

1. INTRODUCTION

A generally accepted definition of quasicrystals is still missing. The
good one will eventually follow from experimental and/or theoretical
investigations. Here a working definition is adopted which makes
quasicrystals a special case of incommensurate crystal structures and
includes the case of the Penrose-like tiling [1]. Embedding of
quasicrystals in a higher dimensional space allows to recover lattice
periodicity and to treat them as classical crystal structures [2]. It
is, however, essential to realize that embedding of a n-dimensional
quasicrystal in a (n+d)-dimensional space does not mean that one can
simply apply the laws of (n+d)-dimensional crystallography. This
because of the restrictions imposed by the existence of the crystal-
like object in the "real" n-dimensional space. In particular, two
crystallographic descriptions of the same object have to be declared
equivalent. For example the oblique and the rectangular lattices are
inequivalent in two dimensions because they belong to different Bravais
classes.   Their distinction make no much sense when considered from the
embedding of a 1-dimensional quasicrystal lattice, because both
represent possible embeddings of a same quasicrystal.
    Furthermore, the recovery of crystallography by embedding on a
lattice, does not imply that the higher dimensional space is Euclidean.

93

A. Amann et al. (eds.),
Fractals, Quasicrystals, Chaos, Knots and Algebraic Quantum Mechanics, 93–109.

The aim of the present paper is to show that alternative embedding in a space with indefinite metrics is not only possible, but that the crystallography one gets is essential for a proper characterization of quasicrystals. This is a surprising result which underlines the view that quasicrystal structures are conceptually important. As a by-product one gets an insight in unexpected relations between Euclidean and non-Euclidean crystallography. Having said that, it should be clear why most of the present paper is restricted to 1-dimensional quasicrystals embedded in a 2-dimensional plane. This does not mean, that the validity of the concepts reported is necessarily limited to this low-dimensional case, but simply that the crystallography of 3-dimensional quasicrystals is so rich that, at present, it cannot be treated properly.

2. SUPERSPACE APPROACH

The geometry of quasicrystal structures has been approached in many different ways and it is not the aim of the present paper to expose and to compare these descriptions [3]. Relevant is a general consensus that the Fourier wave vectors of a quasicrystal are expressible as integral linear combination of a finite number of given ones:

$$\vec{h} = \sum_{i=1}^{n+d} h_i \vec{a}^*_i \qquad h_i \text{ integers.} \tag{1}$$

The full set of allowed Fourier wave vectors generates a Z-module M^* of <u>dimension n</u>, which is the dimension of the space spanned by $\{\vec{a}^*_1, \ldots, \vec{a}^*_{n+d}\}$, i.e. the dimension of the reciprocal space, and of <u>rank n+d</u>, which is the number of free generators of M^*, normally the smallest number of linearly independent ones on the rationals, so that M^* is isomorphic with Z^{n+d}. Accordingly $d \geq 0$ and for $d > 0$ the quasicrystal is an incommensurate crystal, whereas for $d = 0$ it is commensurate and M^* is a (reciprocal) lattice. Considered here is the case $d > 0$.

The Fourier transform of a quasicrystal density $\rho(\vec{r})$ can be written as

$$\rho_{FT(n)}(\vec{k}) = \sum_{\vec{h} \in M^*} \hat{\rho}(h_1, \ldots, h_{n+d}) \delta^{(n)}(\vec{h} - \vec{k}) \tag{2}$$

and $\rho(\vec{r})$ as

$$\rho(\vec{r}) = \sum_{\vec{h} \in M^*} \hat{\rho}(h_1, \ldots, h_{n+d}) e^{2\pi i \vec{h}\vec{r}} . \tag{3}$$

The basic idea of the superspace approach is that the structure is uniquely characterized by the set of complex numbers $\hat{\rho}(h_1, \ldots, h_{n+d})$ defined on Z^{n+d}, allowing thus an embedding on a $(n+d)$-dimensional lattice Σ^*:

$$\vec{a}^*_i \rightarrow a^*_{is} = (\vec{a}^*_i, \vec{a}^*_{iI}) \qquad i=1,2,\ldots,n+d \qquad (4)$$

in a so-called <u>superspace</u> $V_s = V \oplus V_I$, direct sum of the original space V (called external) and an additional d-dimensional one V_I (called internal). In the incommensurate case each $\vec{h}$ of M^* can uniquely be embedded as h_s of Σ^* by

$$\vec{h} = \sum_i h_i \vec{a}^*_i \quad \rightarrow \quad h_s = \sum_i h_i a^*_{is}, \quad \text{given integers } h_i. \qquad (5)$$

Then a (n+d)-dimensional crystal density $\rho_s(r_s)$ is defined by Fourier transformation in (n+d) dimensions

$$\rho_{sFT(n+d)}(k_s) = \sum \hat{\rho}(h_1,\ldots,h_{n+d}) \delta^{(n+d)}(k_s - h_s) \qquad (6)$$

giving

$$\rho_s(r_s) = \sum_{h_s \in \Sigma^*} \hat{\rho}(h_1,\ldots,h_{n+d}) e^{2\pi i h_s r_s}. \qquad (7)$$

As Fourier transforms are invertible, the structural information of the original quasicrystal ρ and that of the crystal embedded ρ_s is the same. That is the strength of the superspace approach, which formally can be characterized by:

$$\rho(\vec{r}) \leftarrow FT(n) \rightarrow \hat{\rho}(h_1,\ldots,h_{n+d}) \leftarrow FT(n+d) \rightarrow \rho_s(r_s). \qquad (8)$$

Note that

$$\rho_s(r_s)\big|_{\vec{r}_I} = \rho(\vec{r}) \qquad (9)$$

i.e. the quasicrystal appears as section of ρ_s, a consequence of the fact that the Z-module M^* is the projection of the reciprocal lattice Σ^*. The 1-to-1 correspondence between ρ and ρ_s does not mean that the latter is uniquely defined by the former. There is a freedom in embedding expressed in eq. (4) by the choice of the internal components $\vec{a}^*_{iI}$, which is a kind of gauge. The relation between ρ_s and ρ recalls that between potentials and fields. Fixing the embedding corresponds to fix the gauge. A given ρ imposes restrictions on the possible ρ_s, and that becomes apparent when one considers among all possible ρ_s the most symmetrical ones (again a well known fact for potentials).

So far, what has been said is common to all incommensurate crystals. Characteristic for a quasicrystal is the existence of a quasicrystal lattice in direct space which can be embedded on the lattice Σ dual to the reciprocal Σ^* one. From Σ the quasicrystal lattice can then be recovered by the so-called cut-projection method. In order to get the corresponding superspace version, consider instead of the general case as in eqs. (6) and (7), densities defined on a lattice only:

$$\rho_s(r_s) = \sum_{m_s \nmid \Sigma} \delta^{(n+d)}(r_s - m_s) \tag{10}$$

with $m_s = \sum_{i=1}^{n+d} m_i a_{is}$, for m_i integers and basis vectors a_{is}

$$\{a_{is} = (\overset{\star}{a}_i, \overset{\star}{a}_{iI})\} \quad \text{dual to} \quad \{a^*_{is} = (\overset{\star}{a}^*_i, \overset{\star}{a}^*_{iI})\}. \tag{11}$$

A characteristic function C (the window) is now defined on Σ taking the value 1 if the lattice point m_s is inside the window and the value 0 otherwise. Suppose, furthermore, that the window contains the space V and satisfies the requirement of a smallest non-zero distance between points of the quasicrystal lattice. The latter is defined by the density

$$\rho(\overset{\star}{r}) = \sum_{\overset{\star}{m}} \delta^{(n)}(\overset{\star}{r} - \overset{\star}{m})C(m_1, \ldots, m_{m+d}) \tag{12}$$

with $\overset{\star}{m} = \sum_{i=1}^{n+d} m_i \overset{\star}{a}_i$. Note that the admissible $\overset{\star}{m}$'s generate in V a Z-module M of dimension n and rank n+d. Note also that the function C, and thus the quasicrystal lattice, depends on the choice of the origin in V_s, a fact not explicit apparent from eqs. (10) and (12). The Fourier transform of $\rho(\overset{\star}{r})$ takes the form (compare with eq. (2))

$$\rho_{FT(n)}(\overset{\star}{k}) = \sum_{\overset{\star}{h} \nmid M^*} \hat{C}(h_1, \ldots, h_{n+d})\delta^{(n)}(\overset{\star}{h} - \overset{\star}{k}) \tag{13}$$

with $\hat{C}$ the Fourier transform of the window. It is the explicit presence of a Z-module structure in both the direct and the reciprocal space which makes incommensurate quasicrystals so similar to crystals and thus fundamental objects. Extension of the above expressions to the general case of a quasicrystal density leads to a structure factor $\hat{S}$ as well. In superspace one then has:

$$\rho_{sFT(n+d)}(k_s) = \hat{S}(k_s)\hat{L}_s(k_s) \tag{14}$$

with $\hat{L}_s(k_s)$ the lattice factor:

$$\hat{L}_s(k_s) = \sum_{h_s \nmid \Sigma^*} \delta^{(n+d)}(k_s - h_s). \tag{15}$$

The general quasicrystal density is expressible as product of structure- lattice- and window-factor

$$\rho_{FT(n)}(\vec{k}) = \hat{S}(\vec{k})\hat{L}(\vec{k})\hat{C}(\vec{k}) \qquad (16)$$

with

$$\hat{L}(\vec{k}) = \sum_{\vec{h} \nmid M^*} \delta^{(n)}(\vec{k}-\vec{h}) \qquad (16a)$$

$$\hat{S}(\vec{h}) = \hat{S}(h_1,\ldots,h_{n+d}) = \hat{S}(h_s) \quad \text{for} \quad \vec{h} \nmid M^* \qquad (16b)$$

and $\hat{C}$ as above. One can equally well avoid to introduce a window (or equivalently one can reduce the window to V itself) and include the factor $\hat{C}(\vec{k})$ in the definition of the structure factor $\hat{S}(\vec{k})$. In the superspace instead of a lattice of points one then gets the typical pattern of (discontinuous) modulation (hyper)lines [2], which by intersection with V produce the same effect as the cut-projection operation considered in the next section.

3. EUCLIDEAN EMBEDDING OF 1-DIMENSIONAL QUASICRYSTALS

3.1 Cut-projection method

The superspace phraseology is here applied to the cut-projection method of Duneau and Katz [4,5,6], based on concepts developed by de Bruijn [7] (see also ref. [8]). Consider the 2-dimensional plane defining the superspace V_s with orthonormal basis e_1, e_2, identifying the space generated by e_1 with the external space V of the quasicrystal, and that by e_2 with the internal space V_I. Correspondingly one has the decomposition:

$$r_s = re_1 + r_I e_2 = (r,r_I) \qquad (17)$$

for any r_s in V_s. The elements of a 2-dimensional lattice Σ with basis

$$a_{1s} = (a_1,a_{1I}) \quad \text{and} \quad a_{2s} = (a_2,a_{2I}) \qquad (18)$$

given by $m_s = m_1 a_{1s} + m_2 a_{2s} = (m_1,m_2)$ are projected on V according to

$$m_s \rightarrow m = m_1 a_1 + m_2 a_2. \qquad (19)$$

The actual positions r_m of a 1-dimensional quasicrystal are selected by cutting V_s according to a window, i.e. a strip containing the line V, taking the values which defines a characteristic function $C(m_1,m_2)$ i.e. as already said, 1 for m inside the strip and 0 for m outside. In particular for a strip of width D limited by lines parallel to V one has:

$$C(m_1,m_2) = \int_D dt \; \delta(te_2 - m_1 a_{1I} - m_2 a_{2I}).$$ (20)

Accordingly, the quasicrystal lattice can be described by the density function

$$\rho(r) = \sum_{m_1,m_2} \delta(r - m_1 a_1 - m_2 a_2) C(m_1,m_2).$$ (21)

The magnitude of D fixes the minimal distance between two points of the quasicrystal lattice. A natural choice is to take $a_{1I} + a_{2I}$ for D, i.e. the projection of the lattice unit cell on V_I. If one takes $D = a_{2I}$ (for $a_{2I} \geq a_{1I}$), then the projection on V of $a_{2s} = (0,1)$ defines a "short" interval $S = |a_2|$ and that of $a_{1s} + a_{2s} = (1,1)$ a "long" one L $= |a_1 + a_2|$, so that the quasicrystal represents a tiling of the line in short and long intervals. The Fourier transform of $\rho(r)$ as in eq. (21) is given by:

$$\rho_{FT(1)}(k) = \sum \delta^{(1)}(k - z_1 a^*_1 - z_2 a^*_2)\hat{C}(z_1,z_2)$$ (22)

where a^*_1 and a^*_2 are the components along V of the basis reciprocal to that given in (18):

$$a^*_{1s} = (a^*_1, a^*_{1I}) \qquad a^*_{2s} = (a^*_2, a^*_{2I})$$ (23)

and

$$\hat{C}(z_1,z_2) = \int_D dt \; e^{-2\pi i(z_1 a^*_{1I} + z_2 a^*_{2I})t}.$$ (24)

In particular, for a symmetric strip centered on V one gets

$$\hat{C}(z_1,z_2) = \frac{\sin\pi D(z_1 a^*_{1I} + z_2 a^*_{2I})}{\pi D(z_1 a^*_{1I} + z_2 a^*_{2I})}.$$ (25)

One sees that from the intensities of the diffraction spots following from eq. (25) one can get information on the window and on the internal components; from the position of these spots information can be obtained on the external components. In eqs. (21) and (22) one recognizes the Z-module structure of M and M^* generated by the quasicrystal atomic positions and by the Bragg spots, respectively.
    Two cases arise:
(a) V is parallel to a line through two lattice points (rational line)
(b) V is parallel to a line through one lattice point only (irrational line).
The line can be expressed by $\alpha + a_{1s} + \beta a_{2s}$. In the case (a) β is rational and the quasicrystal periodic (commensurate), whereas in case (b) β is irrational and the quasicrystal aperiodic, i.e.

incommensurate.

3.2 Canonical parametrization

It is convenient to adopt the following parametrization (denoted as canonical) of the dual lattice basis sets (18) and (23) in terms of real parameters a, a^*, χ, θ and σ (with $aa^* = 1$)

$$a_{1s} = a(1,\chi) \qquad\qquad a_{2s} = a(\theta,-\chi\sigma) \qquad\qquad (26a)$$

$$a^*_{1s} = \frac{a^*}{\theta+\sigma}(\sigma,\frac{\theta}{\chi}) \qquad a^*_{2s} = \frac{a^*}{\theta+\sigma}(1,\frac{-1}{\chi}). \qquad (26b)$$

In terms of these parameters and for appropriate D value (e.g. $D = a_{2I}$ for $a_{2I} > a_{1I}$) the quasicrystal lattice positions are given by

$$r_m = r_o + ma_1 + a_2 \lfloor \frac{-m\ a_{1I}}{a_{2I}} + \beta_o \rfloor \qquad (27)$$

$$= a_1(\alpha_o + m + \theta \lfloor \frac{m}{\sigma} + \beta_o \rfloor)$$

with m integer, α_o and β_o real and $\lfloor....\rfloor$ the greatest integer function. As shown by Levine, Socolar and Steinhardt [9,10] the ratio of the two intervals $S = a_1$ and $L = a_1+a_2$ between neighbouring positions is given by $1+\theta$ and the corresponding relative frequency n_S/n_L by $\sigma-1$. The parameters α_o and β_o represent initial conditions (or a choice of the origin in the superspace) and leave the local isomorphism class of the quasicrystal invariant. For that reason, the influence of the choice of the origin will be disregarded in what follow (taking α_o and β_o both zero).

3.3 Two-dimensional Bravais lattices

In two dimensions there are five Bravais classes of lattices: oblique, rectangular, rhombic (or diamond), square and hexagonal. Embedding on a oblique lattice imposes no conditions; on a rectangular it requires with $a_{1s}\cdot a_{2s} = 0$

$$\chi^2 = \frac{\theta}{\sigma} . \qquad (28)$$

Now χ can be chosen freely (what matters is the projection) and by an appropriate choice of the orientation, eq. (28) can always be satisfied. Embedding on a diamond lattice implies $|a_{1s}| = |a_{2s}|$ i.e.

$$\chi^2 = \frac{\theta^2 - 1}{1 - \sigma^2} . \tag{29}$$

We may suppose $0 < \theta < 1$ and $0 < \sigma < 1$ by an appropriate choice of a_1 and of $\acute{a}^*_1$. Replacing then a_{2s} by $a'_{2s} = a_{1s} + a_{2s}$ leads to $\theta' = 1+\theta$ and $\sigma' = \sigma-1$, i.e. $|\theta'| > 1$ and $|\sigma'| < 1$. Therefore embedding a quasicrystal on a diamond lattice is also always possible. As θ and σ do have structural meaning even after identifying locally isomorphic quasicrystals, the available parameters appearing in eq. (26) are fixed by the choice of a rectangular or of a rhombic embedding. Embedding on a square lattice, or on an hexagonal one, implies additional relations. As both these lattices are isometric (i.e. such that with $|a_{1s}| = |a_{2s}|$) we can adopt without restriction of generality an isometric embedding. We then have for the metrical tensor of the lattice having chosen basis vectors a_{1s} and a_{2s} and ϕ the angle between them

$$g_{11} = a^2_{1s} = g_{22} = a^2_{2s} = a^2(1 + \chi^2) \tag{30}$$
$$g_{12} = g_{21} = a^2(1 + \chi^2) \cos\phi = a^2(\theta - \chi^2\sigma)$$

implying

$$(\sigma - \theta) \cos\phi = \theta\sigma - 1. \tag{31}$$

Accordingly one gets:

Square lattice ($\cos\phi=0$): $\theta\sigma = 1$ (32a)

Hexagonal lattice ($\cos\phi=1/2$): $\sigma = \dfrac{2 - \theta}{2\theta - 1}$ (32b)

or equivalently: $\theta = \dfrac{2 + \sigma}{2\sigma + 1} .$ (32c)

The dual basis for the same hexagonal lattice has $\cos\phi = -1/2$ and generates what we may call the trigonal lattice. One finds

Trigonal lattice ($\cos\phi=-1/2$): $\sigma = \dfrac{2 + \theta}{2\theta + 1}$ (32d)

and correspondingly for θ as a function of σ. The role of these two parameters is interchanged with respect to the first hexagonal lattice case.

We can conclude that the Euclidean embedding of a 1-dimensional quasicrystal lattice leads to three classes: the rhombic (equivalent to the oblique and to the rectangular), the square and the hexagonal ones [11].

Let us remark that the two Z-modules

$$M^* = \{a^*_2, \sigma a^*_2\}, \qquad M = \{a_1, \theta a_1\} \tag{33}$$

of dimension 1 and rank 2 become identical in the square case after an appropriate choice of the units of length. In the hexagonal case, a trigonal M implies an hexagonal M* and vice versa. Furthermore, the Z-modules generated by the internal components behave in the same way. In both cases, σ irrational implies θ irrational.

3.4 Lattice basis transformations

As the hexagonal versus the trigonal case show, different choices of the lattice basis for a same lattice can lead to structurally different quasicrystals (relative size and relative frequency have structurally quite different meaning). Let us, therefore, consider more closely the arithmetic equivalency, i.e. the lattice basis transformations by elements of the arithmetic group Gl(2,Z) (the group of the integral two-by-two matrices with determinant $\pm$ 1). To

$$A = \begin{matrix} n_{11} & n_{12} \\ n_{21} & n_{22} \end{matrix} \notin Gl(2,Z)$$

we associate a lattice basis transformation

$$a'_{1s} = n_{12}a_{1s} + n_{12}a_{2s} = a'(1,\chi') \tag{34}$$

$$a'_{2s} = n_{21}a_{1s} + n_{22}a_{2s} = a'(\theta',-\chi'\sigma')$$

implying the linear fractional transformations:

$$\theta' = \frac{n_{21} + n_{22}\theta}{n_{11} + n_{12}\theta} \qquad \sigma' = \frac{n_{21} - n_{22}\sigma}{-n_{11} + n_{12}\sigma} \tag{35}$$

$$\chi' = \frac{n_{11} - n_{12}\sigma}{n_{11} + n_{12}\theta} \chi \quad \text{and } a' = (n_{11} + n_{12}\theta) a.$$

Let us now consider a matrix A leaving σ invariant. From $\sigma' = \sigma$ one gets the equation

$$n_{12}\sigma^2 - (n_{11} - n_{22}) \sigma - n_{21} = 0, \tag{36}$$

having solutions

$$\sigma = \frac{(n_{11} - n_{22}) \pm \sqrt{Tr^2 - 4 Det}}{2 n_{12}} \tag{37}$$

where $Tr = n_{11} + n_{22}$ denotes the trace of A and Det its determinant. In the same way, the invariance condition for θ leads to the expression

$$\theta = \frac{(-n_{11} + n_{22}) \pm \sqrt{Tr^2 - 4 \ Det}}{2 \ n_{12}} \tag{38}$$

For such σ and θ the transformation A is a symmetry. Both σ and θ belong to the same real quadratic field $Q(d)$, with discriminant $d = Tr^2 - 4 \ Det$, and are conjugated elements. One has

$$\sigma\theta = \frac{n_{21}}{n_{12}} \quad \text{and} \quad \sigma - \theta = \frac{n_{11} - n_{22}}{n_{12}} \tag{39}$$

after the choice the same (positive) square root sign. In particular choosing $n_{12} = 1$ and $n_{11} = 1$ in the case of a square lattice, as $\sigma\theta = 1$ $n_{21} = 1$ and $n_{22} = 0$, one gets $\sigma = (1+\sqrt{5})/2$ and $\theta = (-1+\sqrt{5})/2$,, the generators of the Fibonacci chain. The matrix A expresses the corresponding enflation rule:

$$A = \begin{matrix} 1 & 1 \\ 1 & 0 \end{matrix} \quad A \begin{pmatrix} 1 \\ 0 \end{pmatrix} = \begin{pmatrix} 1 \\ 1 \end{pmatrix} , \quad A \begin{pmatrix} 0 \\ 1 \end{pmatrix} = \begin{pmatrix} 1 \\ 0 \end{pmatrix}$$

4. SELF-SIMILARITY AND ENFLATION RULES

If A is a symmetry for θ and/or for σ, then so also the group K generated by A which contains elements A_+ with determinant +1. Irrational σ implies trace of $A_+ > 2$, and accordingly the group K is of infinite order. Therefore A is not an Euclidean symmetry transformation and the distances are not conserved. Inspection of eq. (35) shows that a symmetry for θ leaves the Z-module M invariant, i.e. self-similar, and K is its self-similarity group. The same can be said for σ with respect to M*.

The situation is slightly different for the quasicrystal lattice. Due to the window function $C(m_1,m_2)$ and the associated existence of a minimal interval, the symmetry is reduced to the semigroup generated by enflations. The inverse operation, the deflation, is beyond the minimal distance no more an allowed transformation. Without claiming completeness, let us now look at the square, hexagonal and trigonal cases.

4.1 The square lattice case

The necessary and sufficient condition for being in this case is $n_{12} = n_{21}$. By total inversion, one can always get $n_{12} > 0$; the case $n_{12} = 0$ being excluded because leading to Tr = 0, and σ (and θ) either rational or not real. The lowest possible value is $n_{12} = n_{21} = 1$ implying $n_{11}n_{22} = 0$ in the Det = -1 case (the Det = +1 case is then also generated). Choosing $n_{22} = 0$ and denoting n_{11} by ν, we have:

$$A_{\nu-} = \begin{matrix} \nu & 1 \\ 1 & 0 \end{matrix} \qquad \text{integer } \nu \geq 1 \tag{40}$$

and correspondingly

$$\sigma_\nu = \frac{\nu \pm \sqrt{\nu^2+4}}{2}, \qquad \theta_\nu = \frac{-\nu \pm \sqrt{\nu^2+4}}{2}. \tag{41}$$

The enflation rule defined by $A_{\nu-}$ is associated with the characteristic equation det $(A_{\nu-} - x) = 0$, which leads precisely to the invariance equation for σ (and θ respectively) [12]:

$$x^2 - \nu x - 1 = 0 \tag{42}$$

and to the enflation rule in terms of

$$C_n = A_{\nu-}^n \, a \qquad \text{for } a = \begin{matrix} 1 \\ 0 \end{matrix} \quad b = \begin{matrix} 0 \\ 1 \end{matrix} \quad . \tag{43a}$$

One gets

$$C_2 = C_1^\nu \, C_0 \tag{43b}$$

as one verifies: $C_0 = a$, $C_1 = a^\nu b$, $C_2 = (a^\nu b)^\nu a$. A general element of the group K has the form:

$$A_{\nu-}^k = \begin{matrix} f_{k+2} & f_{k+1} \\ f_{k+1} & f_{k-1} \end{matrix} \tag{44}$$

with f_k obeying the recursion relation

$$f_{k+1} = f_{k-1} + \nu \, f_k \tag{45}$$

with initial values $f_0 = f_1 = 0$. For $\nu = 1$ one obtains the Fibonacci numbers.

4.2 The hexagonal lattice case

The hexagonal case is more difficult. Again the value $n_{12} = 0$ (or n_{21}) is excluded. We may suppose $n_{12} > 0$, consider $n_{12} = 1$ and parametrize $n_{21} = \lambda + 1$. The Det $= -1$ case implies $\lambda(\lambda+1) = n^2$, which does not has integral solutions. Det $= +1$ leads to the Diophantine equation

$$n^2 = \lambda^2 + \lambda + 2 \tag{46}$$

whose integral solutions build up a matrix $A_{\lambda+}$:

$$A_{\lambda+} = \begin{matrix} \lambda + n & 1 \\ \lambda + 1 & -\lambda + n \end{matrix}$$

which is a symmetry for

$$\sigma_\lambda = \lambda \pm \sqrt{n^2 - 1} \,, \quad \theta_\lambda = -\lambda \pm \sqrt{n^2 - 1} \,. \tag{47}$$

A particular solution is $\lambda = 1$ and $n = 2$ leading to

$$A_{1+} = \begin{matrix} 3 & 1 \\ 2 & 1 \end{matrix} \quad \text{and} \quad \sigma = 1 \pm \sqrt{3}\,, \quad \theta = -1 \pm \sqrt{3}\,, \tag{48}$$

another solution $\lambda = -2$ and $n = 2$ being arithmetically equivalent. The corresponding characteristic equation

$$x^2 - 4x + 1 = 0$$

leads to the enflation rule $Aa = a^3 b^2$ and $Ab = ab$, thus to

$$C_o = a, \quad C_1 = a^3 b^2, \quad C_2 = C_1^4 C_o^{-1} \,. \tag{49}$$

4.3 The trigonal lattice case

As above, we may choose $n_{12} = 1$ and $n_{21} = \lambda + 1$ leading to $n_{22} - n_{11} = 2\lambda$. Again the Det = -1 case is excluded and the self-similarity symmetry transformation is as above, after interchanging n_{11} and n_{22}, i.e. also σ_λ with θ_λ. A particular solution is given by

$$A_{1+} = \begin{matrix} 1 & 1 \\ 2 & 3 \end{matrix} \quad \sigma = -1 \pm \sqrt{3}, \quad \theta = 1 \pm \sqrt{3}. \tag{50}$$

5. MINKOWSKIAN EMBEDDING OF 1-DIMENSIONAL QUASICRYSTALS

The self-similarity symmetries with determinant +1 leave lattices of an indefinite metric plane invariant. Furthermore, as one verifies, the invariance condition (36) for σ implies invariance of the axis e_1, i.e. of the external space V. The same can be said, mutatits mutandi, for θ, e_2 and the internal space V_I. Accordingly, a self-similarity symmetry leaving θ and σ invariant is a <u>crystallographic Lorentz transformation</u> of a plane where V and V_I are the isotropic subspaces leaving invariant the lattice Σ generated by a_{1s} and a_{2s}. The group K_+ of the proper transformations is then the relativistic point group, symmetry (holohedry) of Σ. The self-similarities which are symmetries with negative determinant, if they exist, transform the metric tensor into its negative and are called negautomorphs [13,14].

Let us verify analytically these results by introducing in the plane an indefinite metrics with orthonormal basis vectors ε_o and ε_1 having scalar products $\varepsilon_1^2 = -\varepsilon_o^2 = 1$, $\varepsilon_o \varepsilon_1 = 0$. The embedding of V and V_I being on the light cone we put

$$e_1 = \frac{1}{\sqrt{2}}(\varepsilon_o + \varepsilon_1) \qquad e_2 = \frac{1}{\sqrt{2}}(\varepsilon_o - \varepsilon_1). \tag{51}$$

Accordingly the basis vectors of Σ become

$$a_{1s} = \frac{1}{\sqrt{2}} \left[a (1 + X) \varepsilon_o + (1 - X) \varepsilon_1 \right] \tag{52}$$

$$a_{2s} = \frac{1}{\sqrt{2}} \left[a (\theta - X\sigma) \varepsilon_o + a (\theta + X\sigma) \varepsilon_1 \right] .$$

The corresponding Minkowskian metric tensor is

$$g_M = a^2 X \begin{bmatrix} -1 & \dfrac{\sigma - \theta}{2} \\ \dfrac{\sigma - \theta}{2} & \sigma\theta \end{bmatrix} \tag{53}$$

whereas the Euclidean metric tensor for the same basis is

$$g_E = a^2 (1 + X^2) \begin{bmatrix} 1 & \dfrac{\sigma\theta - 1}{\sigma - \theta} \\ \dfrac{\sigma\theta - 1}{\sigma - \theta} & 1 \end{bmatrix} . \tag{54}$$

The invariance of the Minkowskian metrics with respect to the transformation A is expressed by the condition:

$$A \, g_M \, A^t = g_M \tag{55}$$

(note that components transform contravariantly with respect to the basis vectors). Substituting in eq. (55) the corresponding expressions (53) and (39), one verifies the relations

$$A_+ \, g_M \, A_+^t = g_M \quad \text{if} \quad \det A_+ = +1 \tag{56}$$

and

$$A_- \, g_M \, A_-^t = -g_M \quad \text{if} \quad \det A_- = -1. \tag{57}$$

The surprising result one gets from a crystallographic analysis of 1-dimensional quasicrystal lattices is that square and hexagonal lattices, if adequately embedded in a Minkowskian plane, do have a relativistic holohedry of infinite order.

6. BRAVAIS CLASSES FOR 1-DIMENSIONAL QUASICRYSTAL LATTICES

Euclidean embedding of quasicrystal lattices leads to three Euclidean Bravais lattices, which after identification of local isomorphic quasicrystals represent, on the Z-module level for M and for M*, the natural generalization of the basic properties of the corresponding 2-dimensional Bravais lattices (think at the situation e.g. for Σ and Σ^* square dual lattices). These properties, however, do not reflect enflation rules, if present. This non-trivial structural property of incommensurate quasicrystal lattices is taken into account by Bravais classes of 2-dimensional relativistic lattices. It seems, therefore, natural to associate to a given quasicrystal lattice both the Euclidean as well as the Minkowskian holohedries of the same 2-dimensional lattice (for the most symmetric embedding of the quasicrystal). This allows a formal definition of Bravais classes for quasicrystal lattices by a joint arithmetical equivalence of both types of holohedries, respectively. Testing the validity of these ideas requires, of course, a more detailed investigation of the rhombic lattice case as well. At the present stage of knowledge it is already interesting to mention that the hexagonal and the trigonal lattice cases, which from the Euclidean point of view are arithmetically equivalent, are no more so from the Minkowskian metrical point of view. The trigonal lattice leads to the relativistic Bravais class M_4 characterized by the quadratic form $[\bar{1},\bar{2},2]$, whereas the hexagonal lattice leads to the Bravais class $[1,2,2]$ denoted as $\bar{M}_4$. In the square lattice case (labeled by v) there is only one relativistic Bravais class for each $v \geq 1$, the corresponding Bravais lattice being noted as N_v and characterized by the quadratic form $[\bar{1},v,1]$. This if one disregards possible centering [14]. The notation $[a,b,c]$ stands for the metric tensor $g_{11} = a$, $g_{22} = c$ and $g_{12} = b/2$.

7. RECIPROCAL SPACE ASPECTS

Given a lattice Σ, its dual or reciprocal one Σ^* depends on the metrics. Quite in general, for the identification of the superspace V_s with its reciprocal V^*_s a specific (non-singular) metrics is required. The phase of the Fourier coefficients of a crystal structure represents an Euclidean scalar product of the type $\vec{k} \cdot \vec{r}$. Its extension to a scalar product in superspace, which is of the type

$$k_s \cdot r_s = (\vec{r}, \vec{r}_I) \cdot (\vec{k}, \vec{k}_I) = \vec{k}\vec{r} + \vec{k}_I\vec{r}_I \qquad (58)$$

plays an essential role in the conditions imposed by superspace group symmetry on the crystal structure and leads to selection rules for physical properties, like systematic extinction rules in X-ray diffraction patterns [15,16,17,18].

In the Euclidean case, the lattice basis a^*_{1s}, a^*_{2s}, reciprocal to a_{1s}, a_{2s}, has been explicitly given in eq. (26b). It is straightforward to derive the corresponding reciprocal basis $\hat{a}^*_{1s}, \hat{a}^*_{2s}$ for the Minkowskian metric case. Indeed, from the orientation of the light cone with respect to e_1 and e_2 follows $e_1^2 = e_2^2 = 0$ and $e_1 e_2 = -1$. So instead of eq. (26b) one gets

$$\hat{a}^*_{1s} = \frac{-a^*}{\sigma + \theta} \left(\frac{\theta}{x}, \sigma\right)$$

$$\hat{a}^*_{2s} = \frac{-a^*}{\sigma + \theta} \left(\frac{-1}{x}, 1\right) \tag{59}$$

spanning the lattice Σ^*_M. Comparing the two expressions, one sees that to a reciprocal lattice vector k^E_s of the lattice Σ^*_E it corresponds k^M_s of Σ^*_M with components $(-k_I, -k)$. In the embedding (both Euclidean and/or Minkowskian) the basis is changed according to the increased dimensionality, but the components h_i of the elements, the Z-module and of the lattice are correspondingly the same

$$\vec{k} = \sum_{i=1}^{n+d} h_i \vec{a}^*_i \;\leftrightarrow\; k^E_s = \sum_{i=1}^{n+d} h_i a^*_{is} \;\leftrightarrow\; k^M_s = \sum_{i=1}^{n+d} h_i \hat{a}^*_{is} \tag{60}$$

in the present case n and d being 1. Therefore, the value of the scalar product is independent of the embedding's metrics adopted. For

$$r_s = \sum_{i=1}^{n+d} x_i a_{is}$$

one has in both cases

$$r_s \cdot k^E_s = r_s \cdot k^M_s = \sum_{i=1}^{n+d} x_i h_i \tag{61}$$

8. AN EXAMPLE: THE SQUARE LATTICE CASE

In this last section we simply collect the relevant expressions for the case of a quasicrystal lattice which can be embedded on a 2-dimensional square lattice. We know that, disregarding centring aspects, a positive integer ν label the possibilities. The canonical parameters are

$$\sigma_\nu = \frac{\nu + \sqrt{\nu^2 + 4}}{2}, \qquad \theta_\nu = \frac{-\nu + \sqrt{\nu^2 + 4}}{2} \tag{62}$$

$$x^2_\nu = \frac{\theta_\nu}{\sigma_\nu} \quad \text{thus} \quad x_\nu = \pm\, \theta_\nu. \tag{63}$$

Indeed

$$\theta_\nu \sigma_\nu = 1 \quad \text{and} \quad \sigma_\nu - \theta_\nu = \nu \tag{64}$$

$$\cos \phi_\nu = 0, \qquad \tan \psi_\nu = \theta_\nu ,$$

with ϕ_ν the angle between a_{1s} and a_{2s}, and ψ_ν that between a_{1s} and e_1 (the external space axis). The Euclidean metric tensor is that of a square lattice

$$g_E(\nu) = a^2(1 + \theta_\nu^2) \begin{bmatrix} 1 & 0 \\ 0 & 1 \end{bmatrix} \tag{65}$$

whereas the Minkowskian metric tensor is that of the lattice denoted N_ν

$$g_M(\nu) = 2\,a^2\theta_\nu \begin{bmatrix} \bar{1} & \nu/2 \\ \nu/2 & 1 \end{bmatrix}. \tag{66}$$

The Euclidean holohedry is

$$K^E = \{M_R, M_D\} = 4mm \tag{67}$$

with

$$M_R = \begin{matrix} \bar{1} & 0 \\ 0 & 1 \end{matrix}, \quad M_D = \begin{matrix} 0 & 1 \\ 1 & 0 \end{matrix}.$$

The Minkowskian holohedry (including negautomorphs) is

$$K^M = \{N_R, A_{\nu-}\} \tag{68}$$

with

$$N_R = \begin{matrix} 0 & 1 \\ \bar{1} & 0 \end{matrix}, \quad A_{\nu-} = \begin{matrix} \nu & 1 \\ 1 & 0 \end{matrix}.$$

ACKNOWLEDGEMENTS

Thanks are expressed to the Stichting voor Fundamenteel Onderzoek der Materie of the Nederlandse Organisatie voor Zuiver Wetenschappelijk Onderzoek for partial financial support.

REFERENCES

[1]   Penrose R., Math. Intelligencer 2 (1979) 32-37.

[2]   Janssen T., Acta Cryst. A 42 (1986) 261-271.

[3]   Janssen T. and Janner A., Adv. Phys. (1987). (To appear).

[4]   Duneau M. and Katz A., Phys. Rev. Letters 54 (1985) 2688-2691.

[5]   Katz A. and Duneau M., J. de Physique 47 (1986) 181-196.

[6]   Katz A. and Duneau M., J. de Physique, Colloque 47 (1986) C3 103-113. International Workshop on Aperiodic Crystals, Les Houches.

[7]   Bruijn N.G. de, Proc. Kon. Ned. Ak. Wetenschappen A 84 (1981) 39-52, 53-66.

[8]   Elser V., Acta Cryst. A 42 (1986) 36-43.

[9]   Levine D. and Steinhardt P.J., Phys.Rev. B 34 (1986) 596-616.

[10]   Socolar J.E.S. and Steinhardt P.J., Phys.Rev. B 34 (1986) 617-647.

[11]   Janner A., J. de Physique, Colloque 47 (1986) C3 95-102. International Workshop on Aperiodic Crystals, Les Houches.

[12]   Bombieri E. and Taylor J.E., J. de Physique, Colloque 47 (1986) C3 19-28. International Workshop on Aperiodic Crystals, Les Houches.

[13]   Janner A. and Ascher E., Zeit. Krist. 130 (1969) 277-303.

[14]   Janner A. and Ascher E., Physica 45 (1969) 33-66, 67-85.

[15]   Janner A. and Janssen T., Phys. Rev. B 15 (1977) 643-658.

[16]   Janner A. and Janssen T., Physica A 99 (1979) 47-76.

[17]   Janner A. and Janssen T., Acta Cryst. A 36 (1980) 399-408, 408-415.

[18]   Wolff P.M. de, Janssen T. and Janner A., Acta Cryst. A 37 (1981) 625-636.

N-dimensional crystallographic description of the icosahedral phases; the example of the $Al_{73}Mn_{21}Si_6$ quasiperiodic structure

D. Gratias(*), J.W. Cahn(+), M. Bessiere(*$), Y. Calvayrac(*)
S. Lefebvre(*$), A. Quivy(*) and B. Mozer(+)

(*)CECM/CNRS 15,rue G. Urbain 94407-Vitry/France
(+)National Bureau of Standards, Institute for Materials Science
and Engineering
Gaithersburg MD-20899 USA
($) also at L.U.R.E. bat.209d 91405-Orsay-Cedex/France

Abstract:
X-ray powder diffraction spectra of a well characterized $Al_{73}Mn_{21}Si_6$ icosahedral structure are analyzed using the 6-dimensional space description. The self-correlation function shows essentially two extrema at the nodes and at the body centers of the 6D unit cell. This results rules out models based on simple decoration of a Penrose tiling and strongly suggest that at least two orbits have to be considered in the modelisation.

Diffraction experiments in temperature show a clear change in the thermal expension coefficient at 130K which could be associated with a antiferro magnetic transition.

1 Introduction

Icosahedral phases in aluminium based alloys have been extensively studied by means of X-ray, neutron and electron diffraction[1]. Many spectra have been collected, but no fully satisfying crystallographic model that matches properly the observed diffracted intensities has been proposed. The structural characterization of quasiperiodic phases is a formidable task owing to the numerous parameters which have to be determined in addition to those of standard 3D crystallography[2].

The present paper reports experiments carried out on a reproducibly created and well characterized icosahedral $Al_{73}Mn_{21}Si_6$ ternary alloy which has been investigated by X-ray [3]and neutron diffraction [4]. The crystallographic analysis is based on the following heuristic hypotheses:

i - Non-redundant indexing requires only 6 indices; the 6D space description has therefore been used. (This, however, does not necessarily imply that the actual structure in 6D would fit with atomic 3D surfaces depending only on the perpendicular space variable, $x_\perp$);

ii- The powder diffraction patterns have been indexed using the scheme proposed by Cahn, Shechtman & Gratias[5] assuming a primitive 6D hypercubic lattice; actually, two other equivalent lattices with same multiplicity could have been considered as well : the D_6^+ and D_6^- (dual to each other) lattices defined by,

$$D_6^{(\pm)} = F(2A) + (\tfrac{1}{2},\tfrac{1}{2},\tfrac{1}{2},\tfrac{1}{2},\tfrac{1}{2},\pm\tfrac{1}{2})\, F(2A),$$

where F(2A) designates the face-centered hypercubic 6D lattice with parameter 2A, are consistent with both the icosahedral symmetry and the extinction rules observed in single domain electron diffraction. These lattices lead to projected quasilattices τ times smaller (for D_6^+)or larger(for D_6^-) than the one obtained for the primitive 6Dlattice.

iii - the 6D point group of the structure has been assumed to be m35. It is obvious, even from powder diffraction patterns, that the 6D point group is m35 and not a

A. Amann et al. (eds.),
Fractals, Quasicrystals, Chaos, Knots and Algebraic Quantum Mechanics, 111–119.
© 1988 by Kluwer Academic Publishers.

112

supergroup of m35 (Intensities of equivalent reflections in the B_6 holohedral point group are clearly different). Although the hemihedral non- centrosymmetric 235 group could have been chosen as well, convergent beam electron diffraction[6] and recent neutron contrast experiments[7] have shown that the motif is most likely to be centrosymmetric.

2 -Experimental

A well homogenized master alloy was prepared from ultra pure silicon, 99.99% aluminium and 99.9% manganese by levitation melting in a controlled helium atmosphere. The subsequent rapid quenching was carried out by planar flow casting on a copper wheel under a helium atmosphere. The quenching temperature was 1100°C and the tangential wheel speed $25m.s^{-1}$. The samples obtained were flakes of a few millimeters in size and 30 μm thick.

The X-ray data were obtained using several experimental conditions. The peak intensity measurements at room temperature were carried out using Co K_α radiation on a diffractometer equipped with a graphite monochromator placed in the diffracted beam. To measure the angular positions of the peaks, data were obtained using a Cu K_β beam off a curved LiF monochromator, in order to avoid the broadening due to the usual K_α doublet. Experiments at low temperatures were carried out on a special high precision diffractometer built by Bérar et al.[8] at the Ecole Centrale (Chatenay-France).

The stoichiometry chosen leads to nearly single-phase icosahedral samples; traces of fcc aluminium (1%) and β-crystalline hexagonal phase (2%) were detected. Upon heating to 700 °C the as-quenched alloy transforms nearly completely into β phase (98% β, 2% fcc). The cubic α-phase was prepared from a $Al_{73}Mn_{16}Si_{11}$ alloy by annealing at 500°C (98% α, 2% fcc).

A good fit was found between the observed and theoretical icosahedral positions of the peaks. The largest differences reach 10^{-3} Å^{-1} for two reflections close to the fcc or β phase peaks (see figure 1). No asymmetry of the X-ray lines was observed. The peak width dependence with $(q_{//}, q_\perp)$ has been found to follow quite well a quadratic law $\delta q_{//}{}^2 = \alpha q_{//}{}^2 + \beta q_\perp{}^2$ (with $\alpha = 0.247$ and $\beta = 13.9$) shown on figure 2. as expected from the frozen-in phason strain model studied by Horn et al[9]. whereas no agreement is found with the icosahedral glass model.

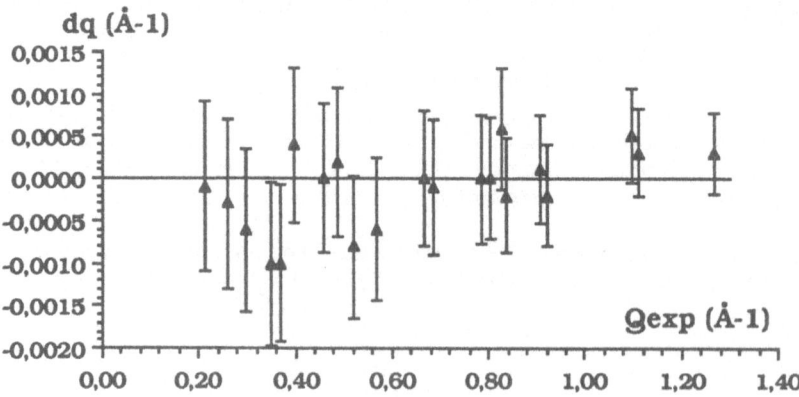

Figure 1 : Differences between the experimental and theoretical icosahedral positions of the reflections; the icosahedral indexing fits better within 10^{-3} Å^{-1} .

X-ray measurements at low temperature (from 294 to 80K) show an isotropic contraction of all samples. The integrated intensities of the icosahedral phase as well as the profile of the reflections are remarkably constant with temperature which implies an unusually small Debye-Waller factor. This could be related to the observed weak variation with temperature of the resistivity measured in a similar alloy by C.

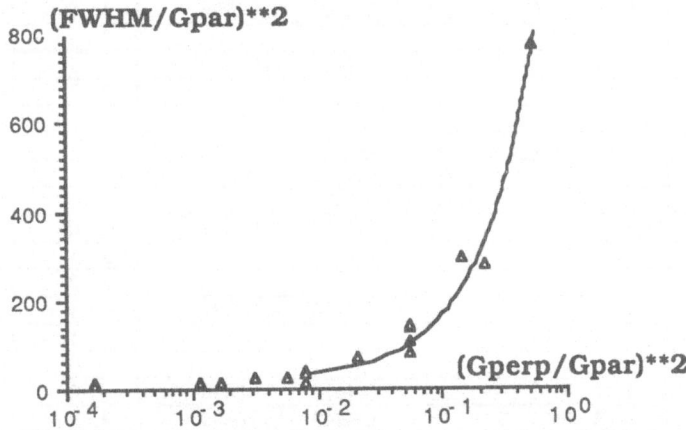

Figure 2 : Variation of the peak width as a function of $(q_{//}, q_\perp)$; a reasonnable fit is obtained with a variation $\delta q_{//}^2 = \alpha q_{//}^2 + \beta q_\perp^2$ corresponding to the frozen-in phason strain type of disorder with $\alpha = 0.247$ and $\beta = 13.9$. This invalidates the icosahedral glass model.

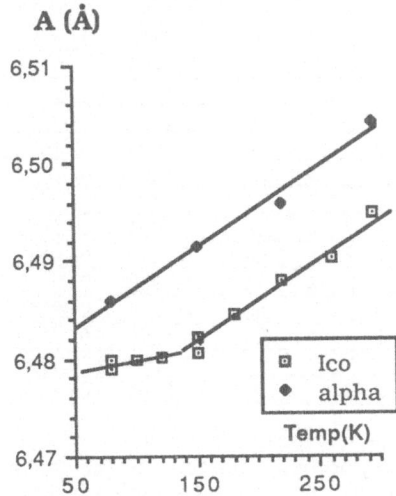

Figure 3 : Variation of the 6D lattice parameter with temperature for the icosahedral and the cubic α phases (see text).

Berger[10]. A possible explanation has been proposed recently by Cyrot & Cyrot-Lackman[11] based on an avalanche effect on electronic coherency loss due to long

N	M	n_1	n_2	n_3	n_4	n_5	n_6	μ	$q(\text{A}^{-1})$	$|f|(e/\text{Å}^3)$	Corresp. α phase			
0	0	0	0	0	0	0	0	1	0.0	1.0316	0	0	0	1.0256
4	4	0	0	1	0	0	1	30	0.1852	0.00752	2	0	0	-0.0201
											2	1	1	0.0036
6	5	0	1	1	0	-1	0	60	0.2146	0.00836				
6	9	0	1	1	0	0	1	20	0.2591	0.07794	3	1	0	0.1046
											2	2	2	0.0767
8	12	0	1	1	0	-1	1	30	0.2989	0.03240	3	2	1	0.0444
											4	0	0	0.0313
12	16	1	1	1	1	-1	1	12	0.3512	0.00957				
12	16	0	1	2	0	0	1	60	0.3512	0.00949				
14	17	0	2	1	0	-1	1	60	0.3676	0.00816				
14	17	0	1	2	1	0	1	60	0.3676	0.00816				
14	21	0	1	2	0	-1	1	60	0.3967	0.02944	1	5	0	-0.0727
											4	3	1	-0.0346
											4	2	2	-0.0291
16	24	1	1	2	0	-1	1	60	0.4236	0.04729	3	4	1	-0.0568
											2	5	1	-0.0926
											4	4	0	-0.0919
18	29	1	1	2	1	-1	1	12	0.4610	0.31898	3	5	0	0.3409
20	32	0	1	2	0	-1	2	30	0.4849	0.26392	5	3	2	0.3217
											6	0	0	0.3325
24	36	1	1	2	1	-1	2	60	0.5181	0.03663				
24	36	0	2	2	0	0	2	20	0.5181	0.03663				
28	44	1	2	2	0	-1	2	60	0.5692	0.03694				
28	44	1	1	3	1	-1	1	12	0.5692	0.03694				
38	61	1	2	3	0	-1	2	60	0.6689	0.09126				
40	64	1	2	3	1	-1	2	60	0.6954	0.04012				
48	76	1	2	3	0	-1	3	120	0.7483	0.02692				
52	84	0	2	3	0	-2	3	30	0.7843	0.25291	8	5	3	0.3408
											10	0	0	0.3529
56	88	1	2	3	1	-2	3	60	0.8059	0.01668				
56	88	0	3	3	0	-1	3	60	0.8059	0.01668				
58	93	1	3	3	0	-1	3	60	0.8267	0.02892				
60	96	1	3	3	-1	-1	3	20	0.8394	0.02855				
60	96	1	2	4	1	-2	2	60	0.8394	0.02855				
70	113	1	2	4	1	-2	3	60	0.9099	0.10689				
72	116	2	2	4	2	-2	2	12	0.9218	0.09148				
72	116	1	3	4	0	-1	3	60	0.9218	0.09148				
102	165	1	4	4	-1	-1	4	20	1.0996	0.13533				
104	168	2	3	5	1	-2	3	60	1.1095	0.07798				
136	220	0	3	5	0	-3	5	30	1.2694	0.08538				

Table 1: X-ray absolute scattering factors $|f|$ of the principal reflections of the icosahedral phase and corresponding reflections of the α phase.
wave length defects (static phasons). The variation with temperature of the 6D lattice parameter, A(T), for both the icosahedral and cubic phases are shown on figure 2.

The cubic phase has an isotropic thermal expansion coefficent $\alpha = 12.75 \ 10^{-6} \ K^{-1}$; the icosahedral phase exhibits a clear change in the slope of A(T) around 130K, from $\alpha = 13.66 \ 10^{-6}K^{-1}$ (similar to the α-phase) above 130K, to $\alpha \approx 2.3 \ 10^{-6}K^{-1}$ below 130K, which is an unusually small value for metallic alloys. The powder spectra show neither additional peaks nor intensity changes after crossing this temperature. This change in slope could be possibly associated with a second-order antiferromagnetic transition.

The integrated intensities (Table 1) have been normalized in absolute units by using a standard Ni_3Fe powder. Five reflections of the standard specimen were carefully measured on the same diffractometer under the same experimental conditions in order to determine the diffractometer proportionality constant. The X-ray normalization is given in units of electron units/Angstrom^3, instead of per unit cell, for the obvious reason of the absence of periodicity in the quasicrystalline samples. Once the normalization is performed, it is easy to calculate the zero term of the table from the concentration and the density of the sample. Assuming a density of $3.62g/cm^3$, a 6D lattice parameter of 6.495Å, the zero term is found to be 1.0316 e/A^3 for the icosahedral phase with composition $Al_{73}Mn_{21}Si_6$ which is very close to the value of 1.025 e/A^3 found for the cubic α structure. The list of the absolute values of the measured structure factors of the icosahedral phase is given in Table 1, where N and M are the powder indices defined in ref. 4, and n_1,n_2,n_3,n_4,n_5,n_6 are the 6D indices of one representative of the orbits considered and μ is its multiplicity. The corresponding structure factors of the α phase are also given for some of the strongest non-degenerate peaks. These structure factors have been calculated with the model of Cooper & Robinson[12]. The α cubic reflections corresponding to a given icosahedral reflection are found from the cut and projection technique applied for a rational cut[13,14] where the golden mean $\tau = \frac{1}{2}(1 + \sqrt{5})$ is approximated by $\frac{p}{q} = \frac{1}{1}$. In this scheme, H,K,L, cubic reflections are obtained from the six indices h/h', k/k', l/l' icosahedral indices by the simple relation:

$$H = q \, h + p \, h' \text{ (and cyclic permutations)} \qquad /1/$$

where p and q are successive integers of the Fibonacci series; the cubic lattice parameter A is given by :

$$A = \frac{2A_0 \, (p\tau+q)}{\sqrt{2(2+\tau)}} \qquad /2/$$

with A_0 as the 6D lattice parameter of the icosahedral phase. The breaking of symmetry from m35 to m3 generally splits single icosahedral orbits into several nonequivalent cubic orbits (a complete description of the cubic/icosahedral phase relation-ships is to be found in [15]). Orbits of medium range q wave vectors with strong intensity in the icosahedral phase correspond to short $q_\perp$ and are easily identified in the α phase with comparable peak positions and intensities. This suggests, as shown in the next section, that the chemical short range order is similar in both phases, as can be expected from a rational approximant of a quasiperiodic structure.

3 -6D Patterson (self-correlation) function

One of the specific problems in incommensurate phases and quasicrystals is the presence of an infinite set of pseudo-translations which result in a partial overlap of the structure with itself. The interatomic distances corresponding to motif atomic distances are not easily distinguishable from those due to the pseudo-translations. 3D Patterson maps show all translations, and can be readily created by a Fourier transform of the indexed intensity data. The decision about which Patterson peaks are due to pseudo-translations and which are true signatures of the elementary

116

interatomic distances within the motif is much clearer in a 6D Patterson map. Such

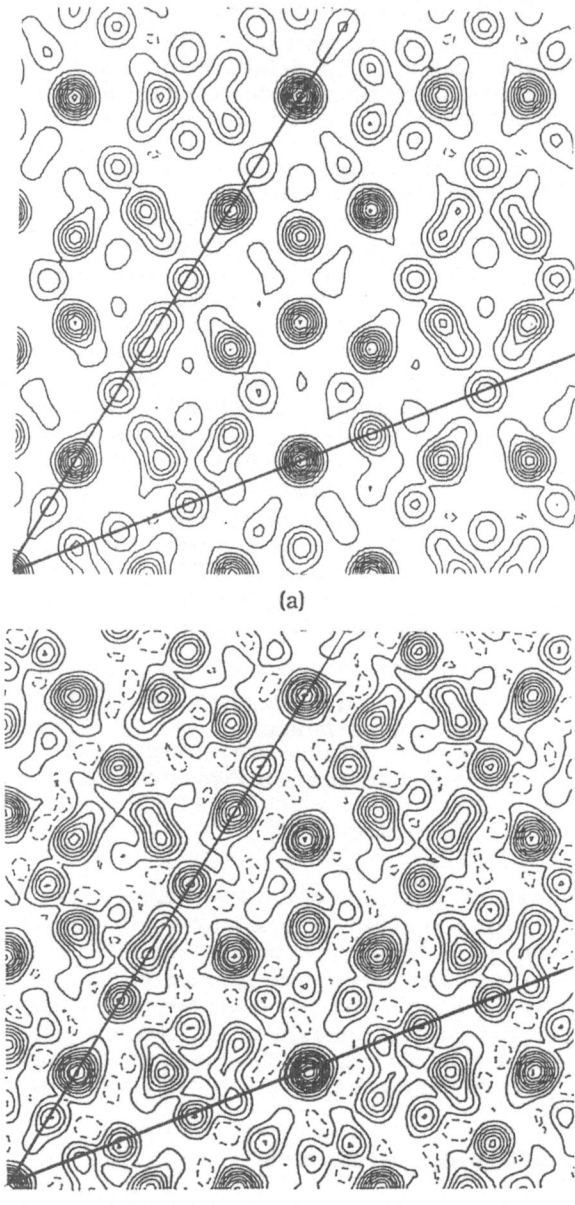

(a)

(b)

Figure 4 : Self-correlation function of (a) the icosahedral phase perpendicular
to a 2-fold axis, (b) the α-phase in the irrational plane $[\tau,1,\tau^2]$ perpendicular to a
pseudo-2-fold axis. The 3-fold and 5-fold direction are shown as full lines with their
corresponding pseudo-axes in the α-phase.

Fourier calculation has been directly performed[16,17] in the 6D space where full periodicity is recovered :

$$\mathbb{P}(\mathbf{R}) = \sum_{(h,h',k,k',l,l')} \mathbb{I}(h,h',k,k',l,l') \, e^{2i\pi \, \mathbf{K}.\mathbf{R}} \qquad \qquad /3/$$

where $\mathbf{K}$ and $\mathbf{R}$ are the reciprocal and direct 6D variables. This yields a 6D periodic mapping of the standard 3D Patterson function that for these alloys is quasiperiodic. The 6D calculation can be displayed in either parallel or orthogonal spaces as well as along rational subspaces of the 6D embedding space.

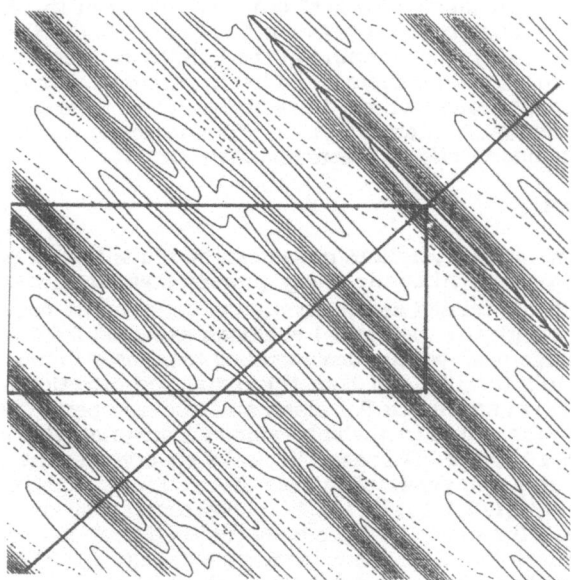

Figure 5 : Self-correlation function of the icosahedral phase displayed on a rational plane of the 6D space (spanned by [1,0,0,0,0,0] and [0,1,1,1,1,-1]) that contains the 5-fold axis in 3D. The full 6D periodicity is recovered. The unit cell is outlined and the traces of the parallel and perpendicular spaces are drawn in full lines.

The 3D pseudo-translation peaks arise from cuts of the fundamental peaks elongated along the orthogonal space and attached periodically to all 6D lattice nodes. When displayed in the 3D physical space, the self-correlation function of the icosahedral phase closely resembles the function obtained from the cubic α phase; both phases have similar short-range atomic order. The typical peaks associated with the double icosahedron of Al-Mn called the Mackay icosahedron (MI) are clearly identified in the quasicrystal because they so closely match those in the crystal. This result experimentally confirms the original idea of Guyot and Audier[18] and the decoration proposed by Elser and Henley[19]. An additional peak is found in the 2-fold projection of the quasicrystal (Figure 4-a) which is not observed in the 2-fold axis map of the crystal, but turns out to be in the irrational $[\tau,1,\tau^2]$ plane which is the pseudo 2-fold plane of the crystal (Figure 4-b). This additional peak is the signature of a stacking of the basic MI's along all the ten 3-fold axes in the quasicrystal, whereas the stacking is only along the four cubic 3-fold axes in the crystal. The display in the perpendicular space shows extended peaks as expected in the 6D models[20] in which the atoms are represented by 3D volumes. This elongated shape is however not surprising since the calculation uses intensities corresponding to short $q_{\perp}$ values and therefore long wave variations in the perpendicular space. A plot along rational 2D planes of the 6D space of course shows exact periodicity. A careful examination within the 6D unit cell revealed no peaks other than the fundamental peak located at the lattice nodes and a secondary peak at the body center of the unit cell(Figure 5). *All peaks observed in the physical 3D space are entirely due to these sole two peaks* giving a surprisingly simple self-correlation function in 6D space; a 6D "CsCl" structure.

The peak at the body centered peak is a little shorter than the one at the origin. This indicates that the window functions (or atomic 3D surfaces) are probably different not just for the different kinds of atoms, but also for different orbits of the same species. Not much chemical order can be seen in the X-ray self-correlation function (not enough contrast between Al, Mn and Si) but the corresponding neutron studies show a strong shortening of the Patterson peak associated with the lattice node which suggests that there could be, in perpendicular space, a spherical shell of heteroatomic distances (which, in neutrons, gives a negative self-correlation function) surrounding an internal sphere of homoatomic distances. The body centered peak corresponds mostly to heteroatomic distances.

4 -Conclusion

A good quality quasicrystalline phase can be made from a rapidly solidified ternary $Al_{73}Mn_{21}Si_6$ alloy. Absolute X-ray powder measurements permit accurate comparisons with the related stable crystalline phases. The peak locations in reciprocal space fit well with the theoretical positions expected from the icosahedral indexing with an average peak width corresponding to some 40 nm correlation length. In order to take full advantage of both the radial and directional information contained in the diffraction spectra, a Patterson analysis was performed (instead of the usual radial distribution function) which exhibits a surprisingly simple internal structure when displayed in the 6D space. All peaks observed in the 3D self-correlation map result from only two peaks of the 6D unit cell, one centered on the origin and the other at the $(1/2,1/2,1/2,1/2,1/2,1/2)$ body center. The comparison between neutron and X-ray spectra shows that the body centered peak is mainly due to heteroatomic Al-Mn distances and that the central peak results from the superimposition of an homoatomic distance surrounded by a spherical shell, in orthogonal space, of heteroatomic distances. The same distances are also identified in the stable crystalline α phase : the short range chemical order of the icosahedral phase is essentially the same as in the crystalline α phase. The basic difference between the two structures is the stacking of the basic atomic clusters periodically connected in the α phase through four among the ten possible icosahedral three-fold axes, and equally connected in average through all the ten three fold axes in the quasiperiodic phase.

Measurements at low temperature show no detectable variations in intensity which implies a very small value of the Debye-Waller term.

Probably the most striking result of the present study is the change in the slope of the 6D lattice parameter versus temperature around 130K ; the cause of this effect is yet unknown. It could be the signature of a second-order magnetic phase transition which would generate no detectable additional peaks and no discontinuities in the intensities. Additional experiments are in progress.

Acknolewdgments

We kindly thank our colleagues J. Bigot, A. Dezellus and S. Peynot for the preparation of the samples; we are extremely gratefull to J. -F. Bérar for the lattice parameter measurements as a function temperature on his specially designed high resolution X-ray diffractometer.

References

[1] see for instance "International Workshop on Aperiodic Crystals" , *J. de Physique* **7** Colloquium 3 (1986)

[2] P. Bak Phys. Rev. Lett **54** 1517 (1985); *Phys. Rev.* **B32** 5764 (1985); *Phys. Rev. Lett* **56** ,861 (1986); *J. de Physique* **7** Colloqium 3 136 (1986)

[3] M. Bessiere et al in preparation

[4] R. Bellissent et al in preparation

[5] J.W. Cahn, D. Shechtman and D. Gratias, *J. Mat. Res.* **1**, 13 (1986)

[6] L. Bendersky and M.J. Kaufman, *Phil. Mag.* **B53** L75 (1986)

[7] C. Janot, J. Pannetier, M. De Boissieu and J.-M. Dubois, *EuroPhys. Lett.* **3** (9) 995 (1987)

[8] J. F. Bérard , G. Calvarin and D. Weigel *J. of Appl. Cryst.* **13**, 201 (1980)

[9] P.M. Horn, W. Malzfeldt, D.P. DiVincenzo,J. Toner and R. Gambino, *Phys. Rev. Lett* **57**,1444 (1986)

[10] C. Berger, PhD *Thesis* Grenoble (1987)

[11] M. Cyrot and F. Cyrot-Lackman, Proc. Int. Conf. LAM6, *Z.Phys.Chem.Neue Folge* **157**,823 (1988)

[12] M. Cooper and K. Robinson, *Acta Cryst.* **20**, 614 (1966)

[13] C.L. Henley, *J. Non Cryst Solids* 75, 91 (1985)

[14] D. Gratias and J.W. Cahn *Scripta Met.* **20**, 1197 (1986)

[15] M. Bessiere et al, in preparation

[16] J.W. Cahn, D. Gratias and B. Mozer submitted to *Phys. Rev. Lett*

[17] D. Gratias , J.W. Cahn and B. Mozer submitted to *Phys. Rev. Lett*

[18] P. Guyot and M. Audier, *Phil. Mag* **B52**, L15 (1985)

[19] V. Elser and C.L. Henley, *Phys. Rev. Lett.* **55**, 2883 (1985)

[20] M. Duneau and A. Katz, *Phys. Rev. Lett* **54**, 2688 (1985); A. Katz and M. Duneau, *J. de Physique* **47**, 181 (1986)

THE GROWTH OF ICOSAHEDRAL PHASE

Veit Elser
AT&T Bell Laboratories
600 Mountain Avenue
Murray Hill, NJ 07974
USA

ABSTRACT. A simple model for the growth of icosahedral phase based on structural information obtained from crystalline phases is defined and simulation results are presented. Growth velocities in the splat-cooling range and 400Å system sizes are attained. The resulting structures are neither crystalline nor quasiperiodic but resemble a glass. Long-range correlations indicate a linear peak width vs. $|\mathbf{G}^\perp|$ relationship with a slope that is within a factor of two of experimental values.

1. INTRODUCTION

Although it has already been three years since the publication of Shechtman's discovery of the icosahedral phase,[1] the question of its structure — when asked in even the most basic terms — remains unanswered.[2] A number of experiments have ruled out proposals based on twinning[3] so that currently two principal contenders remain: (1) the "quasiperiodic" model[4] and (2) the "glass" model.[5-7] Model (1) maintains that the atomic density is a quasiperiodic function consistent with icosahedral symmetry. An icosahedral rotation applied at any point of the structure, when combined with a suitable translation, will bring the whole structure into near incidence with itself. In contrast, model (2) does not invoke quasiperiodicity but merely assumes that atomic clusters with icosahedral symmetry are linked together in a way which propagates a fixed set of icosahedral axes throughout the structure. Although there exist crystalline structures that share this property,[8-10] model (2) requires further that the set of linkages along the various icosahedral axes occur with equal frequency. The term "glass"[6] derives from the use of randomness to achieve this end.

Superficially models (1) and (2) appear to have very little in common. Previously, however, it has been emphasized that from a geometrical point of view these models represent just the extremes of a continuous spectrum.[11] The first step is to consider comparable abstractions of the two models, say, the set of points formed in (1) by the vertices of a quasiperiodic rhombohedral tiling and in (2) by the cluster centers. In both cases the set of points is simply described as the projection of a subset of points forming a six

121

A. Amann et al. (eds.),
Fractals, Quasicrystals, Chaos, Knots and Algebraic Quantum Mechanics, 121–138.

dimensional (6D) lattice. Furthermore, in both cases the subset of 6D lattice points is sufficiently connected (in the sense of near-neighbor bonds) to form a 3D hypersurface. Finally, the orientation of the embedded 3D hypersurface on a macroscopic scale is the same and is uniquely determined by icosahedral symmetry. What sets the two models apart, at least geometrically, is the smoothness property of the 3D hypersurface. At one extreme is the quasiperiodic model which assumes the hypersurface is *maximally* smooth, resembling the perfect order of "atomically smooth" crystal facets. The glass model, at the other extreme, uses random processes to generate very rough hypersurfaces. While the details of the diffraction properties of either extreme do not agree with experiment, it is likely that an intermediate level of roughness does. In spite of the existence of a geometrical reconciliation of models (1) and (2), the debate has remained as lively as ever. This stems almost certainly from the differing *physical* underpinnings of the two models.

The principal physical difference between the "quasiperiodic" (1) and "glass" (2) models is the relative importance of energy and entropy. Model (1) asserts that the ground state (at zero temperature) is quasiperiodic. That this is possible is demonstrated by the existence of "matching rules" for Penrose tiles.[12] The nature of these rules is quite complex so that the question of their implementation by realistic interactions among atoms has not even been addressed. A naive translation of matching rules into physical terms suggests a level of specificity in the local chemical order observed only in organic or strongly covalent materials. The known icosahedral phases occur among the intermetallic compounds which do not share this property. Model (2) on the other hand, was inspired to some extent by the actual atomic arrangements in crystalline intermetallic compounds with known structures.[8-10] There it was observed that icosahedral symmetry is already exhibited by the structure of certain atomic clusters. It is then only necessary to arrange an alternative linkage of these clusters with the property (in the context of the geometrical discussion above) that the 3D hypersurface formed by the cluster centers has an icosahedral orientation with respect to the 6D lattice. To produce this result, model (2) relies heavily on entropy rather than energy. This may be understood in two regimes. First, we may imagine a growing aggregate of linked icosahedral clusters. If clusters are added at the surface such that linkages along the various icosahedral axes are used with equal frequency, then the icosahedral orientation of the hypersurface follows (although it may be very rough). Alternatively, we may imagine a completely formed hypersurface of linked clusters that fluctuates thermally in 6D space. By this we mean that linkages between clusters are broken and reformed differently, all the time maintaining the connectivity of a 3D hypersurface. The directions along which this hypersurface is fluctuating in 6D are orthogonal to the three (physical) dimensions into which it is finally projected. Long-wavelength fluctuations resemble the capillary waves present at the crystalline-solid/vapor interface above the roughening temperature. Due to their similarity with the modes of modulated crystals, they are also referred to as "phasons".[2] The importance of these fluctuations, in the context of model (2), is that they provide an entropic selection mechanism for the orientation of the hypersurface. Whereas energy minimization applied to the set of all possible cluster linkages might result in a periodic crystalline structure (corresponding to a hypersurface with non-icosahedral orientation), it is possible that at higher temperatures the *free* energy minimum occurs at the icosahedral orientation. According to either view of model (2), i.e. as random aggregation or as stabilization by thermal "phason" fluctuations, the icosahedral phase is understood as a *metastable* phase at low temperatures. Thus model (2) predicts a residual configurational entropy that is absent in model (1).

The aim of the present paper is to present the results of computer simulations of the glass model. In Section 2 the structure and linkage of icosahedral atomic clusters is briefly described. In Section 3 a growth model involving clusters is defined and computer simulation results are compared with experiment. The successes of the glass model are summarized in Section 4.

2. STRUCTURE AND LINKAGE OF ICOSAHEDRAL ATOMIC CLUSTERS

A striking pattern in the structure of the large-unit-cell intermetallic phase α(AlMnSi) is the rather large atomic icosahedron shown in Figure 1a.[8-9] Similarly, the Frank-Kasper or tetrahedrally-close-packed phases R(AlCuLi)[13] and $(Al, Zn)_{49}Mg_{32}$[10] contain in their structure the equally large but *different* icosahedron shown in Figure 1b. Although many structural features of aluminum transition-metal compounds and Frank-Kasper compounds are different, such as the degree of close packing, the above instances share some remarkable similarities: (1) both contain large icosahedral clusters; (2) the clusters are centered on the vertices of a bcc lattice and (3) are similarly oriented. But perhaps the most significant similarity is the fact that both form icosahedral phases under conditions of rapid cooling or slightly modified composition.

One way to describe the orientations of the clusters is to relate the directions of their symmetry axes with the symmetry axes of the cubic unit cell. Thus three of the 15 icosahedral 2-fold axes are aligned with <001> while four of the 10 3-fold axes are aligned with <111>. A different approach which immediately suggests the possibility of alternative structures avoids all reference to the cubic unit cell and instead expresses cluster-cluster relationships entirely in terms of a set of icosahedral basis vectors, $e_1^{\parallel}$, ..., $e_6^{\parallel}$, intrinsic to each cluster. These are vectors of length a_R that point from the center to the six vertices on one hemisphere of the icosahedron. In the actual

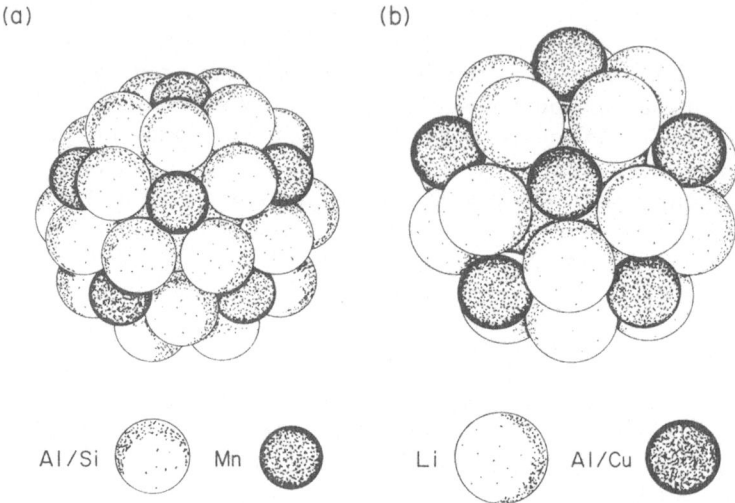

(a)   (b)

Al/Si   Mn   Li   Al/Cu

Figure 1:   Icosahedral atomic clusters taken from the structures of (a)
α(AlMnSi) and (b) R(AlCuLi). Only the second shell of atoms is
shown. Each cluster also contains an internal icosahedron of (a)
Al/Si and (b) Al/Cu. In addition, (b) contains a central Al/Cu
atom whereas the center of (a) is vacant.

clusters they correspond to the directions of the Mn atoms (Fig. 1a) and
Al/Cu atoms (Fig. 1b). The relationships of adjacent clusters are shown in
Figure 2. Sums of icosahedral basis vectors originating at the center of one
cluster arrive at the center of a neighboring cluster. Figure 2a shows how the
cubic lattice constant a is related to the length of a sum of four basis vectors:

$$a = |e_1^{\parallel} + e_3^{\parallel} + e_4^{\parallel} + e_5^{\parallel}| = (4 + 8/\sqrt{5})^{\frac{1}{2}} a_R \ . \tag{1}$$

Similarly, the sum of three basis vectors and the cube corner-to-body-center
distance are also simply related (see Fig. 2b) since

$$|e_1^{\parallel} + e_2^{\parallel} + e_3^{\parallel}| = (3 + 6/\sqrt{5})^{\frac{1}{2}} a_R = \frac{1}{2}\sqrt{3} \, a \ .$$

The two cluster-cluster relationships, along 2-fold axes (Fig. 2a) and 3-fold
axes (Fig. 2b) will be referred to as 2-fold and 3-fold "linkages". The
possibility of a high multiplicity of alternate structures (formed by clusters) is
a consequence of the large number of possible linkages per cluster, only a few
of which are used in the crystalline bcc compounds. The set of points
generated by arbitrary and repeated application of these linkages to a given
point automatically form a subset of projected 6D lattice points.[2] Physical

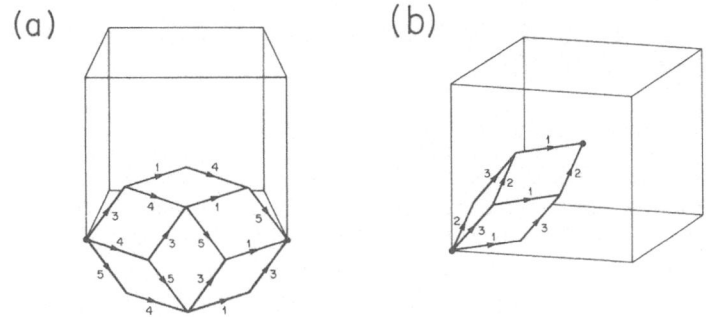

Figure 2: Linkage of icosahedral atomic clusters in the bcc structure expressed in terms of icosahedral basis vectors. (a) 2-fold linkage (b) 3-fold linkage.

considerations greatly constrain this subset. For example, combinations of linkages should not be chosen which result in overlapping clusters. Since these linkages are definitely not chemical bonds, the physical or "mechanical" integrity of a structure also requires that several linkages are impingent on any one cluster. Finally, the necessary presence of atoms filling the space between clusters is not an insurmountable problem. It is always possible to return to a description with space filling tiles whose vertices correspond to cluster centers.[14] Due to the physical considerations above, the number of kinds of tiles is small (perhaps four suffice) and their "decoration" with atoms is largely determined by the structure of the clusters themselves.

There are local and global experimental probes that give strong evidence that the structural modification described above is a good model of icosahedral phase. EXAFS measurements show that the local atomic environments in the crystalline phases $\alpha(\text{AlMnSi})$[15] and $R(\text{AlCuLi})$[16] are very similar to their counter-parts in corresponding icosahedral phases — and that this similarity does not extend to other, simpler crystalline phases without clusters. Diffraction, on the other hand, can in principle discover the spatial ordering of clusters if indeed this description applies to icosahedral phase. The scale of the 6D hypercubic lattice can be obtained by indexing the icosahedral phase diffraction pattern. When this scale is expressed in terms of the length of icosahedral basis vectors, one finds $a_R=4.60\text{Å}$[17] and $a_R=5.05\text{Å}$[18] for AlMnSi and AlCuLi icosahedral phases, respectively. These values should be compared with $a_R=4.61\text{Å}$[9] and $a_R=5.05\text{Å}$[18] obtained from equation (1) and the known cubic lattice constants.

3. COMPUTER SIMULATIONS OF THE GROWTH OF ICOSAHEDRAL PHASE

3.1 Definition of the Model

The growth algorithm described below is similar to the one applied previously to the study of decagon packings in the plane.[11] There are some important differences, however. In particular, the number of parameters in the model has been reduced to three: the interfacial thermal gradient h, the growth velocity v, and a ratio of cohesive energies ϵ_2/ϵ_3. As before, the two concerns which motivate the form of the model are (1) the desireability of generating a highly connected, homogeneous network of linked clusters and (2) the inclusion of thermal fluctuations in the formation process.

The growth geometry is a cone with the nucleus, a single cluster, placed at its apex. Using cylindrical coordinates (z, ρ, ϕ), the cone interior is given by $z > 2\rho$. Growth proceeds in the z-direction and is implemented by a linear temperature field $T(z) = h(z - z_0)$ where the zero temperature isotherm at z_0 moves at a constant velocity $\dot{z}_0 = v$. In the region $z < z_0$ there are N_0 clusters which are "frozen" while in the region of active growth, $z > z_0$, there are N_T clusters. The number N_T fluctuates and grows linearly with the area of the circular interface at $z = z_0$.

In order to promote the formation of highly connected structures, negative cohesive energies are assigned to the cluster-cluster linkages. Since there are two kinds of linkages, along 2-fold and 3-fold axes, the ratio of the two cohesive energies, ϵ_2/ϵ_3, constitute a third parameter in the model. In general, the assignment of cohesive energies is somewhat arbitrary since it is only required to *parametrize* the total energy of a small number of local cluster environments. Before actual total energy calculations are performed (a formidable task considering the number of atoms) it is impossible to know how well a parametrization by ϵ_2 and ϵ_3 works and what the true values of these cohesive energies are. The approach adopted here was to let ϵ_2 and ϵ_3 define the scale of energy (relative to which h is measured) according to $\epsilon_2 + \epsilon_3 = -2$. Then the ratio ϵ_2/ϵ_3 was varied until well-connected cluster networks already appeared at relatively large values of the growth velocity v. The optimal value obtained in this way was $\epsilon_2/\epsilon_3 = 2$. This value was used in all subsequent simulations.

The elementary growth and thermalization processes involve two operations. (1) A new cluster may be linked to an already existing cluster in the region $z > z_0$ if (i) it is linked to at least one other cluster, and (ii) if its distance from clusters to which it is not linked is greater than the length of the 2-fold linkage. (2) A cluster in the region $z > z_0$ is removed probabilistically according to the Metropolis criterion: a random number r, uniform in $(0, 1)$, is chosen and if $r < \exp(-E/T)$ the cluster is removed. In the above expression, $T = T(z)$ is the local temperature at the cluster center and $E = \frac{4}{3}n_2 + \frac{2}{3}n_3$ is the (positive) change in cohesive energy upon breaking n_2 2-fold linkages and n_3 3-fold linkages. An elementary process is defined to be the application of both (1) and (2) to a single cluster chosen at random from the N_T clusters in the region $z \to z_0$.

After each elementary process, the isotherms are advanced according to $z_0 \to z_0 + v/N_T$. The unit of time τ, implicit in this definition of v is the time required for one elementary process. A rough estimate of τ goes as follows. In order for a cluster to "dissolve" and perhaps reappear somewhere else nearby, it is necessary for the 12 atoms at the icosahedral vertices to diffuse a distance on the order of a_R. This estimate uses the minimum number of atoms required to form an icosahedral object since the likelihood of short range chemical order suggests that not all the atoms are diffusing independently. Alternatively, the Mn atoms in Fig. 1a and the Al/Cu atoms in Fig. 1b constitute a kind of backbone for their respective clusters. Thus $\tau \sim 12 a_R^2/D$ where D is the liquid-phase diffusion constant. Using the typical value $D \sim 10^{-4}$ cm^2/sec, and $a_R = 5$Å, we have $\tau \sim 3 \times 10^{-10}$ sec. It is useful to express experimental growth velocities[19] in terms of the characteristic velocity $v_0 = a_R/\tau \sim 200$ cm/sec. The highest velocities, attained by splat cooling, are typically $v \sim (0.02)v_0$. For chill-cast samples of AlCuLi icosahedral phase, $v \sim (5 \times 10^{-5})v_0$. Finally, for controlled growth of large AlCuLi single quasicrystals using the Bridgman technique, v can be as small as $(5 \times 10^{-7})v_0$. For comparison, velocities on the order of $(0.001)v_0$ appear to be a practical lower limit for computer simulations such as described here.

The remaining parameter of the model, the interfacial thermal gradient h, is much less accessible to experimental measurement. The term "interfacial" is included to express the fact that h does not measure the externally imposed temperature gradient, say in a Bridgman apparatus. Rather, h is interpreted as a variable which couples directly to the diffusivity or thickness of the interface. For the situation we are mostly interested in, namely the formation of well-connected cluster networks under even high growth velocities, the interface should be thick (small h). A thick interface allows sufficient time for a well-connected network to develop even when the zero temperature isotherm is rapidly advancing. Although the interface becomes quite thick as h vanishes, it is also truly diffuse in the sense that the high connectivity develops gradually as a function of z. One measure of this is the average coordination c (by linkages to other clusters) of a cluster as a function of its distance beyond the zero temperature isotherm, $z - z_0$. With 2-fold and 3-fold linkages treated equally, the variation of c for $h = 0.5$ and

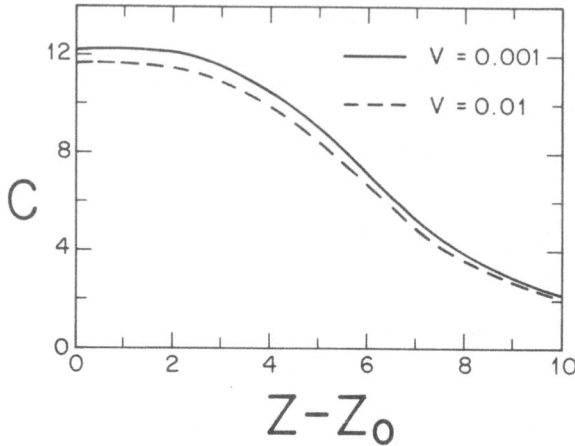

Figure 3: Total coordination per cluster, c, as a function of the distance, $z-z_0$, beyond the zero temperature isotherm. The unit of length is a_R.

the two velocities $v/v_0 = 0.01, 0.001$ is shown in Figure 3. For both velocities the decay of c from its large value at $z-z_0 \sim 2a_R$ to $c \sim 2$ occurs gradually over a distance $8a_r$ or roughly 40Å. An interfacial width of this magnitude, although not ruled out by experiment, is at variance with the results of microscopic growth simulations of one-component, Lennard-Jones systems.[20] It should be remarked, however, that the intermetallic phases being considered here are considerably more complex. Since representative Frank-Kasper crystal structures have unit cell sizes nearly one order of magnitude larger than simple monatomic systems, it is quite conceivable that interfacial widths are also larger by about the same factor.

Apart from the observation that well-connected cluster networks are easily formed when h is small, there is an additional reason for believing that a small h describes the correct physics. A characteristic property of icosahedral phases is their ability to nucleate and grow at rates much faster than crystalline phases of similar composition. This property has been ascribed to an anomalously low value of the solid/liquid surface tension.[21] A possible microscopic explanation of the low surface tension is simply that the structure of the interface is diffuse, so that thermal fluctuations may easily transform significant amounts of material from one phase to the other. On the other hand, the interface should not be too diffuse since then it is unlikely that the effects of anisotropy would manifest themselves in dendritic growth[22] and faceting.[23,24] The value $h=0.5$ which gives an interfacial width of roughly 3-4 cluster diameters is probably a good compromise. No attempt was made to measure the effect of anisotropy; in all simulations the icosahedral 5-fold axis was aligned with the growth direction.

Two types of measurement, local and global, were performed on each conical sample generated by the growth algorithm. The local measurements include the density of clusters and the coordination number distribution of 2-fold and 3-fold linkages. A convenient dimensionless expression for the density is the equivalent packing fraction of spheres having diameters equal to the length of the 3-fold linkage.

Global measurements investigate the nature of the embedding of the abstract 3D hypersurface in 6D space. The correlation function of the hypersurface relates directly to diffraction line shapes and shifts. For each cluster center coordinate $\mathbf{x}^{\|}$, the growth algorithm also generates the corresponding 3D coordinate $\mathbf{x}^{\perp}$; the pair $(\mathbf{x}^{\|}, \mathbf{x}^{\perp})$ form a 6D lattice point in the usual way.[2] A complete description of diffraction properties is provided by the distribution of

$$\Delta\mathbf{x}^{\perp}(\mathbf{R}) = \mathbf{x}^{\perp}(0) - \mathbf{x}^{\perp}(\mathbf{R})$$

where $\mathbf{x}^{\perp}(0)$ and $\mathbf{x}^{\perp}(\mathbf{R})$ are $\mathbf{x}^{\perp}$ coordinates of two clusters separated in physical space by $\mathbf{R}$. For technical reasons relating to the form of the computer code, only the first and second moments of $\Delta\mathbf{x}^{\perp}(\mathbf{R})$, effectively summed over $\mathbf{R}$, were obtained. Specifically, the procedure used was to assume the *ansatz*

$$\mathbf{x}_\alpha^\perp = \delta\mathbf{x}_\alpha^\perp + \sum_{\beta=1}^{3} A_\alpha^\beta \mathbf{x}_\beta^\| + b_\alpha$$

where A_α^β and b_α are constants and $\delta\mathbf{x}_\alpha^\perp$ varies from cluster to cluster. A_α^β and b_α are then varied so that the sum

$$(\Delta x^\perp)^2 = \frac{1}{N_0} \sum |\delta\mathbf{x}^\perp|^2$$

is minimized (N_0 is the total number of clusters). The matrix A_α^β and vector b_α then have the interpretation of giving respectively the orientation and position of the best (least squares) approximating hyperplane. Nonzero values of A_α^β correspond to departures from icosahedral symmetry. The variance, $\Delta x^\perp$, measures the roughness of the hypersurface. Values of the physically relevant quantities A_α^β and $\Delta x^\perp$ were tabulated at four stages of growth: $z_0/a_R = 20, 40, 60, 80$ ($z_0 = 0$ corresponds to the apex of the cone geometry). Consequently, the scaling of $\Delta\mathbf{x}^\perp(\mathbf{R})$ with $|\mathbf{R}|$ can be investigated by comparing these measurements.

130

(a)

$n_2 \backslash n_3$	3	4	5	6	7	8
3	0	0	0	0	1	0
4	0	1	1	3	3	0
5	1	2	5	10	12	0
6	1	3	9	20	12	0
7	0	2	7	5	0	0

(b)

$n_2 \backslash n_3$	3	4	5	6	7	8
3	0	0	0	0	0	0
4	0	0	0	1	1	0
5	0	1	2	6	14	0
6	0	1	5	23	27	1
7	0	1	6	9	0	0

Table 1: Distribution of cluster coordinations, (n_2, n_3), by 2-fold and 3-fold linkages, (a) $v=0.01$ (b) $v=0.001$. All values are rounded to the nearest percent.

3.2 Results

The results described below were obtained from measurements of 10 samples at each of the two growth velocities, $v=0.001$ and $v=0.001$ (the unit of velocity, v_0, is understood). Each sample had a conical geometry with height $80a_R \sim 400$Å and contained roughly 14000 clusters. Approximately 25 hours of cpu time was required to grow one sample at the low velocity.

An examination of the distribution of coordinations (n_2, n_3) of clusters by n_2 2-fold and n_3 3-fold linkages shows that the structures generated by our growth process are indeed highly connected. The frequencies of coordinations, rounded to the nearest percent, are given in Tables 1a $(v=0.01)$ and 1b $(v=0.001)$. The most frequently occurring coordination is $(n_2, n_3)=(6,6)$ for $v=0.01$ and $(6,7)$ for $v=0.001$. The predominant effect of decreasing the growth velocity is an increase in the amount of 3-fold coordination. This is reflected in the values of the average coordinations:

$$(\overline{n}_2, \overline{n}_3) = (5.62, 5.82), \quad v=0.01$$

$$(\overline{n}_2, \overline{n}_3) = (5.86, 6.25), \quad v=0.001 \ .$$

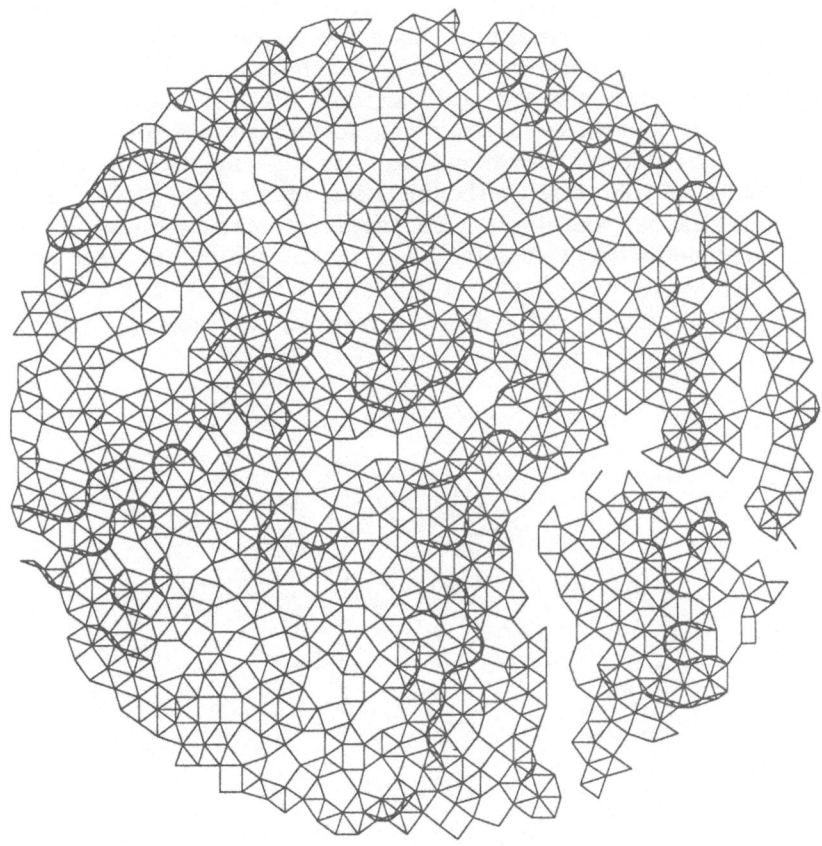

Figure 4:   Cross-section of a cluster network grown at low velocity,
             $v = 0.001$. A tear has formed in the lower right hand corner.

A comparison of the packing fractions $f = 0.590$ for $v = 0.01$ and $f = 0.615$ for $v = 0.001$ with the packing fraction of "random loose packings",[25] $f_{RLP} = 0.60 \pm 0.02$, suggests that our cluster packings are close to being mechanically stable even when only the shorter 3-fold linkages are considered. A "random loose packing" describes the situation of ball bearings dumped into a container which is not shaken. Random *close* packings, produced by shaking the container, have $f_{RCP} \sim 0.64$. The maximum packing fraction that utilizes only the icosahedral 3-fold contacts is the value attained by the bcc lattice packing, $f_{bcc} \cong 0.68$. Using a construction involving a complicated "acceptance domain" for the $\mathbf{x}^{\perp}$ coordinates, Henley[26] has shown how in principle a deterministic, quasiperiodic cluster packing may be generated. Unfortunately, his packing fraction $f \cong .5535$ is rather low.

Although the average and local connectedness of our cluster packings is quite good, it is also necessary to check the degree of *global* connectedness since large scale properties are crucially dependent on this.[11] In particular, the average coordination is not much affected if the structure contains a low density of internal surfaces across which no linkages are present. In the 6D geometry, such defects correspond to "tears"[11] in the 3D hypersurface: opposite sides of the tear, although near in physical space, are widely separated in the x^1 coordinate. To check for tears, circular slices of the packing were periodically examined during the simulation. An example of the final slice at $z_0 = 80a_R$ is shown in Figure 4. The thickness of the slice is $(1+3/\sqrt{5})a_R$; vertices and edges correspond to clusters and linkages,

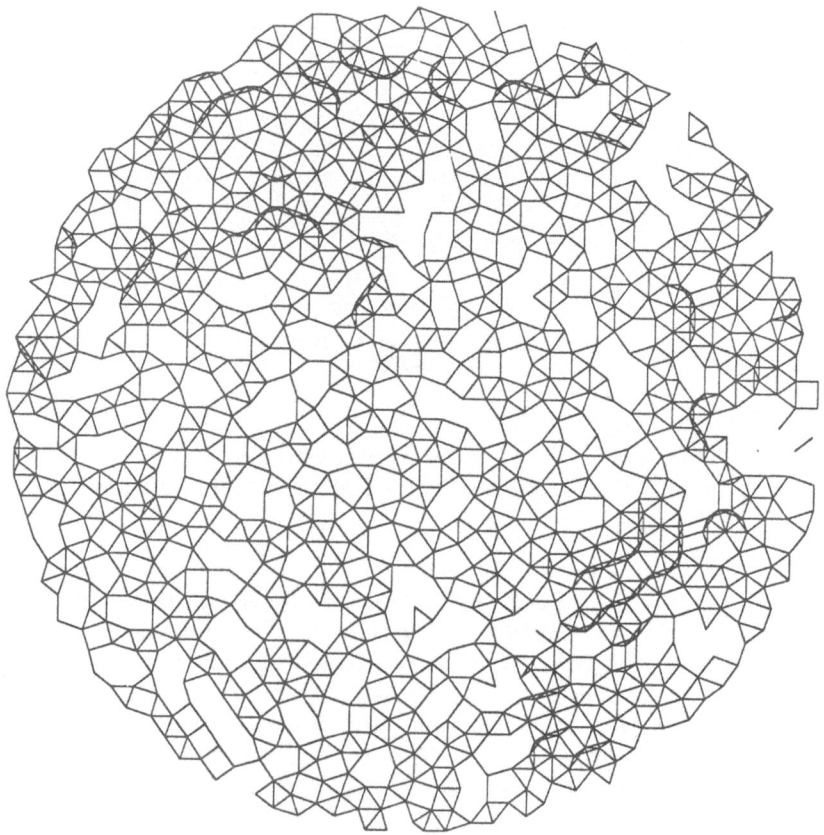

Figure 5: Same as Fig. 4 but with $v=0.01$. Two short tears have formed at the surface.

respectively. Of the 10 samples generated at the low velocity (v=0.001), two developed long tears such as the one in Fig. 4. Of the remaining eight samples, five did not develop tears at all and three had very short tears that may have grown in length had the simulation been continued. In all cases the tears originated at the surface of the cone. Essentially the same statistics apply to the structures grown at the higher velocity (v=0.01), although three samples developed long tears. The final slice of a rapidly grown sample containing a short tear is shown in Figure 5. Occasionally, a short tear present in an earlier slice was found to have mended in a later slice. The observation that tears do not originate spontaneously in the bulk of the sample suggests that their density is controlled by boundary effects.

The departure from planarity of the hypersurface, $\Delta x^{\perp}$, for various system sizes $L = z_0/a_R$ is shown in Figure 6. The apparent *linear* rise in $\Delta x^{\perp}$ with L, although unexpected, is a welcome result. An equilibrium model of fluctuating hypersurfaces[2] predicts that the second moment of the $\Delta x^{\perp}(\mathbf{R})$ distribution is *finite* in the limit $|\mathbf{R}| \to \infty$. Since this would imply that $\Delta x^{\perp}(L)$ approaches a constant for large L, the data seems to rule out the equilibrium model. The mechanism responsible for the increase in $\Delta x^{\perp}$ is not understood at present.

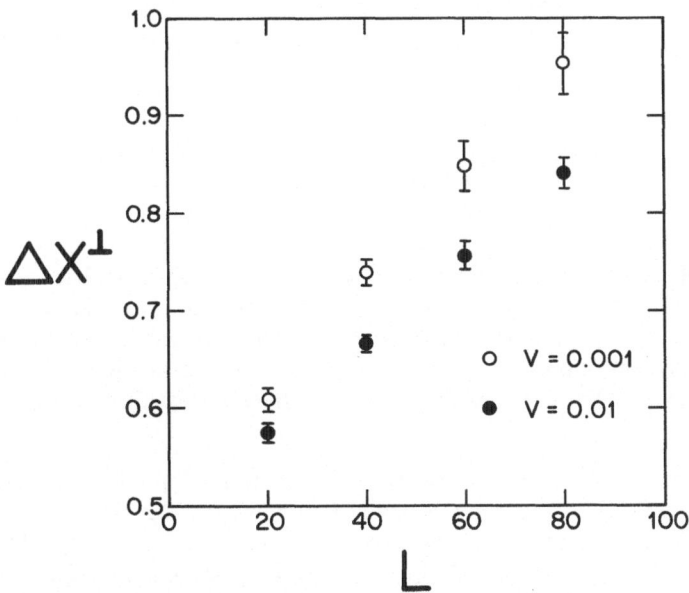

Figure 6: Root-mean-square fluctuation of the hypersurface, $\Delta x^{\perp}$, as a function of system size, L.

A linear increase in $\Delta x^\perp$ with system size suggests that fluctuations in the hypersurface's average orientation, A, are very important. A reliable estimate of the sample average, $<A>$, required as many as 10 samples and showed that, except for $<A_3^3>$, all matrix elements are consistent with zero. The index 3 denotes the growth direction so that the small values

$$<A_3^3>=0.011 \pm 0.001, \; v=0.01$$

$$<A_3^3>=0.010 \pm 0.002, \; v=0.001$$

represent a growth induced anisotropy. Although a nonzero average for A does shift the positions of Bragg peaks $(\mathbf{G}^\parallel, \mathbf{G}^\perp)$ by $\mathbf{G}^\parallel \rightarrow \mathbf{G}^\parallel + \delta\mathbf{G}$ where $\delta\mathbf{G} = A^{tr}\mathbf{G}^\perp,$[27,28] the fluctuating part of A is just as important. The magnitude of the shift produced by fluctuations is

$$<|\delta\mathbf{G}|^2> = (\mathbf{G}^\perp)^{tr} \, M \, \mathbf{G}^\perp$$

where

$$M = <A^{tr} A> - <A^{tr}><A> \; .$$

Averaging over 10 samples showed that within statistical errors the matrix M was a multiple of the identity so that

$$\delta G_{rms} \equiv [<|\delta\mathbf{G}|^2>]^{\frac{1}{2}} \simeq \Delta A \, |\mathbf{G}^\perp| \tag{2}$$

where

$$(\Delta A)^2 = \frac{1}{3} \, \mathrm{Tr} M \; .$$

Values of ΔA at the four system sizes and two growth velocities are given in Figure 7. This data conveys three interesting results: (1) ΔA is considerably larger than $<A_3^3>$; (2) the decay of ΔA with L is rather slow; (3) a lower growth velocity *increases* ΔA. The last two remarks are consistent with the behavior of $\Delta x^\perp$. Unfortunately, their explication is likewise a subject of speculation.

If equation (2) is to serve as an explanation of peak broadening, it is necessary to establish the length scale L for which ΔA is to be evaluated. Two effects that restrict the translational correlation length are (1) tears and (2) strain in the relaxation of the rigid linked structure. Either of these effects, or a combination,[29] will decompose a macroscopic sample into regions which contribute incoherently to the diffracted intensity. This is the sense in which the average in equation (2) is to be understood. Since tearing appears

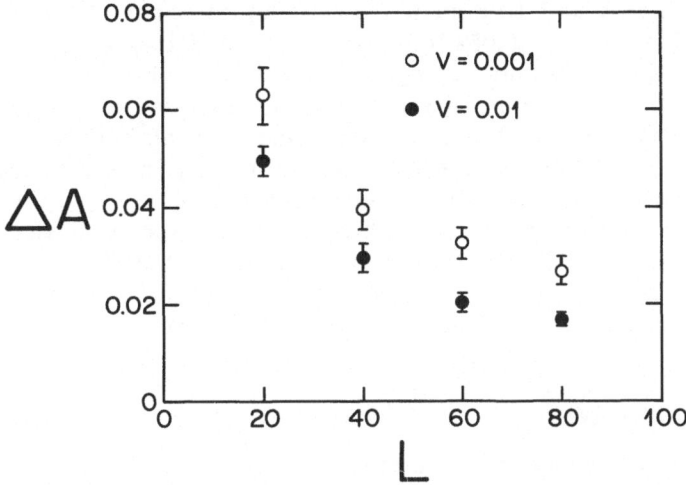

Figure 7: Root-mean-square fluctuation of the uniform phason strain amplitude, ΔA, as a function of system size, L.

to be associated with the boundary in these simulations, one expects that a correlation length determined by tears is close to the length scale of inhomogeneities during growth. Thus for melt-spun samples having morphological features as small as $\sim 400 \text{Å}$ in size,[30] a reasonable guess is $L \sim 80$. Slowly cooled AlCuLi samples exhibit extended defects separated by roughly twice this length.[31] Consequently, the tearing phenomenon alone suggests that $L \lesssim 160$.

Comparison with experimental peak widths is further complicated by considerable asymmetry in the peak shapes. It has been argued that asymmetry is the result of a superposition of peaks from regions having different average phason amplitude $<A>$.[27] To the extent that these effects are small, experiments have extracted the half-width-at-half-maximum (HWHM) of diffraction peaks, δG_{HWHM}, using fits to symmetric gaussians. Horn et al.[32] were the first to notice the approximate constancy of $\delta G_{\text{HWHM}}/|\mathbf{G}^\perp|$. Their data from furnace-annealed AlMnSi films gives a value $\Delta A_{\text{exp}} \simeq 0.044$.[33] Probably the strongest evidence for the linear δG_{HWHM} vs. $|\mathbf{G}^\perp|$ behavior is the data of Heiney et al.[34] obtained from single dendrites of chill-cast AlCuLi. Their two samples gave different slopes corresponding to $\Delta A_{\text{exp}} \simeq 0.067, 0.051$. These values are about a factor of two larger than the simulation results at $L \sim 80$. Although the agreement is improved if the effective length scale is smaller, say $L \sim 40$, a change in the experimentally inaccessible parameters h and ϵ_2/ϵ_3 may also produce the desired result. Chen et al.[19] also report measurements of AlCuLi peak widths. Chill-cast and Bridgman grown samples appear to give essentially the same value, $\Delta A_{exp} \simeq 0.059$ in agreement with Heiney et al. Interestingly, however, their melt-spun samples appear to give a *smaller* value, $\Delta A_{exp} \simeq 0.022$. A change in this direction as a function of growth velocity is consistent with the simulation results. Budai et al.[28] have been able to extract $<A>$ for AlMn icosahedral phase formed by implanting Mn in an Al matrix. Although the symmetry of the anisotropy is different, the magnitude of the largest matrix element, 0.0065, is comparable with the simulation result $<A_3^3> \simeq 0.01$. Moreover, the measurements of Budai et al. also show that if the peak widths are interpreted as random shifts, then their magnitude, i.e. ΔA, is significantly larger than the average shift, $<A>$.

4. CONCLUSIONS

There are two independent arguments that form the basis of all hopes that a glassy structure of linked atomic clusters is the correct model of icosahedral phase:

(1) Complementary evidence from EXAFS and diffraction shows that the icosahedral phase in both the aluminum transition-metal and Frank-Kasper systems is a subtle modification of particular crystalline structures. A natural choice for what such a modification might involve is provided by the only structural element common to both systems: icosahedral atomic clusters.

(2) Randomness is intrinsic to the glass model and as such can potentially account for the large peak widths and metastability of icosahedral phases in a natural way.

In the present study, computer simulations of a growth model based on (1) were used to make the expectations of (2) more explicit. These simulations are more realistic than previous work[5,6] in that the construction of the linked cluster network involves a fair amount of annealing. A diffuse growth interface provided the means of producing a highly connected network at even high growth velocities. Essentially the same mechanism was invoked by Bendersky and Ridder[21] to explain the high (homogeneous) nucleation rate of icosahedral phase. The principal result of the simulations was the observation of large phason fluctuations that persisted for large system sizes. The random excitation of long-wavelength phasons produces a peak broadening linear in $|G^1|$ as proposed phenomenologically by Lubensky et al.[27] The simulation gives a slope for this relationship that is within a factor of two of experimental values. The growth induced, average phason amplitude was also measured. Again, there is rough agreement with the peak shift data of Budai et al.[28]

In spite of the simplicity of the model, the occurrence of long-wavelength phason fluctuations that seem to evade self-averaging is mysterious and suggestive of glassy behavior. The elucidation of this phenomenon in the context of the present growth model (or a simpler one) would be most welcome.

Correspondence and discussions with J. Cahn, H. S. Chen, C. Henley, and F. Spaepen are gratefully acknowledged.

REFERENCES

1. D. Shechtman, I. Blech, D. Gratias, and J. W. Cahn, Phys. Rev. Lett. **53**, 1951 (1984).

2. An excellent survey of the literature is given by C. L. Henley, Comments Cond. Mat. Phys. **13**, 59 (1987).

3. J. W. Cahn, MRS Bulletin (March/April), 9 (1986).

4. D. Levine and P. J. Steinhardt, Phys. Rev. Lett. **53**, 2477 (1984).

5. D. Shechtman and I. Blech, Metall. Trans. **16A**, 1005 (1985).

6. P. W. Stephens and A. I. Goldman, Phys. Rev. Lett. **56**, 1168 (1986); Phys. Rev. Lett. **57**, 2331 (1986).

7. V. Elser, Phys. Rev. Lett. **54**, 1730 (1985).

8. P. Guyot and M. Audier, Philos. Mag. **B52**, L15 (1985).

9. V. Elser and C. L. Henley, Phys. Rev. Lett. **55**, 2883 (1985).

10. P. Ramachandrarao and G. V. S. Sastry, Pramana **25**, L225 (1985).

11. V. Elser, in *Proceedings of the XVth International Colloquium on Group Theoretical Methods in Physics*, Vol. 1, eds. R. Gilmore and D. H. Feng (World Scientific Press, Singapore, 1987).

12. D. Levine and P. J. Steinhardt, Phys. Rev. **B34**, 596 (1986).

13. E. E. Cherkashin, P. L. Kripyakevich, and G. I. Oleksiv, Sov. Phys. Crystallogr. **8**, 681 (1964).

14. C. L. Henley, unpublished.

15. M. A. Marcus, H. S. Chen, G. P. Espinosa and C. -L. Tsai, Solid State Comm. **58**, 227 (1986).

16. Y. Ma, E. A. Stern, and F. W. Gayle, unpublished.

17. V. Elser, Phys. Rev. **B32**, 4892 (1985).

18. M. A. Marcus and V. Elser, Phil. Mag. **B54**, L101, (1986).

138

19. H. S. Chen, A. R. Kortan, and J. M. Parsey, Jr., unpublished.

20. J. Q. Broughton, G. H. Gilmer, and K. A. Jackson, Phys. Rev. Lett. **49**, 1496 (1982).

21. L. A. Bendersky and S. D. Ridder, J. Mat. Res. **1**, 405 (1986).

22. R. J. Schaefer, Scripta Met. **20**, 1187 (1986).

23. B. Dubost, J. M. Lang, M. Tanaka, P. Sainfort, and M. Audier, Nature **326**, 4 (1986).

24. W. Ohashi and F. Spaepen, to appear in Nature (1987).

25. G. D. Scott and D. M. Kilgour, J. Phys. D**2**, 863 (1969).

26. C. L. Henley, Phys. Rev. B **34**, 797 (1986).

27. T. C. Lubensky, J. E. S. Socolar, P. J. Steinhardt, P. A. Bancel, and P. A. Heiney, Phys. Rev. Lett. **57**, 1440 (1986).

28. J. D. Budai, J. Z. Tischler, A. Habenschuss, G. E. Ice, and V. Elser, Phys. Rev. Lett. **58**, 2304 (1987).

29. Ordinary elastic strain may mend a tear and produce instead a dislocation. Dislocations are discussed in: D. Levine, T. C. Lubensky, S. Ostlund, S. Ramaswamy, P. J. Steinhardt, and J. Toner, Phys. Rev. Lett. **54**, 1520 (1985).

30. J. L. Robertson, M. E. Misenheimer, S. C. Moss, and L. A. Bendersky, Acta. Metall. **34**, 2177 (1986).

31. C. H. Chen, J. P. Remeika, G. P. Espinosa, and A. S. Cooper, Phys. Rev. B **35**, 7737 (1987).

32. P. M. Horn, W. Malzfeldt, D. P. DiVincenzo, J. Toner, and R. Gambino, Phys. Rev. Lett. **57**, 1444 (1986).

33. The quoted values of ΔA_{exp} follow from the assumption $\delta G_{HWHM} = [(\log 4)/3]^{\frac{1}{2}} \delta G_{rms}$ (valid for symmetric gaussian peak shapes).

34. P. A. Heiney, P. A. Bancel, P. M. Horn, J. L. Jordan, S. LaPlaca, J. Angilello, and F. W. Gayle, to appear in Science (1987).

ON THE ELECTRONIC STRUCTURE OF CALAVERITE

B. Krutzen
Research Institute for Materials
Faculty of Science, University of Nijmegen
Toernooiveld, 6525 ED Nijmegen
The Netherlands

ABSTRACT. The problem of the electronic structure of calaverite is exposed and the current state of research is pointed out.

The gold mineral calaverite ($Au_{1-x}Ag_xTe_2$) attracted attention of mineralogists and crystallographers already from at least 1878 [1]. In a paper of G. van Tendeloo e.a. [2] it was finally understood to be an incommensurate modulated structure. Very accurate X-ray diffraction data were obtained in the summer of 1986 by W. Schutte e.a. [3]. From these data, with the help of the superspace refinement program REMOS by Yamamoto [4], the crystal structure could be identified: The basisstructure has spacegroup C2/m (figure 1). The material is displacively modulated with wave vector $\vec{q} = -0.4076\vec{a}^* + 0.4479\vec{c}^*$, a direction of polarisation along the b-axis and an amplitude of 0.36 Å. The displacements are roughly on the Te-sites only. The superspace group is C2/m (-1/2,0,1/2) (1$\bar{s}$).

An interesting feature is the fact that the refinement reliability factor could be improved significantly by allowing for silver substitution. The result is a substitution modulation with the same $\vec{q}$-vector as for the displacive modulation. This can be explained with simple chemical arguments [5] and may be verified by an appropriate band structure calculation and suitable experiments. The driving force for the modulation is Te-Te bonding. (Compare with $[S_2]^{2-}$ and $[Se_2]^{2-}$ pair formation in 3D transition metal pyrites [6].) The Au-atoms are forced into a mixed valence situation (Au^+, Au^{3+}).

The crystal structure itself is consistent with this picture. The Au-Te distances show linear (Au^+) and square (Au^{3+}) coordination of gold atoms (figure 2). The silver substitution is found to be maximal on Au^+-sites and vanishes at Au^{3+}-sites. The substitution stabilises and pins the modulation wave.

A verification through simulation with a band structure calculation is in progress. This band structure calculation should be fully relativistic since preliminary crystal field calculations showed that spin-orbit coupling is essential for the effect of the driving mechanism mentioned above. For this purpose a version of the ASW

139

A. Amann et al. (eds.),
Fractals, Quasicrystals, Chaos, Knots and Algebraic Quantum Mechanics, 139–141.
© 1988 by Kluwer Academic Publishers.

formalism [6] based on the Dirac equation has been implemented. The results for a fourfold supercell approximation will be published in the near future.

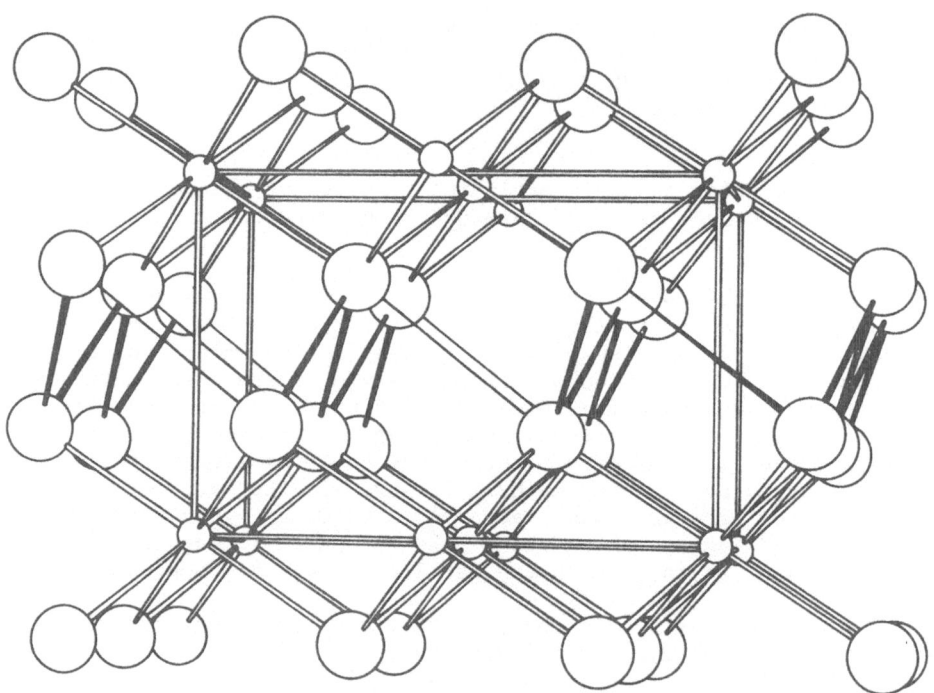

Figure 1.   The basisstructure of calaverite. The large spheres represent Te-, the small spheres Au-atoms. The a- and c-axis are in the plane of the paper along the horizontal and vertical direction respectively. The b-axis is perpendicular to the paper. Note the Te-chains along the b-axis (bold connections). The Au-atoms are on the edges of the nearly orthorombic conventional cell (γ = 90.03°) and on the face centered positions in the ab-planes. In the real structure the Te-Te distances in each chain are modulated.

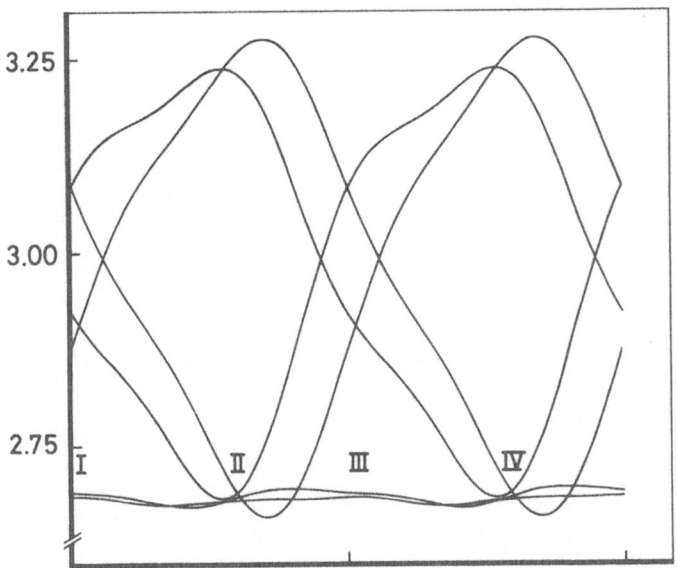

Figure 2. The distances (in Å) of a Au-atom to its 6 neighbouring Te-atoms as a function of the phase of the modulation wave. Region I shows linear (Au$^+$) and region II square (Au^{3+}) coordination. The silver substitution modulation has a vanishing amplitude in region II and is maximal in region I.

REFERENCES

[1] Schrauf, *Zeit. Krist.* 2 (1878), 211.
[2] Tendeloo, G.v. et al., *J. Sol. Stat. Chem.* 50 (1983), 321.
[3] Schutte, W., Boer, J. de, to be published in *Acta Cryst.* B.
[4] Yamamoto, A., *Acta Cryst.* A38 (1982), 87 (REMOS version 82.0).
[5] Groot, R. de, private communication.
[6] Folkerts, W. et al., *J. Phys.* C 20 (1987), 4135.
[7] Williams, A. et al., *Phys. Rev.* 19 (1979), 6094.

Chaotic Behavior of Classical Hamiltonian Systems

H.–D. Meyer
Theoretische Chemie, Physikalisch–Chemisches Institut
Universität Heidelberg
Im Neuenheimer Feld 253
D–6900 Heidelberg, West–Germany

ABSTRACT. The characterization of regular and irregular motion in classical Hamiltonian systems is briefly reviewed. It is shown that the investigation of the Liapunov exponents yields the most detailed information on the irregular (i.e. chaotic) behavior. Comments on the corresponding quantal systems are provided.

1. INTRODUCTION

In this article I want to review on what is known of the chaotic behavior in classical systems, or — more precisely — what distinguishes chaotic from regular motion. After that I shall present some numerical results to illustrate the concepts.

We shall see that the notion chaos is well defined in classical mechanics. In the field of quantum mechanics, on the other hand, there are indications of irregular behavior [1] but a precise definition of the notion quantum chaos is lacking. We feel that it is helpful to precisely know what is meant by "chaos" in classical systems if one wants to investigate irregular behavior in quantal systems.

The paper is restricted to the discussion of classical systems but from time to time I will comment on the corresponding quantal ones. The paper is limited to the study of Hamiltonian systems which are <u>conservative</u>, obey <u>time–reversibility</u> and for which the motion can take place only in a <u>bounded</u> region of phase–space. The consideration of time–independent Hamiltonians only is not a restriction. Any time–dependent Hamiltonian H can be replaced by a time–independent one, H', by just adding one artificial degree of freedom, i.e.

$$H(x_1, ..., x_N, p_1, ..., p_N, t) \rightarrow$$
$$H'(x_1, ..., x_{N+1}, p_1, ..., p_{N+1})$$
$$= H(x_1, ..., x_N, p_1, ..., p_N, x_{N+1}) + p_{N+1} . \qquad (1.1)$$

To simplify the discussion we shall further assume that the Hamiltonian is <u>analytic</u> in all coordinates and momenta. This excludes the popular billiard systems.

One may define "irregular" or "chaotic" as not being regular. Hence I shall start with discussing regular systems. The third section is then devoted to the discussion of the celebrated KAM–theorem. After that we introduce the stability matrix and the Liapunov exponents and finally show some numerical results on the determination of the irregular part of the phase–space of a model system.

A. Amann et al. (eds.),
Fractals, Quasicrystals, Chaos, Knots and Algebraic Quantum Mechanics, 143–157.
© 1988 by Kluwer Academic Publishers.

2. REGULAR SYSTEMS

The well known Hamilton equations of motion read

$$\dot{x}_i = \partial H / \partial p_i$$
$$\dot{p}_i = -\partial H / \partial x_i \qquad i = 1, ..., N \tag{2.1}$$

where we have assumed that the system has N degrees of freedom. Introducing the phase–space point γ

$$\gamma = (x_1, x_2, ..., x_N, p_1, p_2, ..., p_N)^T \tag{2.2}$$

where T denotes the transpose and introducing the symplectic matrix J

$$J = \begin{bmatrix} 0 & 1 \\ -1 & 0 \end{bmatrix} \tag{2.3}$$

where 1 and 0 denote the n × n unit and zero matrices, respectively, one may write the equations of motion very compactly as

$$\dot{\gamma} = J \frac{\partial H}{\partial \gamma}. \tag{2.4}$$

Let $G(\gamma)$ be a single valued analytic function. One calls G a constant of motion if $G(\gamma)$ remains constant along each trajectory. G is a constant of motion if and only if its Poisson bracket with the Hamiltonian vanishes identically, i.e.

$$0 = \{G, H\}$$

$$= \sum_{j=1}^{N} \left[\frac{\partial G}{\partial x_j} \frac{\partial H}{\partial p_j} - \frac{\partial G}{\partial p_j} \frac{\partial H}{\partial x_j} \right]$$

$$= \left[\frac{\partial G}{\partial \gamma} \right]^T J \frac{\partial H}{\partial \gamma} \tag{2.5}$$

where we have given the definition of the Poisson bracket for the convenience of the reader. The Poisson bracket is a canonical invariant, i.e. its value is independent of the particular set of generalized coordinates and momenta used to evaluate it. We now can define integrable systems.

Definition

A Hamiltonian is called to be <u>integrable</u> if there exist N single valued analytic functions, G_1, ..., G_N, which are

i) functional independent, i.e. $\frac{\partial G_1}{\partial \gamma}$, ..., $\frac{\partial G_N}{\partial \gamma}$ are linearly independent vectors a.e.

ii) constants of motion, i.e. $\{G_n, H\} = 0$

iii) in involution, i.e. $\{G_n, G_m\} = 0$

The Hamiltonian itself is usually included in the set of constants of motion $\{G_n\}$. The requirement *ii*) is hence usually included in *iii*). In order to interpret the condition *iii*) let us assume that the system is invariant under rotations in physical space. The angular momenta L_x, L_y and L_z are then constants of motion. They are, however, not in involution because

$$\{L_x, L_y\} = L_z \neq 0 \tag{2.6}$$

and similar for the other Poisson brackets. The rotational symmetry gives raise to only two constants of motion which are in involution. These two constants are e.g. L^2 and L_z.

The above Definition leads to the following Theorem.

Theorem (Liouville)

If a Hamiltonian system is integrable then the equations of motion can be solved by quadratures.

We shall demonstrate this property for a system of one degree of freedom. (Systems of one degree of freedom are always integrable since the total energy, H, is always a constant of motion). Assume that H is of the standard form

$$H(x,p) = p^2/2m + V(x). \tag{2.7}$$

Invoking $H(x,p) = E$ and $\dot{x} = p/m$ one arrives at

$$\frac{dx}{dt} = \sqrt{2(E-V(x))/m} \tag{2.8}$$

or

$$t(x) = \int_{x_0}^{x} [2(E-V(x'))/m]^{-1/2} \, dx' \tag{2.9}$$

which yields the desired trajectory $x(t)$ by inverting $t(x)$. Integrable Hamiltonian systems are the prototypes of regular systems.

In classical mechanics one is, of course, not restricted to perform the calculations in cartesian coordinates. One may choose any conical set of generalized coordinates and momenta. Of particular importance are the so called action—angle variables. This set of coordinates is characterized by the fact that the generalized momenta, the actions I_j, are constants of motion. Hence a Hamiltonian in action—angle variables is integrable. The other direction also holds.

Theorem (Liouville—Arnold)

> If a system is integrable, then there exists a canonical transformation to action angle variables
>
> $$(x_1, \ldots, x_N, p_1, \ldots, p_N) \rightarrow (\phi_1, \ldots, \phi_N, I_1, \ldots, I_N)$$
>
> such that the Hamiltonian, expressed in the new variables, depends only on the actions
>
> $$H(x, p) \rightarrow H(I).$$
>
> The canonical transformation is periodic in the angles ϕ_j with a periodicity of 2π, i.e.
>
> $$\gamma(\phi, I) = \gamma(\phi + 2\pi 1, I) \tag{2.10}$$

Assume that the Hamiltonian has been transformed to action—angle variables. The time evolution is then given by

$$\dot{I}_j = -\partial H/\partial \phi_j = 0 \tag{2.11}$$

$$\dot{\phi}_j = \partial H/\partial I_j \equiv \omega_j \tag{2.12}$$

The frequencies $\omega_j = \omega_j(I)$ are constant along the trajectory; the integration of the equations of motion becomes trivial

$$I_j(t) = I_j(0) \tag{2.13}$$

$$\phi_j(t) = \phi_j(0) + \omega_j t . \tag{2.14}$$

Hence we have solved the time—evolution once for all! At this point I would like to make a comment on quantum mechanics. Transforming the classical Hamiltonian to action—angle variables correspond to diagonalizing the quantal Hamiltonian. If one has diagonalized the quantal Hamiltonian then the time evolution of the wavefunction is given by

$$\psi(t) = \sum_{j=1}^{\infty} a_j \, \psi_j \exp\left(-iE_j t/\hbar\right) \tag{2.15}$$

where

$$a_j = \langle \psi_j | \psi(0) \rangle \tag{2.16}$$

and

$$H\psi_j = E_j \psi_j \tag{2.17}$$

Similar to the classical case, after finding the eigenbasis the time evolution becomes trivial! There are again constant of motion and phases which vary linearly in time. A closer analysis gives the following correspondence

$$\hbar |a_j|^2 \qquad \longleftrightarrow \qquad I_j \tag{2.18}$$
$$\frac{1}{i} \log (a_j |a_j|^{-1}) \qquad \longleftrightarrow \qquad \phi_j(0) \tag{2.19}$$
$$E_j/\hbar \qquad \longleftrightarrow \qquad \omega_j \tag{2.20}$$

The correspondence between transforming the classical Hamiltonian to action–angle variables and diagonalizing the quantal Hamiltonian becomes even more apparent if one recalls that the generator for the transformation to action–angle variables is the action integral $S(x, I)$. From the equations

$$\phi_j = \partial S / \partial I_j \tag{2.21}$$
$$p_j = \partial S / \partial x_j \tag{2.22}$$

one may determine the canonical transformation $(x, p) \rightarrow (\phi, I)$. The action integral satisfies the Hamilton–Jacobi equation

$$H(x_1, x_2, ..., x_N, \frac{\partial S}{\partial x_1}, ..., \frac{\partial S}{\partial x_N}) = E . \tag{2.23}$$

This partial–differential equation is of great formal similarity to the time–independent Schrödinger equation

$$H\psi = E\psi \tag{2.24}$$

For instance, employing the WKB ansatz for the wavefunction

$$\psi(x) = \exp [\frac{i}{\hbar}(S_0(x) + \hbar S_1(x) + ...)] \tag{2.25}$$

and requiring that ψ satisfied (2.24) one finds that S_0 has to obey the Hamilton–Jacobi equation.

Since an integrable system has N constants of motion, the trajectory cannot wander all over phase–space but is restricted to remain on an 2N–N=N dimensional hypersurface. This hypersurface has the topology of a torus [2]. The torus is called an "invariant torus" because it remains invariant under the time evolution of the system. For a system of two degrees of freedom a torus is depicted in Fig. 1.

148

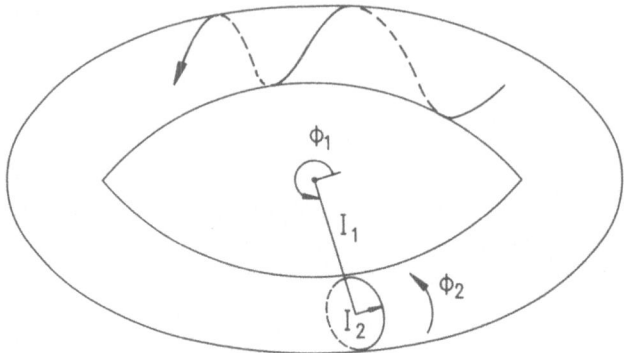

Figure 1 Invariant torus of a system with two degrees of freedom. Different tori are specified by different actions I_1 and I_2 which can be visualized as the radii of the torus. The angle ϕ_1 and ϕ_2 define a point on the torus. A trajectory winding around the torus is indicated.

The torus is characterized by the actions which are the radii and it is parameterized by the angles. One must not forget that the torus of Fig. 1 lies in a 4–dimensional euclidian space rather than in the 3–dimensional world. Different tori do not intersect similar as different trajectory do not cross in phase–space. I finally remark that it is the invariant torus which corresponds to the wavefunction and not the trajectory, i.e.

$$T_I(\phi) \longleftrightarrow \psi_m(x) \tag{2.26}$$

The torus T and the wavefunction ψ are specified by a set of N action variables I or quantum numbers m, respectively. Both quantities depend on a set of N coordinates, ϕ or x, respectively.

Since we know that the phase–space point γ depends periodically on the angle ϕ we may expand the transformation $\gamma = \gamma(\phi, I)$ into a Fourier series

$$\gamma(\phi, I) = \sum_m \gamma_m(I) \, e^{im \cdot \phi} \tag{2.27}$$

where m denotes a vector of N integers. The trajectory $\gamma(t)$ is hence given by

$$\gamma(t) = \sum_m \gamma_m(I) \exp(i(m \cdot \omega)t + im \cdot \phi(0)) . \tag{2.28}$$

The Fourier transform of any function of the trajectory $f(\gamma(t))$ — e.g. $f(\gamma) = x_1$ —

has a discrete stick spectrum. The sticks appear at the frequencies

$$\Omega = \mathbf{m} \cdot \boldsymbol{\omega} \tag{2.29}$$

We have now completed our discussion of regular systems and it is useful to compile our findings:

Regular systems

> The system is integrable. There exist N independent constants of motion which are in involution.
> The classical motion is restricted to an N-dimensional hypersurface.
> The transformation to action—angle variables is possible; the Hamilton—Jacobi equation has a solution.
> The invariant tori exist.
> The Fourier spectrum of $f(\gamma(t))$ is discrete.
> All Liapunov exponents are zero.

The last point, the vanishing of all Liapunov exponents, will be discussed in section 4. Defining irregular systems as not being regular we arrive at the following list.

Irregular systems

> The system is non—integrable. There exist only n_c constants of motion with $1 \le n_c < N$.
> The classical motion is restricted to an $2N-n_c$ dimensional hypersurface.
> The transformation to action—angle variables is not possible; the Hamilton—Jacobi equation has no solution.
> Invariant tori do not exist.
> The Fourier spectrum of $f(\gamma(t))$ is continuous for almost all functions f.
> There are $n_L = N - n_c$ positive Liapunov exponents.

If the number of constants of motion, n_c, is equal to one (the Hamiltonian itself is always a constant of motion) then one speaks of "full dimensional chaos". Otherwise, if $1 < n_c < N$ one speaks of "low dimensional chaos". Quantum mechanical measures of chaos as the statistical properties of energy levels [1,6,9] can only detect full dimensional chaos. If the constants of motion are known from the outset then one may proceed as in the following example. Assume that the total angular momentum J is a constant of motion. One then investigates the statistical properties of the energy levels for each fixed value of (J^2, J_z) separately. If the constants of motion are not known from the outset then one may not be able to detect low dimensional chaos quantum mechanically. By investigating the spectrum of Liapunov exponents, on the other hand, there is no problem in studying low dimensional chaos classically. The subtlety of low dimensional chaos does not exist for systems with two degrees of freedom.

3. THE KAM THEOREM

Is a classical system either regular or chaotic or do there exist mixed forms? This question was answered by the celebrated KAM Theorem [3,4] named after Kolmogorov, Arnol'd and Moser. Moser and Arnol'd proved the theorem in 1962 on the basis of suggestions by Kolmogorov in 1954.

Assume that one starts with an integrable Hamiltonian and adds some perturbation to it

$$H(\phi, I) = H_0(I) + \epsilon V(\phi, I) \tag{3.1}$$

What happens if we turn on ϵ? Fermi, for instance, assumed that all tori are immediately destroyed and that the motion becomes completely chaotic. The situation is not that bad. Kolmogorov, Arnol'd and Moser have shown that if one turns on ϵ then the invariant tori are deformed but almost all tori still exist if ϵ is smaller than some critical value ϵ_0. If ϵ surpasses ϵ_0 then a some of the tori are destroyed but others remain. When ϵ becomes sufficiently large then eventually almost all tori are destroyed and the motion becomes completely chaotic. Hence the typical situation is that there exist regions in phase space which are filled with invariant tori and other regions in which no tori exist. We therefore may decompose the phase space Γ as

$$\Gamma = \Gamma_R \cup \Gamma_I, \quad \Gamma_R \cap \Gamma_I = \emptyset \tag{3.2}$$

where Γ_R and Γ_I denote the regular and irregular parts of the phase space. A similar decomposition holds for the energy surface Γ_E

$$\Gamma_E = \Gamma_{E,R} \cup \Gamma_{E,I}, \quad \Gamma_{E,R} \cap \Gamma_{E,I} = \emptyset. \tag{3.3}$$

A convenient measure of the degree of irregularity is the relative weight of the irregular part of the energy shell.

$$q(E) = V(\Gamma_{E,I}) / V(\Gamma_E)$$
$$= \int d\gamma \, \chi(\gamma) \, \delta(H(\gamma) - E) / \int d\gamma \, \delta(H(\gamma) - E) \tag{3.4}$$

Here V denotes the volume (Liouville measure) and its precise meaning is given in the second line of eq. (3.4). $\chi(\gamma)$ is the characteristic function on the irregular part of the phase space

$$\chi(\gamma) = \begin{cases} 1 & \text{if } \gamma \in \Gamma_I \\ 0 & \text{if } \gamma \in \Gamma_R \end{cases} \tag{3.5}$$

The quantity $q(E)$ has proven to be very important in the correlation of classical chaos with the statistical properties of the distribution of energy levels [1b,5–9].

It is important to note that a system may be neither regular nor irregular, a trajectory, however, is always either regular or chaotic. If a trajectory is on a torus then it has to stay on it for all times. If a trajectory is not on a torus then it cannot jump onto one because time reversibility would otherwise require that this trajectory can leave a torus. By running trajectories one therefore can decide for each point γ out of the phase space Γ whether it belongs to Γ_R or to Γ_I. If the trajectory which is started at the phase space point γ is regular then $\chi(\gamma) = 0$ and $\gamma \in \Gamma_R$, otherwise $\chi(\gamma) = 1$ and $\gamma \in \Gamma_I$. The task is now to distinguish regular from irregular trajectories. How to do this is discussed in the following section.

4. LIAPUNOV EXPONENTS

Regular and irregular trajectories show different stability properties. As we shall see, these stability properties can be characterized by the Liapunov exponents (LE). The full set of LE gives, in fact, the most detailed information on the chaotic behavior of a trajectory. The theory of LE was developed by Oseledec [10]. We use here, however, a different approach. The first extensive use of the LE in a numerical study is due to Benettin et al [11,12].

The stability of a trajectory with respect to changes in the initial conditions is expressed by the stability matrix $\mathbf{M}$

$$[\mathbf{M}(\gamma(0),t)]_{ij} = \frac{\partial \gamma_i(t)}{\partial \gamma_j(0)}. \tag{4.1}$$

The stability matrix $\mathbf{M}$ depends on the time t and on the phase space point $\gamma(0)$ which specifies the initial conditions of the trajectory $\gamma(t)$. Each phase space point γ has its own stability matrix and hence its own set of LE. The matrix $\mathbf{M}$ maps the infinitesimal changes of the initial conditions onto the thereby produced changes in the final conditions. To visualize the meaning of M(t) we place an infinitesimal droplet on the phase space and let it evolve in time. Because we assume the droplet to be infinitesimal it can be deformed only by a linear transformation, i.e. it can be deformed only into an ellipsoid. (see Fig. 2).

$t=0$ $t=t_1$ $t=t_2$

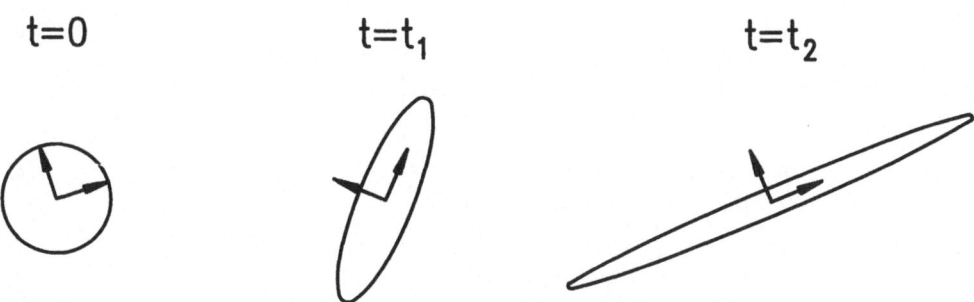

Figure 2 Time evolution of an infinitesimal droplet in phase space. As time increases the length of N principal axes of the ellipsoids shrinks and the other N axes grows.

We decompose the stability matrix $\mathbf{M}$ into a product of a diagonal matrix $\mathbf{D}$ and two orthogonal matrices $\mathbf{U}$ and $\mathbf{V}$.

$$\mathbf{M} = \mathbf{V} \, \mathbf{D} \, \mathbf{U}^{\mathrm{T}} \; . \tag{4.2}$$

This so called "singular value decomposition" [13] can be performed for any real matrix. By construction we find

$$d_j \, \mathbf{v}_j = \mathbf{M} \, \mathbf{u}_j \tag{4.3}$$

where $\mathbf{v}_j$ and $\mathbf{u}_j$ denote the column vectors of $\mathbf{V}$ and $\mathbf{U}$ and where d_j are the elements of the diagonal matrix $\mathbf{D}$. The figure 2 is just a graphical representation of the above equation. The principal axes of the ellipsoid have lengths d_j and directions $\mathbf{v}_j$.

The matrix $\mathbf{M}$ is symplectic because the transformation $\gamma(0) \rightarrow \gamma(t)$ is canonical [14,15]. From this follows that the eigenvalues d_j and the vectors $\mathbf{u}_j$ and $\mathbf{v}_j$ can be arranged such that

$$d_j \geq 1 \tag{4.4a}$$
$$d_{j+N} = d_j^{-1} \tag{4.4b}$$
$$\mathbf{u}_{j+N} = -\mathbf{J} \, \mathbf{u}_j \tag{4.4c}$$
$$\mathbf{v}_{j+N} = -\mathbf{J} \, \mathbf{v}_j \tag{4.4d}$$

holds for $j = 1, ..., N$. In particular one finds $\det(\mathbf{M}) = 1$ which is just Liouville's theorem. The differential equation for $\mathbf{M}$ follows via the chain rule from eqs. (2.4, 4.1).

$$\dot{\mathbf{M}} = \mathbf{J} \frac{\partial^2 \mathbf{H}}{\partial \gamma^2} \mathbf{M} \tag{4.5}$$

with the initial condition

$$\mathbf{M}(0) = 1 \; . \tag{4.6}$$

The differential equations for $\mathbf{U}$, $\mathbf{V}$ and $\mathbf{D}$ read [12]

$$\dot{d}_j = h_{jj} \, d_j \tag{4.7}$$
$$(\mathbf{U}^{\mathrm{T}}\dot{\mathbf{U}})_{ij} = \frac{1-\delta_{ij}}{d_j^2 - d_i^2} \, d_i d_j (h_{ij} + h_{ji}) \tag{4.8}$$
$$(\mathbf{V}^{\mathrm{T}}\dot{\mathbf{V}})_{ij} = \frac{1-\delta_{ij}}{d_j^2 - d_i^2} \, (h_{ij} \, d_j^2 + h_{ji} \, d_i^2) \tag{4.9}$$

where

$$h_{ij} = (\mathbf{V}^{\mathrm{T}} \mathbf{J} \frac{\partial^2 \mathbf{H}}{\partial \gamma^2} \mathbf{V})_{ij} \tag{4.10}$$

From eq. (4.7) we observe that the eigenvalues d_j have the tendency to behave exponentially in time. This suggest the following definitions of the Liapunov functions (LF) and Liapunov exponents

$$\lambda_j(t) = t^{-1} \log d_j(t) \tag{4.11}$$

$$\overline{\lambda}_j = \lim_{t \to \infty} \lambda_j(t) \tag{4.12}$$

The limes, i.e. the LE $\overline{\lambda}_j = \overline{\lambda}_j(\gamma)$, can be shown to exist almost everywhere [10,15]. From eqs. (4.7, 4.11) now follows

$$\lambda_j(t) = t^{-1} \int_0^t h_{jj}(t') \, dt' . \tag{4.13}$$

Hence the LE $\overline{\lambda}_j$ is the time average of the bounded function $h_{jj}(t)$.
The set of LE is not independent. Because of eq. (4.4) we find

$$\overline{\lambda}_j \geq 0 \tag{4.14a}$$

$$\overline{\lambda}_{j+N} = -\overline{\lambda}_j \tag{4.14b}$$

for $j = 1, ..., N$. Hence the first N LE determine the spectrum of the LE. If one is interested in the maximal LE only then one can avoid finding the eigenvalues d_j. The consideration of the euclidian norm of $\mathbf{M}$ is sufficient. We define

$$\lambda(t) := t^{-1} \log ((2N)^{-1/2} \|\mathbf{M}(t)\|)$$

$$= (2t)^{-1} \log ((2N)^{-1} \sum_{j=i}^{2N} d_j^2(t)) \tag{4.15}$$

and

$$\overline{\lambda} := \lim_{t \to \infty} \lambda(t) = \max_j \{\overline{\lambda}_j\} , \tag{4.16}$$

Inspecting eq. (4.13) one supposes the LE to be non zero. Why should the time average of h_{jj} vanish? I will now show that the regular systems impose conditions on the motion such that all LE vanish! Using action angle variables one may immediately integrate the stability matrix yielding [15]

154

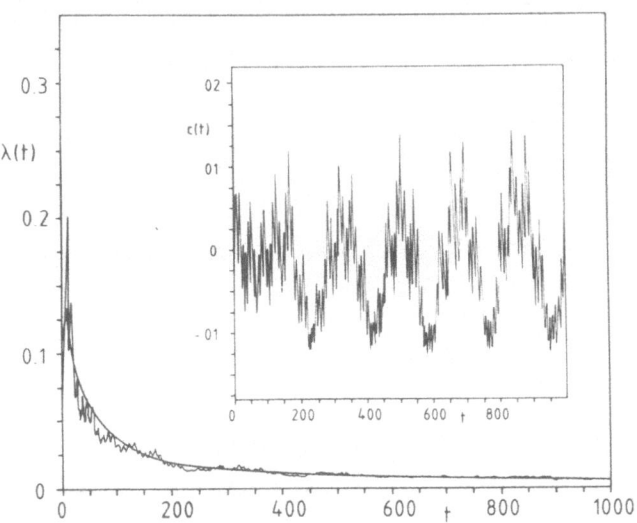

Figure 3 Liapunov function $\lambda(t)$ [eq. (4.15)] of a regular trajectory. The smooth interpolating line represents the function (4.18) with some best fit value for $\|\partial \omega / \partial I\|$. The function $c(t)$ (see eq. (4.19)), which is given by the difference between the two curves multiplied by t, is shown in the inset.

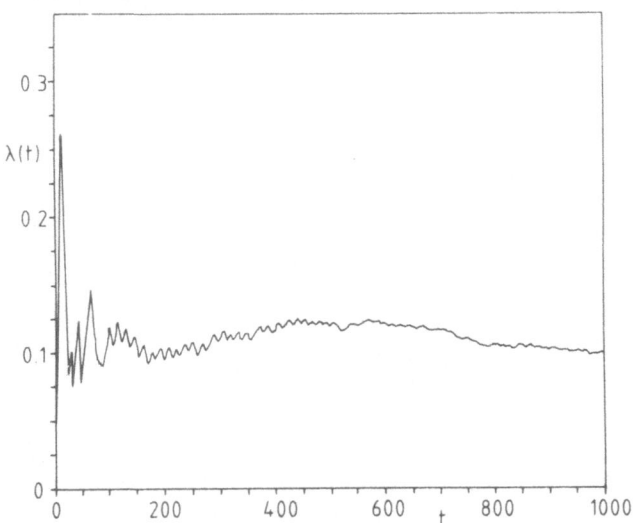

Figure 4 Liapunov function of a irregular trajectory.

$$M(t) = \exp(J \frac{\partial^2 H}{\partial \gamma^2} t) = \begin{bmatrix} 1 & \frac{\partial \omega}{\partial I} t \\ 0 & 1 \end{bmatrix}. \tag{4.17}$$

The LF (4.15) takes the appearance

$$\lambda(t) = (2t)^{-1} \log \left(1 + (2N)^{-1} \| \frac{\partial \omega}{\partial I} \|^2 t^2 \right) \tag{4.18}$$

which has a vanishing limes. Hence for integrable systems all LE vanish! The chaotic systems, on the other hand, are characterized by non—vanishing LE. There are generically as many positive LE as there are missing constants of motion, i.e. $n_L = N - n_c$ where n_L and n_c denote the numbers of positive LE and constants of motion, respectively. The investigation of the spectrum of LE yields the most detailed information on the chaotic behavior of a trajectory. The number of positive LE gives the dimensionality of the chaos. This allows for the identification and characterization of low dimensional chaos.

We now discuss numerically calculated LF. Fig. 3 shows the LF of a regular trajectory. The calculation was performed in cartesian variables rather than in action angle variables. Hence one does not follow the analytic form (4.18) exactly but one follows it in the mean. In fact, one can show [15] that

$$\lambda^{(1)}(t) = \lambda^{(2)}(t) + c(t)/t \tag{4.19}$$

holds where $\lambda^{(1)}$ and $\lambda^{(2)}$ denote the LF of the same trajectory computed in different sets of generalized coordinates and momenta. The function $c(t)$ is bounded and hence

$$\overline{\lambda}^{(1)} = \overline{\lambda}^{(2)}. \tag{4.20}$$

The LE is independent of the particular coordinate system used.

The LF depicted in Fig. 3 was calculated for a Hamiltonian which consists of two harmonic oscillators coupled by a quartic term in the coordinates [16].

$$H = \frac{1}{2}(p_x^2 + p_y^2 + x^2 + y^2) + 4kx^2y^2 \tag{4.21}$$

This Hamiltonian has the scaling property that the trajectories essentially depend on kE only rather than on k an E separately [15]. The LF shown in Fig. 4 was computed for the same Hamiltonian and the same coupling ($kE = 0.2$) but for a different trajectory as in Fig. 3. Obviously the LF does not vanish for $t \to \infty$, the trajectory is chaotic. Even if one follows the trajectory only a finite time (for about 150 oscillations in the examples given above) one can quite safely distinguishes regular from irregular trajectories.

Following these ideas one may compute $q(kE)$, the relative volume of the energy shell filled with irregular trajectories. The energy shell is divided into a large number of small cells. Within each cell a trajectory is started and according

156

to its LE it is decided whether the cell belongs to $\Gamma_{E,R}$ or $\Gamma_{E,I}$. The details are given on ref. (15). For the model Hamiltonian (4.21) the final result is shown in Fig. 5. For weak coupling (kE $\leq$ 0.05) the system remains regular. Then the transition region begins and for kE $\geq$ 1.2 one arrives at an essentially totally chaotic system.

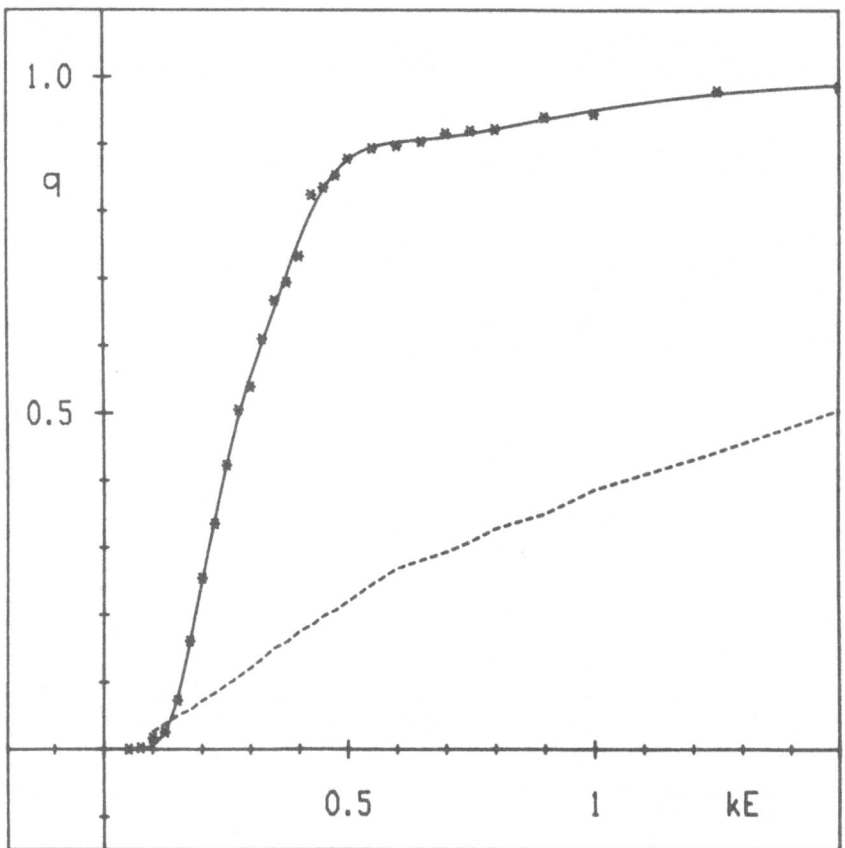

Figure 5 The stars (*) represent the computed values of q(kE). the full line is a fit to these numbers. The dashed line depicts λ_{av}(kE), the average value of all nonvanishing LE.

REFERENCES

[1]   See the contributions to this workshop by a) O. Bohigas, b) L.S. Cederbaum and c) R. Jost.

[2]   V.I. Arnold, <u>Mathematical Methods of Classical Mechanics</u>, Springer, New York (1978)

[3]   G. Casati and J. Ford, <u>Stochastic Behavior in Classical and Quantum Hamiltonian Systems</u>, Springer, Berlin (1979). These proceedings contain a translation of the original article of A.N. Kolmogorov.

[4]   A.J. Lichtenberg and M.A. Lieberman <u>Regular and Stochastic Motion</u>, Springer, New York (1983)

[5]   M.V. Berry and M. Robnik, J. Phys. A 17, 2413 (1984)

[6]   H.–D. Meyer, E. Haller, H. Köppel and L.S. Cederbaum, J. Phys. A 17, L831 (1984)

[7]   Th. Zimmermann, H.–D. Meyer, H. Köppel and L.S. Cederbaum, Phys. Rev. A 33, 4334 (1986)

[8]   Th. Zimmermann, H. Köppel, E. Haller, H.–D. Meyer and L.S. Cederbaum, Physica Scripta 35, 125 (1987)

[9]   Th. Zimmermann, L.S. Cederbaum, H.–D. Meyer and H. Köppel, J. Phys. Chem. 91, 4446 (1987)

[10]   V.I. Oseledec, Tr. Mosk. Mat. Obsch. 19, 179 (1968) [trans. Mosc. Math. Soc. 19, 197 (1968)]

[11]   G. Benettin, L. Galgani and J.M. Strelcyn, Phys. Rev. A 14, 2338 (1976)

[12]   G. Benettin, C. Froeschle and J.P. Scheidecker, Phys. Rev. A 19, 2454 (1979)

[13]   G. Dahlquist and A. Björck, <u>Numerical Methods</u>, Prentice–Hall, Englewood Cliffs (1974)

[14]   H. Goldstein, <u>Classical Mechanics</u>, Second edition, Addison–Wesley, Reading (1980)

[15]   H.–D. Meyer, J. Chem. Phys. 84, 3147 (1986)

[16]   R.A. Pullen and A.R. Edmonds,.J. Phys. A 14, L447 (1981)

Statistical Properties of Energy Levels and Connection to Classical Mechanics

L. S. Cederbaum, Th. Zimmermann, H. Köppel, and H.–D. Meyer
Theoretische Chemie, Physikalisch–Chemisches Institut
Im Neuenheimer Feld 253
D–6900 Heidelberg
Federal Republic of Germany

ABSTRACT. Basic ideas and methods of the statistical analysis of energy spectra are reviewed. Complex nuclear, atomic and molecular spectra show typical fluctuation patterns which can be modeled with the aid of random matrices. Calculations on model systems and theoretical results show the existence of two universality classes of spectral fluctuations which depend on whether there are strong or only weak couplings between the various degrees of freedom. This correspondence also provides a link between quantum spectral statistics and classical dynamics. We illustrate statistical methods by analyzing model spectra of coupled oscillators paying particular attention to the connection between quantum and classical results.

1. INTRODUCTION

Traditionally, a spectrum of energy levels is given an interpretation by assigning to each energy level a set of quantum numbers. In doing this one assumes, e.g., that all couplings between the various degrees of freedom can be neglected. This approximation has turned out to be often very useful, and many experimental spectra can be understood quite well in terms of these approximate quantum numbers, in particular, when considering only low excitations. For reasons to become more obvious later such spectra will be called *regular*. However, it is clear that there will be regimes where these approximations fail owing to a stronger interaction among the various degrees of freedom. In such cases the spectrum will lose its simple structure. Due to strong mixing of zeroth–order states the only (trivial) remaining good quantum number will be the energy itself. Such spectra will be called *irregular*.

The first example has been encountered some 30 years ago in nuclear physics. With the aid of shell model calculations one could achieve a detailed line–by–line understanding of low lying excitations of nuclei. With increasing energy, however, the density of states swells enormously. Clearly these spectral regimes cannot be fully reproduced by simple model calculations, and it became obvious, that a characterization of individual lines then was not useful anymore[1]. Instead, a statistical description of such complicated spectra becomes more appropriate. Wigner was the first to consider the statistical distribution of spacings between adjacent energy levels[2]. He gave arguments that for a given spectrum one

A. Amann et al. (eds.),
Fractals, Quasicrystals, Chaos, Knots and Algebraic Quantum Mechanics, 159–173.
© *1988 by Kluwer Academic Publishers.*

can distinguish between the average behaviour of the level density, which is system dependent, and fluctuations around the mean, which should not depend on specific properties of the system. This was confirmed by the analysis of a number of suitable experimental spectra which gave very similar level spacing distributions, the spacings being given in units of local mean spacing[3,4]. In particular it was found that energy levels of irregular spectra seem to repel each other, i.e. the spacing distribution always peaked at non—zero spacings and vanished for very small ones[5]. Assuming a degree of level repulsion proportional to the spacing, Wigner derived a simple heuristic distribution which agrees with empirical spacing histograms, but this gave no hint how to explain the observed universality. The phenomenon of level repulsion, however, can be understood within the context of the 'non—crossing rule'[6].

The next important step in order to arrive at a better understanding of spectral fluctuations was the idea to make statistical assumptions about a typical Hamiltonian that gives rise to an irregular spectrum. Taking an arbitrary basis set to represent a Hamiltonian, one arrives at a hermitian matrix. If the systems is 'complex' enough, one may assume that the matrix elements will exhibit some kind of random behaviour. This led to the construction of *ensembles* of random matrices, first proposed by Wigner[7], with the aim to calculate spectral fluctuation measures by averaging over all members of the ensemble instead of performing a spectral average. Random matrix ensembles were defined using very few assumptions, namely the incorporation of basic space—time symmetries, invariance under change of basis and independence of the matrix elements. The most simple and successful ensemble became the *Gaussian orthogonal ensemble*, GOE[8]. It consists of real symmetric matrices whose elements are independent and identically distributed Gaussian random variables. The GOE is uniquely defined by requiring independence of the matrix elements and invariance of the ensemble under orthogonal transformations. This mathematical simplicity allows to obtain analytical results not only for the (ensemble averaged) level spacing distribution, but also for other statistics, e.g. the spectral rigidity (see below)[8]. The comparison with statistical properties of experimental (irregular) nuclear spectra showed a remarkably good agreement between empirical and random matrix fluctuations[3,4,9-11]. The conclusion is that random matrix theory provides suitable models to describe spectral fluctuations. However, it does not constitute a proof for the observed universality of spectral statistical properties, since beyond symmetry requirements no further *physical* considerations have entered the construction of the ensembles. The class of systems that this universality pertains to is not defined, in other words, the notion of a 'complex' system remains unclarified. Further experimental evidence for the universality of spectral fluctuations in complex spectra has been given by, mostly more recent, statistical investigations of atomic[12,13] and molecular[14-21] spectra, where also random matrix—type fluctuations have been found.

In this article we want to report about new progress that has been made in the field of statistical analysis of energy spectra. The investigation of experimental spectra is limited by the relatively small samples of reliable spectral data and is always subject to 'impurities' like missing or spurious lines. In the last years there has thus been increasing interest in the study of simple theoretical model systems where large sequences of energy levels become available for a thorough statistical analysis. This gave new insight into the problem of characterizing 'complexity' and of giving a theoretical foundation for the application of random matrix theory. In particular the inspection of the semiclassical limit uncovered an intimate connection between *classical* dynamics and quantum spectral statistics[22-26]. Another important subject became the question of how a given system evolves from a regular–type spectrum (the statistics of which will be discussed below) to an irregular, 'complex' spectrum as energy and/or coupling strength are increasing. It was found that there is a continuous transition between the two universal regimes of regular– and irregular–type spectral fluctuations and that intermediate spectra reflect properties that are specific to the system under consideration.

2. BASIC STATISTICAL MEASURES

A number of statistical measures has been developed to characterize a spectral sequence. In the following we shall introduce the most important ones. A convenient and easy to calculate measure is the frequency distribution $P(S)$ of spacings S between neighbouring energy levels. $P(S)dS$ gives the probability of finding a spacing in the range $[S, S+dS]$. For GOE sequences $P(S)$ is very closely approximated by the Wigner distribution, (see, e.g., ref. 8)

$$P(S) = \frac{\pi}{2} S e^{-\pi S^2/4} \tag{1}$$

where S is given in mean spacing units. It is instructive to compare this with the result for an uncorrelated level sequence, which may be constructed, e.g., by ordering a set of random numbers generated from a uniform probability distribution. For this case one finds a Poisson distribution,

$$P(S) = \exp(-S) \tag{2}$$

Interestingly, this distribution peaks for vanishing S, i.e., uncorrelated levels tend to cluster. The Wigner distribution, in contrast, vanishes for $S=0$, a situation, which is commonly characterized as *level repulsion*. It is an indication of correlations between GOE levels at a scale length of approximately one mean spacing.

To account for level correlations on larger scales (often denoted as 'spectral rigidity') the so–called Δ_3–statistic was introduced. It is defined by[27]:

$$\Delta_3(L)$$
$$= \left\langle \frac{1}{L} \min_{A,B} \left\{ \int_E^{E+L} \left[N(E') - AE' - B \right]^2 dE' \right\} \right\rangle_E \tag{3}$$

$N(E)$ is the number of levels below energy E and $\langle .. \rangle_E$ denotes averaging over a suitable energy range. $\Delta_3(L)$ gives the average least–squares deviation of $N(E)$ from the best straight line fitting it over an interval of length L. $\Delta_3(L)$ and $P(S)$ are independent statistics and give complementary information in the sense that they are mainly sensitive to long and short range spectral correlations, respectively. The GOE ensemble average of $\Delta_3(L)$ can be calculated analytically and is for $L \gg 1$ given by

$$\bar{\Delta}_3(L) = \pi^{-2} \ln L - 0.007. \tag{4}$$

The result for an uncorrelated level sequence is given by

$$\bar{\Delta}_3(L) = L/15. \tag{5}$$

The linear behaviour in the latter case merely reflects the rule that the variance of the number of uncorrelated random events is proportional to the expected mean number of events. The logarithmic dependence of $\bar{\Delta}_3(L)$ on L in the GOE case demonstrates the presence of strong long–range level correlations. It implies for instance that statistical deviations from the expected mean number of levels contained in a fixed energy range of length L are very small. Therefore one speaks of strong spectral rigidity or of a 'semicrystalline nature' of GOE spectra.

Before proceeding, we have to mention here a somewhat technical, but very important point. In dealing with statistical properties one is only interested in fluctuations around mean values. In our case this means that we are not interested in the average variation ('secular behaviour') of the spectral density with respect to energy, but only in density fluctuations. The mean level density is strictly system dependent and can often be calculated by, e.g., semiclassical methods. Before performing any statistical analysis one therefore scales the spectrum to have constant mean level density. This procedure is commonly termed *deconvolution*. Technically this can be done by using fit formulas to the integrated level density $N(E)$ (which is a staircase), e.g. cubic splines, see, e.g. refs. 4 and 14. Usually one finds a quite sharp distinction between spectral fluctuations and the secular behaviour, so statistical results do not depend on details of the deconvolution. Note that N(E) for the deconvoluted sequence fluctuates around a straight line.

As a further statistical measure we mention the frequency distribution $\rho(I)$ of line intensities I, where $\rho(I)dI$ gives the probability to find an intensity in the range $[I, I+dI]$. For the GOE one finds a Gaussian distribution for transition amplitudes, for the corresponding squares this gives the Porter–Thomas distribution:[4,28]

$$\rho(I) = (2\pi I)^{-1/2} e^{-I/2}. \tag{6}$$

I is in mean intensity units.

3. EXPERIMENTAL SPECTRA

The validity of the GOE as a model for spectral fluctuations has of course to be checked by comparison with spectra from physical systems. So far this has been done for a number of experimental spectra[9-21], as well as for many numerically obtained spectra of model systems[25,29-48]. GOE fluctuations have first been found for resonances in neutron scattering cross sections of various nuclei, and by now all three statistics introduced above, as well as higher correlation properties, have been shown to agree there very well with the corresponding GOE predictions[9-11]. Similar findings, though not with the same precision, have been reported for atomic spectra[12,13]. In molecular spectra the first signatures of GOE fluctuations have been found in the optical absorption spectrum of NO_2[14,21], later also in stimulated emission pumping spectra of acetylene[15-20] and in spectra of methylglyoxal[18].

The statistical analysis of experimental spectra faces severe difficulties. Due to the large number of dynamical degrees of freedom, these spectra often exhibit very high level densities. It is thus very difficult to extract a level sequence which is suitable for a statistical analysis. Besides the demands on spectral resolution, such a sequence has to be complete and pure. Complete means that no lines should be missing. Pure means that all lines have to be of the same symmetry species. The reason for the latter is that different symmetries define different eigenvalue problems, so the corresponding sequences are independent and their superposition leads to an effective loss of correlations between levels. Experimentally these requirements are difficult to fulfill. For a review of these topics see, e.g., ref. 49.

4. LEVEL STATISTICS AND SEMICLASSICAL MECHANICS

In the following we shall concentrate on the statistical analysis of numerically obtained model spectra. The above mentioned quality requirements for spectra are of course now easy to fulfill, and the number of levels available can now be large enough to allow for good statistical significance. The study of model systems has in recent years particularly enlightened the relevance and interpretation of spectral fluctuation properties. Before we proceed with the discussion of a model in the context of molecular dynamics, we, therefore, briefly report the basic findings of these investigations. (For a more detailed review of this topic see, e.g., refs. 6 and 26.)

The basic insight obtained from the study of model systems is the finding of an intimate connection between level statistics and characteristics of the corresponding *classical* dynamics (if the system has a classical analogue). To explain this in some more detail, we make a short excursion into basic results from classical and semiclassical mechanics. For a more detailed discussion of classical mechanics we also refer to another contribution to this workshop by Meyer[50]. Semiclassical quantization is an important tool for both interpretation and approximate calculation of quantum properties of Hamiltonian systems. In its standard formulation, however, it is only applicable, when the classical system is *integrable*. These systems are distinguished by the existence of f constants of motion (for f degrees of freedom). All trajectories then are quasiperiodic and wind around f-dimensional tori in $2f$-dimensional phase space[51]. The constants of motion may be chosen in a coordinate independent manner as action integrals over topologically different closed paths on those tori. As already recognized by Einstein[52], this has the important consequence that it is just this class of systems that can easily be quantized semiclassically by requiring the f actions to be integer multiples of $\hbar$ (plus some constant); this is referred to as EBK–quantization[53]. Using this quantization scheme Berry and Tabor[26] were able to prove that for such systems in the semiclassical limit energy levels locally appear to be uncorrelated, giving for the spacing statistic a Poisson distribution, Eq. (2). Less formally, this follows by recognizing that at high energies successive energy levels are characterized by very different values of their quantum numbers (n_1, ..., n_f), and hence are effectively uncorrelated. We mention that harmonic oscillators form an important exception from that rule, because their oscillator frequencies do not depend on energy and this leads to strong level correlations over the entire spectrum[26]. Classical motion of integrable systems is often called *regular*, which is the motivation to use the same expression also for the corresponding quantum spectra.

Irregular motion, on the other side, occurs when, e.g. adding a sufficiently strong nonintegrable perturbation to an integrable Hamilton function. All constants of motion apart from energy then are destroyed and trajectories cover densely the whole accessible $(2f-1)$–dimensional energy shell in phase space. It has been conjectured[22,23] that the quantum counterparts of these systems should exhibit random matrix type spectral fluctuations. This conjecture has been made more precise[25] and has been verified numerically for a considerable number of systems[25,29-48]. Also theoretical arguments support this[54-56] (especially for the Δ_3–statistic, which for regular and irregular systems has been shown to behave asymptotically as given by eqs. 5 and 4, respectively[56]), although there is still no complete and rigorous formal proof and the precise conditions of this correspondence are not known yet.

To summarize, we now have that regular spectra, i.e. those which allow complete line assignments by quantum numbers, are locally uncorrelated. Irregular spectra, i.e. those which do not possess any good quantum numbers apart from energy, exhibit GOE–type spectral statistics. So, interestingly, the basic two types of spectral fluctuations introduced in the beginning are both realized in nature. Furthermore, we now can give the notion of a 'complex' spectrum a more precise meaning, namely by the association with an irregular behaviour of the classical analog.

Of course there are also intermediate systems, where, e.g., the nonintegrable perturbation is only weak. Here, there is a mathematical theorem, namely the KAM theorem[51], which states that some tori of the integrable system persist though slightly deformed. The energy shell in phase space then consists of subdomains containing regular trajectories and others filled with irregular ones. (For $f \geq 3$ degrees of freedom the latter are all connected.) For an increasingly strong perturbation an increasing number of tori is destroyed until eventually the whole motion is irregular. One may now argue[57] that for the regular domains semiclassical (torus–) quantization is still applicable, giving a set of locally uncorrelated energy levels (regular spectrum). Irregular subdomains, on the other hand, give energy levels with random matrix–type spectral fluctuations (irregular spectrum). Semiclassically these sequences will be independent and their superposition gives the total spectrum.

The level statistics of such intermediate systems can now simply be calculated by a weighted superposition of GOE– and Poisson–type statistics. The weights are given by the classical volumes of the corresponding phase space domains on the energy shell. For the most simple case of one irregular region with relative weight q and a regular one with weight $1-q$, for the level spacing distribution one obtains[58]:

$$P(q;S) = \exp\left[-(1-q)S - \frac{\pi}{4}q^2S^2\right] \times$$
$$\times\left[1 - q^2 + \frac{\pi}{2}q^3S - (1-q)^2 R(qS)\right] \tag{7}$$
$$R(z) = 1 - \exp(\pi z^2/4)\,\mathrm{erfc}(\sqrt{\pi}\,z/2),$$

where erfc denotes the well known complementary error function. For $S=0$ this spacing distribution gives $P(q,0)=1-q^2$, i.e. for all finite values of q there is level clustering. The Δ_3–statistic is given by[59]:

$$\bar{\Delta}_3(q,L) = \bar{\Delta}_3^{(\text{regular})}((1-q)L) +$$
$$+ \bar{\Delta}_3^{(\text{irregular})}(qL), \tag{8}$$

As q varies between zero and one, these formulas interpolate between regular (eqs. 2, 5) and irregular (Eqs. 1, 4) spectral fluctuations. If classically there are several distinct irregular phase space domains, semiclassically one has to mix a corresponding number of independent spectra. The resulting spectrum then will appear more regular.

The validity of this approach has been investigated for a number of model systems and the semiclassical approach discussed above is quite successful in explaining spectral statistics. In the next section this will be discussed in detail for two examples of coupled oscillators.

5. LEVEL STATISTICS IN MODEL SYSTEMS

The connection between quantum energy level statistics and corresponding classical dynamics has been studied for a considerable number of model systems. In all cases the study of the quantum spectra confirms the association of regular and irregular classical motion with regular and irregular quantum spectral statistics, respectively. In particular, the classical transition from regular to irregular motion is likewise reflected by a transition of the corresponding spectral fluctuations from Poisson— to GOE—type statistics.

Prototype systems which have been studied extensively are the so—called 'quantum billiards', i.e. a free particle enclosed by hard walls[25,29-31,34,35,41], and nonlinearly coupled oscillators[32,33,36-40,42-44]. By introducing a system parameter that controls the shape of the walls or the coupling strength, respectively, usually a whole system family is considered. Classically, these families typically comprise systems with regular dynamics (e.g. for highly symmetric walls or zero coupling strength) and others with irregular dynamics (irregularly shaped walls or strong coupling). For fixed system parameter both types of behaviour may also occur in different energy ranges. On variation of the parameter values or energies between these two regimes one finds a corresponding continuous transition of dynamical properties between regular and irregular behaviour. We remark here that billiard systems deserve some caution because of the nonanalytic potential energy, which, e.g., has the consequence that the KAM theorem is not valid. Coupled oscillators are much more representative for realistic physical systems.

In the following we want to discuss these observations in some detail for two model systems of coupled harmonic and quartic oscillators, respectively. The Hamiltonians are

$$\mathcal{H}_1 = \frac{1}{2}(p_x^2 + p_y^2 + x^2 + y^2) + 4kx^2y^2 \tag{9}$$

and

$$\mathcal{H}_2 = \frac{1}{2}(p_x^2 + p_y^2 + x^4 + y^4) - kx^2y^2 . \tag{10}$$

We first turn to the quantum spectra. For several values of the coupling parameters k we calculated energy level sequences by expanding $\mathcal{H}_1$ or $\mathcal{H}_2$ in a basis of optimum harmonic oscillator states and subsequently diagonalizing the truncated matrix. By exploiting the geometrical symmetries of $\mathcal{H}_1$ and $\mathcal{H}_2$ one obtains level sequences of definite symmetry. For all technical details we refer to refs. 36, 38 ($\mathcal{H}_1$) and 43 ($\mathcal{H}_2$).

Let us first discuss the level spacing distribution. Taking all levels within suitable energy intervals we calculated spacing histograms to which we fitted the distribution $P(q,S)$, Eq. (7). This gives parameter values $q = q_{qm}$, that, in general, will depend on the coupling parameter k and on the energy range the levels are taken from. Figs. 1, a–c, show typical spacing histograms for the Hamiltonian $\mathcal{H}_2$, Eq. (10), with $k = 0$, 0.3, 0.6, respectively. For increasing coupling strength k the empirical spacing distribution evolves from a Poisson to a Wigner distribution, shown as dashed lines in Figs. 1a and c, respectively. The full line in Fig. 1b is the best–fit distribution $P(q,S)$, Eq. (7), with $q_{qm} = 0.72$. Fig. 1 shows that there is a transition from regularity to irregularity as the coupling of the two degrees of freedom increases.

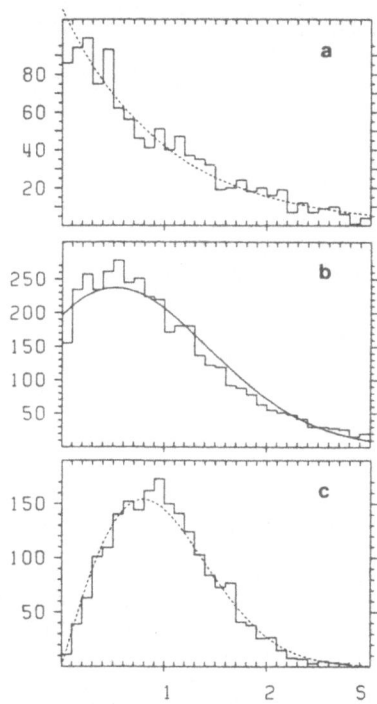

Figure 1. Level spacing histograms for the Hamiltonian, Eq. 10. (a) $k = 0$, constructed from 1142 levels. The dashed line gives the Poisson distribution, Eq. 2. (b) $k = 0.3$, constructed from 4014 levels. The full line gives the best–fit distribution, Eq. 7, with $q_{qm} = 0.72$. (c) $k = 0.6$, constructed from 2028 levels. The dashed line indicates the Wigner distribution, Eq. 1.

In Fig. 2 the symbols give the fitted q_{qm} values for the first Hamiltonian $\mathcal{H}_1$, Eq. (9); different symbols denote different coupling parameters k. For this system the spacing distribution for fixed k varies continuously with respect to energy. Choosing kE as abscissa, where E is the center of the energy interval used to calculate q_{qm}, the q_{qm} values for all coupling parameters k fit together to a smooth curve. These data again indicate a transition from Poisson– to GOE–statistic for increasingly strong coupling.

The choice of kE as a measure for the coupling strength is motivated by a scaling property of the classical Hamilton function $\mathcal{H}_1$, Eq. (9),

$$k \, \mathcal{H}_1(k;\, p_x,\, p_y,\, x,\, y)$$
$$= \mathcal{H}_1(1;\, \sqrt{k}\, p_x,\, \sqrt{k}\, p_y,\, \sqrt{k}\, x,\, \sqrt{k}\, y). \tag{11}$$

This means that the classical dynamics, apart from a simple scaling of the coordinates and momenta, depends solely on kE. Fig. 2 demonstrates, that this

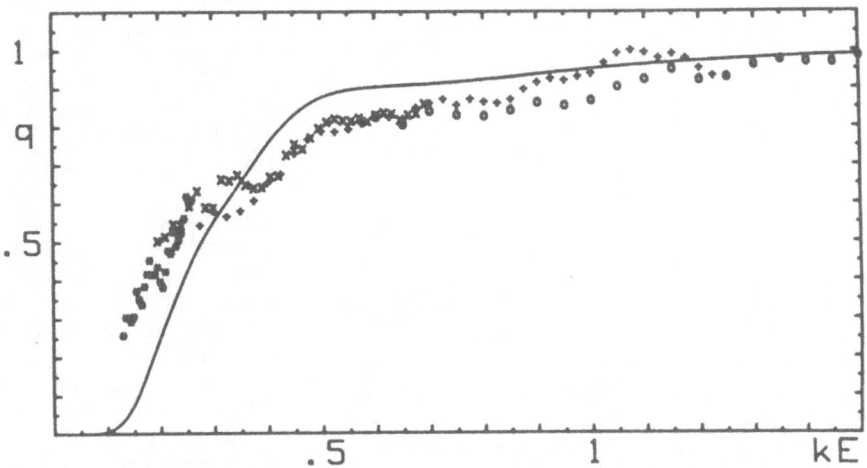

Figure 2. Classical irregular fraction q_{cl} (full line) and quantal best—fit values q_{qm} for the Hamiltonian, Eq. 9. Different symbols represent different coupling parameters k: *, 0.001, × 0.003, + 0.005; o 0.01, respectively.

classical property is reflected by the corresponding quantum system, in the sense that the spacing distribution essentially depends on kE. The analogous statement also holds for $\mathcal{H}_2$, Eq. (10). There, the potential energy is a homogeneous function in the coordinates, and by the principle of mechanical similarity[60] the classical dynamics then (apart from scaling) is independent of energy and only depends on k. Due to the finite value of $\hbar$ such classical scaling properties strictly cannot be translated into the quantum system. However, from the correspondence principle one expects that classical properties come into play as $\hbar$ becomes small, i.e. in the semiclassical limit. (For scaling systems, like $\mathcal{K}_2$, Eq. (10), this is equivalent to the high energy limit.) The above results demonstrate that the spacing distribution reflects scaling properties of the classical dynamics, even in the quantum regime where the semiclassical limit is not fully reached.

To analyze the connection between quantum statistics and classical dynamics more deeply, the classical relative fractions q_{cl} of irregular subdomains on the energy shell is shown in Fig. 2 as a solid line for the first example[50,61]. Due to the classical scaling properties, q_{cl} depends solely on kE. The result for q_{cl} (an analogous one is obtained for the second example[43]) illustrates that these systems classically undergo a transition from regular to fully irregular motion. For both systems the quantal fitted values q_{qm} tend to follow the corresponding classical curves for q_{cl}. There are however systematic deviations, namely $q_{qm} > q_{cl}$ for weak and the reverse tendency for strong coupling. For a better comparison fig. 3 shows the collected data, together with the results of a similar investigation of a billiard system[41], by displaying q_{qm} with respect to q_{cl}. All data points fit well together in the tendencies just stated.

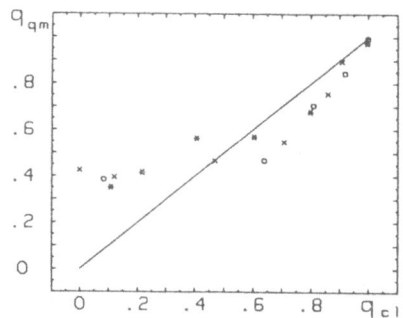

Figure 3. Fitted quantal values q_{qm} vs. corresponding classical irregular fractions q_{cl}. Circles, crosses and stars represent different systems; see text.

In order to understand the deviations towards larger values $q_{qm} > q_{cl}$ for weak coupling, we analyzed very long level sequences of the uncoupled system $\mathcal{K}_2$, Eq. (10), $k = 0$. The spectrum shows an extremely slow degree of convergence to the semiclassically expected Poisson limit[43]. Fitting $P(q,S)$, Eq. (7), to such level spacing histograms gives values $q_{qm} > q_{cl} = 0$. This will of course still hold for weakly coupled systems that classically are dominated by regular motion.

For stronger coupling one has to regard that in case of two degrees of freedom, as in our examples, the energy shell consists of several unconnected irregular subdomains of considerable weights[50,51,61], and each of them independently contributes a GOE–type level sequence. The total spectrum then is a mixture of more than two sequences and will appear more regular due to an effective decrease of level correlations. Fitting the spacing distribution by $P(q,S)$, Eq. (7), which assumes only one irregular region, thus gives values q_{qm} which are smaller than the corresponding classical irregular fraction q_{cl}. This is indeed observed in Fig. 3.

We now turn to the spectral rigidity. For numerical calculations of $\bar{\Delta}_3(L)$ we consider only the system $\mathcal{K}_2$, Eq. (10), for which we have the largest number of eigenvalues and which, due to its classical scaling property, allows a better comparison with semiclassical predictions. Fig. 4a and b show $\bar{\Delta}_3(L)$ for the integrable system, $k = 0$, and for the completely irregular system, $k = 0.6$, respectively. The full lines give the theoretical results for uncorrelated and GOE spectra. The findings agree very well with the theoretical curves for not too large values of L. The saturation of the empirical $\bar{\Delta}_3(L)$ for large L has been explained by a semiclassical theory for the spectral rigidity[56]. It is an effect of additional long range level correlations that are still present for finite $\hbar$. The value L_{max} up to which $\bar{\Delta}_3(L)$ approximately lies on the theoretical curves can be calculated semiclassically[56].

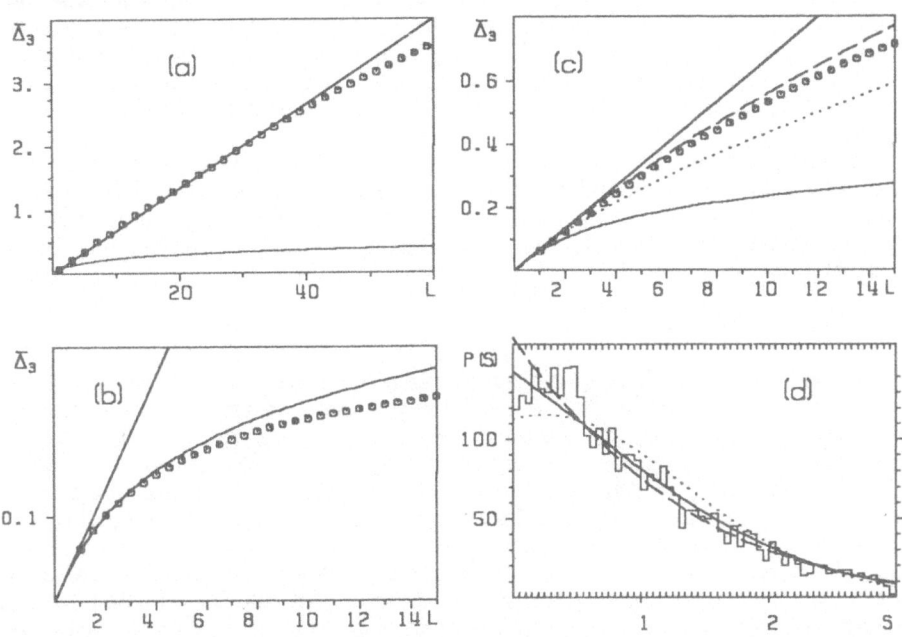

Figure 4. (a) $\bar{\Delta}_3(L)$ for the Hamiltonian, Eq. 10, and $k = 0$. Full lines are $L/15$ and GOE prediction (cf. Fig. 2), circles give $\bar{\Delta}_3(L)$. (b) Same as (a) for $k = 0.6$. (c) Same as (a) for $k = 0.2$. The dotted line gives the semiclassical prediction with one irregular fraction q_{cl}; the dashed line the same with three irregular fractions; see text. (d) Level spacing histogram for the same system as in (c). Full line, best—fit distribution $P(q_{qm}; S)$, $q_{qm} = 0.50$. Dotted and dashed line, semiclassical predictions analogous to (c); see text

The results for an intermediate system, $k = 0.2$, are shown in Fig. 4c. As one expects, the data points lie in between the two limiting universal curves (full lines). Taking Eq. (8) for $\bar{\Delta}_3(L)$ of mixed sequences and setting $q = q_{cl} = 0.64$, one arrives at the dotted curve. However, the actual behaviour of $\bar{\Delta}_3(L)$ for $L \ll L_{max}$ is clearly different. In order to achieve better agreement one has to regard the classical phase space structure in more detail. The classical calculations for $k = 0.2$ revealed several distinct irregular regions[50, 61], each of which independently contributes a GOE—type level sequence. Taking into account the three most important irregular subdomains of weights 0.26, 0.19, and 0.19, one arrives at the dashed curve (using Eq. (8) extended accordingly[43]). This now gives a remarkable agreement between a pure semiclassical prediction and a corresponding quantum calculation (for $L < L_{max}$).

In the same way, of course, one can also extend Eq. (7) to account for the influence of the partitioning of the classical irregular fraction on the level spacing distribution[58]. The result (for $k = 0.2$) is presented in Fig. 4d, where the full line gives the best–fit of Eq. (7) to the histogram with $q = q_{qm} = 0.50$, the dotted line gives the same with $q = q_{cl} = 0.64$, and the dashed line accounts for the three independent irregular regions (in analogy to Fig. 4c). The latter curve agrees best with the spacing histogram for $S > 0.5$, but does not reproduce the actual degree of level repulsion. The dotted line clearly fails, and the fit curve lies in between the two classical predictions trying to match the small– and large–S behaviour of the histogram.

In summary we find for long range spectral correlations, namely $P(S)$, $S > 0.5$, and $\bar{\Delta}_3(L)$, that semiclassical theories well describe spectral fluctuations. $P(S)$ for small S shows deviations, namely the probability for very small spacings is smaller than expected. To explain this finding we first observe that by the uncertainty relation

$$\Delta t \cdot \Delta E \approx \hbar$$

one can associate spectral long and short range correlations with short and long time behaviour, respectively. If one considers for instance the time evolution of a wave packet, one has from the Ehrenfest theorem (see, e.g. ref. 62), that for short times it will follow a classical trajectory. For longer times this relation breaks down due to the spreading of the wave packet. Together, it becomes obvious that semiclassical arguments are valid on large spectral scales, but fail on small ones. In this picture it is clear that for longer times the wave packet will tunnel through dynamical barriers that separate the different irregular phase space domains. The details of the classical phase space structure thus will not be reflected by short range spectral correlations. Level repulsion, as observed e.g. in Fig. 4d in terms of finding fewer small spacings than expected for $\hbar \to 0$ (dashed line), can be interpreted in terms of level interactions due to such tunneling. In the spirit of this discussion we argue that the spacing distribution effectively probes the total irregular fraction, whereas $\bar{\Delta}_3(L)$ sensitively responds to the partitioning of the energy shell into several independent subregions. For a more rigorous discussion of the above arguments see refs. 6, 43, 56 and 63.

6. DISCUSSION

The study of simple model systems has confirmed the usefulness of a statistical description of energy levels and has contributed to a deeper understanding of its physical meaning. Theoretical and numerical statistical investigations indicate two universal prototypes of spectral fluctuations. There are locally uncorrelated spectra where levels tend to cluster, and which are called regular. On the other hand there are irregular random matrix type spectra, where the levels are highly correlated and show level repulsion. Remarkably, by looking at the semiclassical limit $\hbar \to 0$, these two types of fluctuations can be associated with two basic types of classical (bound) motion, namely regular and irregular motion, respectively. Regular motion is quasiperiodic and can be quantized semiclassically by standard methods[53]. Corresponding regular quantum spectra thus can be characterized by a set of quantum numbers. Irregular trajectories are only constrained by energy conservation. They have the specific property that small perturbations grow exponentially in time. Irregular motion is therefore also called chaotic. The study of classical chaotic systems has attained much attention in the past decades (see, e.g. refs. 51, 64 and 65), and the inspection of the corresponding quantum systems is of relevance, see, e.g. refs. 24, 66 and 67. We have shown here that there is an intimate connection between classical chaos and statistics of quantum energy levels.

It is now possible to state more clearly what is meant by a 'complicated' spectrum or by a complex system, namely by associating these notions with a classically chaotic counterpart. In contrary to what one might tend to believe, even two degrees of freedom are sufficient to produce these phenomena. The connection between level statistics and classical dynamics also provides insight into how quantum spectral fluctuations evolve from regular to irregular behaviour for increasing coupling strength between zeroth–order states. In the classical transition from regular to fully chaotic motion an increasing fraction of the energy shell is covered by irregular trajectories. In the semiclassical picture the corresponding quantum transition from regular to irregular spectral statistics proceeds in a very similar way, namely an increasing fraction of levels becomes distributed irregularly. The spectral statistics then follow from an independent superposition of regular and irregular sequences that are associated with classical regular and irregular phase space domains, respectively.

The specific way in which statistical properties evolve between the two universal regimes of fluctuations depends of course on the given system. It is the route from regular to irregular behaviour which contains the most specific information on the system. For instance we have demonstrated, that this transition reflects classical scaling properties of the Hamiltonian. The study of the transition deserves further interest concerning such questions as, e.g., whether different coupling mechanisms between zeroth–order states manifest themselves differently in spectral statistics. As an example from molecular spectroscopy one may think of underlying couplings between vibrational modes alone or vibronic coupling between electronic states. The models discussed in this work are related to the former coupling case. Vibronic models, which do not possess an obvious classical analogue, are investigated elsewhere[68].

The arguments which link spectral statistics and classical dynamics, apply strictly only in the semiclassical limit. Only then it is really possible to define a local level statistic, since the level density then tends to infinity. Actual spectra usually cover a regime where this limit is not fully reached and where genuine quantum effects come into play. These effects, such as tunneling between separate irregular phase space domains, have the consequence that the classical phase space structure is not reflected in all details by short range spectral fluctuations. In particular it is found that the spacing distribution mainly probes the total classical irregular phase space fraction. The simple one–parameter family of distributions $P(q,S)$, Eq. (7), is therefore a suitable working tool to characterize the degree of irregularity of a spectrum. The spectral rigidity $\Delta_3(L)$, on the other hand, mainly accounts for long range fluctuations and reflects rather sensitively the classical phase space structure.

REFERENCES

1   F. J. Dyson, J. Math. Phys. 3, 140(1962).
2   E. P. Wigner, Ann. Math. 53, 36(1951).
3   *Statistical Theories of Spectra: Fluctuations* edited by C. E. Porter, Academic Press, New York, 1965.
4   T. A. Brody, J. Flores, J. B. French, P. A. Mello, A. Pandey, and S. S. M. Wong, Rev. Mod. Phys. 53, 385(1981).
5   I. I. Gurevich and M. I. Pevsner, Nucl. Phys. 2, 575(1957).
6   M. V. Berry, in ref. 64, p. 123
7   E. P. Wigner, Can. Math. Congr. Proc., Univ. of Toronto Press, Toronto, Canada, p. 174 (1957).
8   M. Mehta, *Random Matrices and the Statistical Theory of Energy Levels*, Academic Press, New York, 1967.
9   R. U. Haq, A. Pandey, and O. Bohigas, Phys. Rev. Lett. 48, 1086(1982).
10 O. Bohigas, R. U. Haq, and A. Pandey, Phys. Rev. Lett. 54, 1645(1985).
11 J. F. Shriner, Jr., G. E. Mitchell and E. G. Bilpuch, Phys. Rev. Lett. 59, 435(1987).
12 N. Rosenzweig and C. E. Porter, Phys. Rev. 120, 1698(1960).
13 H. S. Camarda and P. D. Georgopulos, Phys. Rev. Lett. 50, 492(1983).
14 E. Haller, H. Köppel, and L. S. Cederbaum, Chem. Phys. Lett. 101, 215(1983).
15 E. Abramson, R. W. Field, D. Imre, K. K. Innes, and J. L. Kinsey, J. Chem. Phys. 80, 2298(1984).
16 S. Mukamel, J. Sue, and A. Pandey, Chem. Phys. Lett. 105, 134(1984).
17 R. L. Sundberg, E. Abramson, J. L. Kinsey, and R. W. Field, J. Chem. Phys. 83, 466(1985).
18 L. Leviandier, M. Lombardi, R. Jost, and J. P. Pique, Phys. Rev. Lett. 56, 2449(1986).
19 J. P. Pique, Y. Chen, R. W. Field and J. L. Kinsey, Phys. Rev. Lett. 58, 475(1987).
20 R. Jost, this issue.
21 G. Persch, W. Demtröder, Th. Zimmermann, H. Köppel and L. S. Cederbaum, to be published in Ber. Bunsenges. Phys. Chem.; Th. Zimmermann, H. Köppel, L. S. Cederbaum, G. Persch and W. Demtröder, to be published.
22 G. M. Zaslavskii, Zh. Eksp. Teor. Fiz. 73, 2089(1977), [Sov. Phys. JETP 46, 1094(1977)].
23 M. V. Berry and M. Tabor, Proc. R. Soc. London, Ser.A 356, 375(1977).
24 G. M. Zaslavskii, Phys. Rep. 80, 157(1981).
25 O. Bohigas, M. J. Giannoni, and C. Schmit, Phys. Rev. Lett. 52, 1(1984).
26 O. Bohigas and M.–J. Giannoni, in *Mathematical and Computational Methods in Nuclear Physics* edited by J. S. Dehesa, J. M. G. Gomez, and A. Polls, Lecture Notes in Physics Vol. 209, Springer Verlag, Heidelberg, 1984.
27 F. J. Dyson and M. L. Mehta, J. Math. Phys. 4, 701(1963).
28 C. E. Porter and R. G. Thomas, Phys. Rev. 104, 483(1956).
29 S. W. McDonald and A. N. Kaufman, Phys. Rev. Lett. 42, 1189(1979).
30 G. Casati, F. Valz–Griz, and I. Guarneri, Lett. Nuovo Cimento 28, 279(1980).
31 M. V. Berry, Ann. Phys. (N.Y.) 131, 163(1981).
32 V. Buch, M. A. Ratner and R. B. Gerber, Mol. Phys. 46, 1129(1982).
33 V. Buch, R. B. Gerber, and M. A. Ratner, J. Chem. Phys. 76, 5397(1982).
34 O. Bohigas, M. J. Giannoni, and C. Schmit, J. Physique Lett. 45, L1015(1984).
35 M. Robnik, J. Phys. A 17, 1049(1984).

36 E. Haller, H. Köppel, and L. S. Cederbaum, Phys. Rev. Lett. 52, 1665(1984).
37 T. H. Seligman, J. J. M. Verbaarschot, and M. R. Zirnbauer, Phys. Rev. Lett. 53, 215(1984).
38 H.-D. Meyer, E. Haller, H. Köppel, and L. S. Cederbaum, J. Phys. A 17, L831(1984).
39 T. H. Seligman and J. J. M. Verbaarschot, J. Phys. A 18, 2227(1985).
40 T. H. Seligman, J. J. M. Verbaarschot, and M. R. Zirnbauer, J. Phys. A 18, 27(1985).
41 T. Ishikawa and T. Yukawa, Phys. Rev. Lett. 54, 1617(1985).
42 E. Caurier, B. Grammaticos, Europhys. Lett. 2, 417(1986).
43 Th. Zimmermann, H.-D. Meyer, H. Köppel, and L. S. Cederbaum, Phys. Rev. A 33, 4334(1986).
44 Th. Zimmermann, H. Köppel, E. Haller, H.-D. Meyer, and L. S. Cederbaum, Physica Scripta 35, 125(1986).
45 D. Wintgen and H. Friedrich, Phys. Rev. Lett. 57, 571(1986).
46 D. Delande and J. C. Gay, Phys. Rev. Lett. 57, 2006(1986).
47 G. Wunner, U. Woelk, I. Zech, G. Zeller, T. Ertl, F. Geyer, W. Schweizer and H. Ruder, Phys. Rev. Lett. 57, 3261(1986).
48 O. Bohigas, this issue.
49 Th. Zimmermann, L. S. Cederbaum, H.-D. Meyer and H. Köppel, J. Phys. Chem. 91, 4446(1987).
50 H.-D. Meyer, *Chaotic Behaviour of Classical Hamiltonian Systems*, this issue.
51 A. J. Lichtenberg and M. A. Liebermann, *Regular and Stochastic Motion*, Springer Verlag, Heidelberg, 1983.
52 A. Einstein, Verh. dt. Phys. Ges. 19, 82(1917).
53 I. C. Percival, Adv. Chem. Phys. 36, 1(1977).
54 P. Pechukas, Phys. Rev. Lett. 51, 943(1983).
55 T. Yukawa, Phys. Rev. Lett. 54, 1883(1985).
56 M. V. Berry, Proc. R. Soc. London, Ser. A400, 229(1985).
57 I. C. Percival, J. Phys. B 6, L229(1973).
58 M. V. Berry and M. Robnik, J. Phys. A 17, 2413(1984).
59 A. Pandey, Ann. Phys. (N. Y.) 119, 170(1979).
60 L. D. Landau and E. M. Lifshitz, *Mechanics*, Pergamon, Oxford, 1969.
61 H.-D. Meyer, J. Chem. Phys. 84, 3147(1986).
62 A. Messiah, *Quantum Mechanics*, North–Holland Publ. Co., Amsterdam, 1970.
63 M. V. Berry, in *The Wave–Particle Dualism*, edited by S. Diner, D. Fargue, G. Lochak, and F. Selleri, Reidel, Dordrecht, 1984.
64 *Chaotic Behaviour of Deterministic Systems*, edited by G. Iooss, R. H. G. Helleman, and R. Stora, North–Holland, Amsterdam 1983.
65 H. G. Schuster, *Deterministic Chaos*, Physik Verlag, Weinheim 1984.
66 *Chaotic Behavior in Quantum Systems*, edited by G. Casati, Plenum, New York, 1985.
67 *Quantum Chaos and Statistical Nuclear Physics*, edited by T. H. Seligman and H. Nishioka, Lecture Notes in Physics Vol. 263, Springer Verlag, Heidelberg, 1986.
68 Th. Zimmermann, L. S. Cederbaum and H. Köppel, *Statistical Properties of Energy Levels in Non–Born–Oppenheimer Systems*, to be published in Ber. Bunsenges. Phys. Chem.

QUANTUM SUPPRESSION OF CLASSICAL CHAOS AND MICROWAVE IONIZATION OF HYDROGEN ATOM

Giulio Casati
Dipartimento di Fisica dell'Universita'
Via Celoria 16,
20133 Milano, Italy

ABSTRACT. We discuss the problem of excitation and ionization of an hydrogen atom under a linearly polarized, monochromatic, microwave field. We show that, in spite of the suppression of classical chaotic diffusion produced by quantum interference effects, strong excitation and ionization may take place. This result, which is a quantum manifestation of classical chaos, leads to new unexpected results and explains the presence of a large ionization peak at frequencies much below those required for the conventional one-photon photoelectric effect.

1. In the early seventies, experimental results [1] on ionization of highly excited hydrogen atoms (principal quantum number n ~ 60) in microwave fields of frequency $\omega/2\pi$ ~ 10GHz and peak intensity ϵ~ 10V/cm revealed a surprising high ionization probability given that, in such a situation, ionization would require absorption of about 100 photons.

Already in 1978 Delone et al. [2] suggested a diffusive ionization mechanism in order to explain the high ionization probability experimentally observed. In the same year Leopold and Percival [3], in consideration of the high quantum numbers involved in the experiments, suggested that classical mechanics could give an accurate description

175

A. Amann et al. (eds.),
Fractals, Quasicrystals, Chaos, Knots and Algebraic Quantum Mechanics, 175–187.
© 1988 by Kluwer Academic Publishers.

and indeed they showed that a numerical solution of classical equations of motion gives a quite good agreement with experimental data. More recent experimental results [4] with microwave frequency $\omega_0 = \omega n^3 < 1$ ($\omega/2\pi \sim 10$ GHz, $30 < n < 70$) were also shown to be in quite good agreement with classical mechanics.

On the other hand it is quite obvious that a real understanding of the experimental results can be provided only by a quantum theory. At the same time, this theory must be able to assess the validity and the limits of the classical description. In this report we will briefly present a quantum theory of hydrogen atom excitation which already appeared in several papers [5 - 11]. We refer the interested reader to these papers.

From a general physical viewpoint, this problem lies at the intersection of several lines of contemporary research. First of all, transition to chaotic motion in the corresponding classical model plays an important rôle in the excitation process. However, the hydrogen atom is a quantum object and therefore it is necessary to study in detail the modifications introduced by quantum mechanics. This leads us directly to the question of manifestations of chaos in quantum mechanics or, as commonly termed, "quantum chaos". Quite often this question has been indicated as controversial. As already stressed in several occasions we would like to repeat once more that the question of quantum chaos is not controversial at all. As a matter of fact, classical ergodic theory provides a fairly complete classification of statistical properties of classical dynamical systems. Among these, algorithmic complexity theory [14] shows that systems with positive metric entropy possess the property that almost all orbits are random, unpredictable, uncomputable. These systems are referred as deterministically chaotic despite the seeming contradiction of these terms. The subject of "quantum chaos" is the investigation of the quantum dynamics of such systems. This is a quite exciting, well defined and important problem both for the foundations of quantum statistical mechanics as well as for many applications. For example, as it is well known, a necessary (not sufficient) condition for chaotic motion in classical systems is continuous spectrum of the Liouville operator on the energy surface. On the other hand, bounded, conservative, finite particle systems possess discrete spectrum no matter whether the corresponding classical systems are chaotic or not. As a consequence the quantum motion is always almost periodic and there is no room for any statistical behaviour beyond ergodicity. This do not implies that classical chaos is irrelevant for such systems; actually the contrary is true: for example the energy levels distribution of classically chaotic systems belongs to different universality classes depending on whether the dynamics possesses the time-reversal property or not[23].

Even more interesting is the study of quantum systems under time-periodic perturbations since, for these systems, the spectrum of the quantum motion may be, in principle, continuous. The hydrogen atom under a microwave field provides a physical example for which real laboratory experiments are feasible which may unveil the manifestations of classical chaos on the quantum evolution of such a fundamental object. In particular, the important phenomenon of quantum suppression of classical chaos discussed in the previous talk by Guarneri (which is essential for the understanding of the present analysis) may find here its first experimental verification. At the same time we will gain new insight on multiphoton processes and collisionless ionization of atoms and molecules in laser fields.

2. Let us consider the Hamiltonian

$$H = p^2/2 - 1/r + \epsilon z \cos\omega t \tag{1}$$

where ϵ and ω are the field strength and frequency, and the z-coordinate is measured along the direction of the external field. Here and in the following, we use atomic units. To facilitate conversion to physical units, we recall that for principal quantum number $n_0 = 100$ the frequency $\upsilon = \omega/2\pi = 10$GHz corresponds to $\omega_0 = \omega n_0^3 = 1.51998$ and $\epsilon_0 = \epsilon n_0^4 = 0.1$ corresponds to $\epsilon = 5.14485$ V/cm. According to this choice of units, $\hbar = 1$.

An essential step for the understanding of this problem has been the consideration of the one-dimensional model. Indeed this model gives a good approximate description of excitation of atoms initially prepared in very extended states along the field direction (parabolic quantum numbers $n_1 \gg n_2$, $n_1 \gg m$)[12]; such states can be prepared in laboratory experiments [13,21]. Even more important is the fact that the simplified one-dimensional model turns out to reproduce the essential qualitative features of the real problem as has been theoretically shown and numerically verified [10,11].

We would like also to remark that in building up our theory we were mainly guided by numerical experiments on modern supercomputers which played a more important rôle than real laboratory experiments. In this respect the one-dimensional model can be much more easily, and therefore more precisely, numerically handled.

Let us consider therefore the one-dimensional Hamiltonian

$$H = p^2/2 - 1/z + \epsilon z \cos\omega t, \quad z > 0 \tag{2}$$

A simplified description of the classical evolution which, however, retains all the essential features of the problem has been given in the previous talk by Guarneri. The classical motion can be approximately described by the area preserving map

$$\overline{N} = N + k \sin\phi$$
$$\overline{\phi} = \phi + 2\pi \omega (-2\omega\overline{N})^{-3/2} \tag{3}$$

which gives the variation of the quantities $N = E/\omega = -1/(2n^2\omega)$ and $\phi = \omega t - s\theta$ after each unperturbed Kepler period and where $k = 0.822\pi\epsilon/\omega^{5/3}$. If we linearize this map around the initial value $N_0 = -1/(2n_0^2\omega)$ we obtain the well-known standard map or kicked rotator map

$$\overline{N} = N + k \sin\phi$$
$$\overline{\phi} = \phi + T\overline{N} \tag{4}$$

where $T = 6\pi\omega^2 n_0^5$.

As it is well known [15], the qualitative behaviour of map (4) crucially depends on the value of parameter $K=kT= 49\ \epsilon_0\ \omega_0^{1/3}$. If $K<1$ the N motion is bounded by invariant curves and $|\Delta N| < K^{1/2}$. Viceversa if $K>1$, namely for $\epsilon_0 > \epsilon_c$ where

$$\epsilon_c = 1/(49\omega_0^{1/3}) \tag{5}$$

a diffusion process will take place in N-space leading to the unlimited average linear growth

$$<N^2> \approx D\ t \tag{6}$$

and to the Gaussian distribution

$$f(N,t) \approx (\pi k^2 t)^{-1/2}\ exp\ (-(N-N_0)^2/k^2 t) \tag{7}$$

In eqs. (6) and (7), t is time measured in number of orbital periods and $D=k^2/2$ is the diffusion coefficient.

As a consequence, under condition (5), strong excitation and ionization takes place in the hydrogen atom. As explained in ref. [11], the approximation involved in deriving mapping (3) is good for $\omega_0>1$. However, it is possible to generalize mapping (3) to frequencies $\omega_0 <1$ [11]. We will not enter in these details here. For the purpose of the present paper it is sufficient to remark that expression (5) is valid for $\omega_0> 1$ and that a correction to this formula can be obtained which is valid down to $\omega_0 \approx 0.5$. For still smaller frequencies ω_0, we are not able to compute the critical classical chaos border; however this border will approach the critical value for ionization in static field (dotted line in Fig. (3)).

3. The analysis of the quantum version of map (4) shows that quantum mechanics suppresses the classical diffusive process [16,17] and leads to the exponential localization of diffusive excitation [18-20]. This is a <u>dynamical version of Anderson localization</u> which is very well known in

solid state physics [18]. According to analytical estimates [19,20] and to numerical data [19] the quantum excitation process reaches a steady-state distribution which can be satisfactory described by the formula

$$f(N) \approx (1/2l)(1+2|N|/l) \exp(-2N/l) \tag{8}$$

where l is the localization length which, in semiclassical conditions, is [19,20]

$$l \approx D \approx k^2/2 = 3.33 \, \epsilon^2/\omega^{10/3} \tag{9}$$

Notice that exponential localization takes place in the variable N, namely in the number of absorbed photons; only in the vicinity of the initially excited level there may be appropriate exponential localization on unperturbed levels also.

The above theoretical predictions have been numerically checked [9,11] for different values of ω_0, ϵ_0, n_0. For each case we numerically determined the photonic localization length l and plotted it in Fig. 1 versus the field intensity ϵ. As is clearly seen from the figure, the numerical data are in good agreement with the theoretical expression (9) for a wide range of ϵ.

4. The phenomenon of photonic localization described above shows that actual suppression of strong excitation and ionization in the hydrogen atom takes place if the localization length is small. Instead if the localization length l is larger than the number of photons required for ionization $N_I = 1/(2n_0^2\omega)$ then probability will flow into the continuous part of the spectrum and strong excitation will take place as in the classical case [5,8,9,11]. Therefore the condition $l=N_I$ leads to the critical threshold value

$$\epsilon_q \approx \omega_0^{7/6}(6.6 \, n_0)^{-1/2} \tag{10}$$

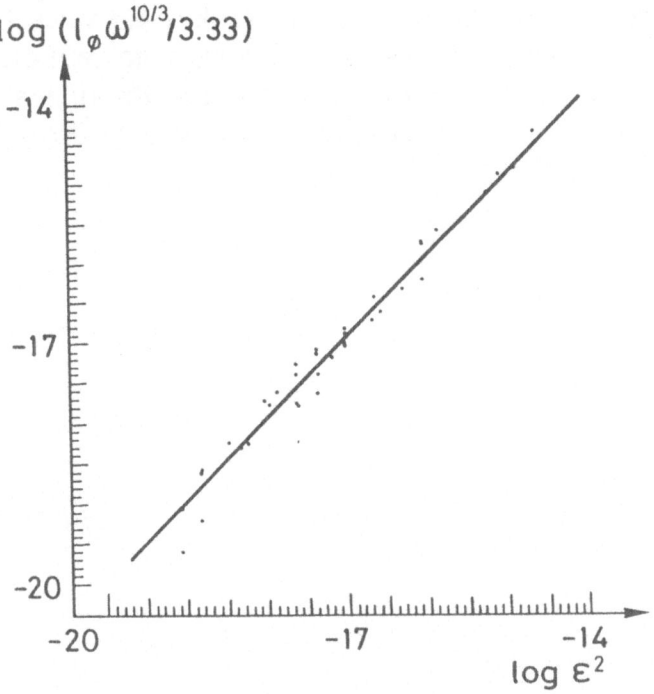

$\log (l_{\emptyset}\omega^{10/3}/3.33)$

$\log \varepsilon^2$

Fig.1 The rescaled photonic localization length as a function of the field strength in logarithmic scale. The dots are obtained from quantum numerical integration with different parameter values while the straight line is drawn according to the theoretical expression (9).

This is the critical delocalization border described in refs. [5,8]. In order to have strong ionization in the hydrogen atom one needs a field intensity larger than both the threshold value ε_c for transition to classical chaos and the threshold value ε_q for transition to quantum delocalization.

The two critical borders, classical and quantum, are reported in Fig. 2 in the ε_0, ω_0 plane for a fixed frequency $\omega/2\pi \approx 9.9$ GHz. The dots are experimental results [4] and give the threshold field value ϵ_0 for 10% ionization. The ϵ_0 values for 10% ionization must be very close to the critical border for ionization and therefore our theory predicts that experimental points must be close to the ionization border evidentiated by the dashed area in Fig. 2.

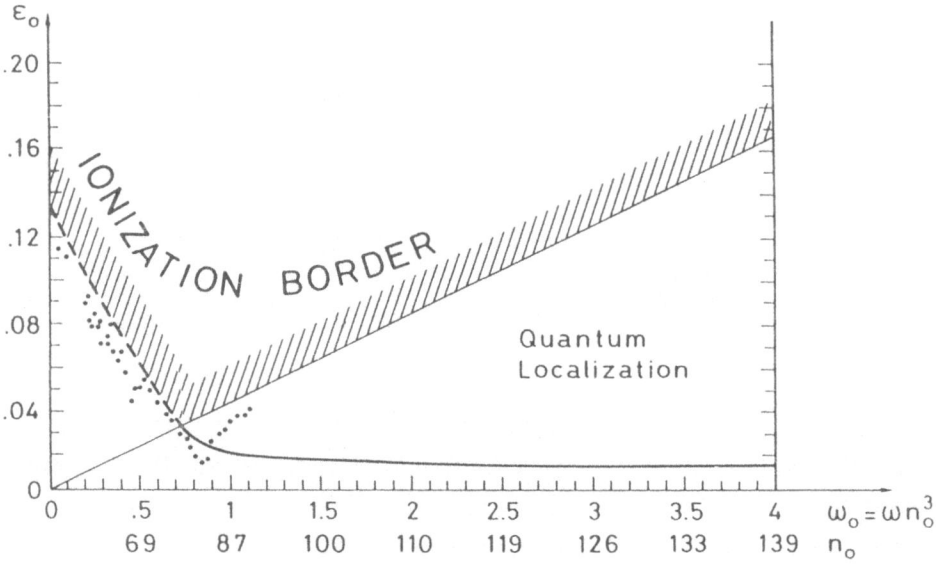

Fig. 2 - Ionization threshold as a function of the initially excited unperturbed state n_0 at fixed frequency $\omega \approx 9.9$ GHz. The dots are experimental data obtained in ref. [4] for the threshold field value for 10% ionization. The thin curve is the classical border for transition to chaos given by expression (5). As explained in the text the dotted line is just a smooth continuation of the curve from $\omega_0 \approx 0.5$ up to the value 0.13 for ionization in static field. The gross line is the quantum delocalization border (10) which for fixed $\omega = 9.9$ GHz writes
$\epsilon_q \approx (\omega^{1/6}/\sqrt{6.6})$. $\omega_0 = 0.0417 \, \omega_0$.

Notice that for $\omega_0 < 1$ the ionization border is determined by the classical chaos border. This is the reason why good agreement has been found between experimental data and the result of numerical computations on the classical model. The most interesting region lies in the frequency range $\omega_0 > 1$ where according to our theoretical predictions one should observe the effect of quantum suppression of classical chaos and of diffusive excitation. This can be done either by increasing the initially excited state n_0, which appears to be difficult, or by increasing the microwave frequency ω. Experiments with larger ω are now under preparation [22].

It is necessary to mention that the experimental data reported in Fig. 2 were obtained on full three-dimensional atoms, which are actually two-dimensional since, due to axial symmetry, the z-component m of the angular momentum is an integral of the motion. On the other hand, our theoretical predictions refer to a one-dimensional model. However, as we have recently shown [10], the energy excitation process in the two-dimensional atom can be well described by the one-dimensional model. The main reason of this fact is that the l-motion is very slow and does not significantly influence the N-motion which is therefore again described by the approximate map (3).

5. From the previous discussion it appears that the quantum excitation process is determined by a sort of competition between to opposite mechanisms: the diffusion process that would be predicted on classical grounds and the "localizing" effect produced by quantum interference. This picture allows to predict new interesting effects such as the strong ionization peak at frequencies much below the conventional one-photon threshold, shown in fig. 3. Here we plot the quantum ionization probability as a function of microwave frequency $\omega_0 = \omega n_0^3$ for a given initially excited state with principal quantum number n_0 and for a fixed peak microwave intensity $\epsilon_0 = \epsilon n_0^4$. Here we chose $n_0 = 66$, $\epsilon_0 = 0.05$ and the ionization probability, which is defined as the total probability above the level n=99, $W_i = \sum_{n>99} |c_n|^2$, is plotted after $\tau = 40\omega_0$ microwave periods, i.e., after 40 orbital periods of the electron with n=66.

184

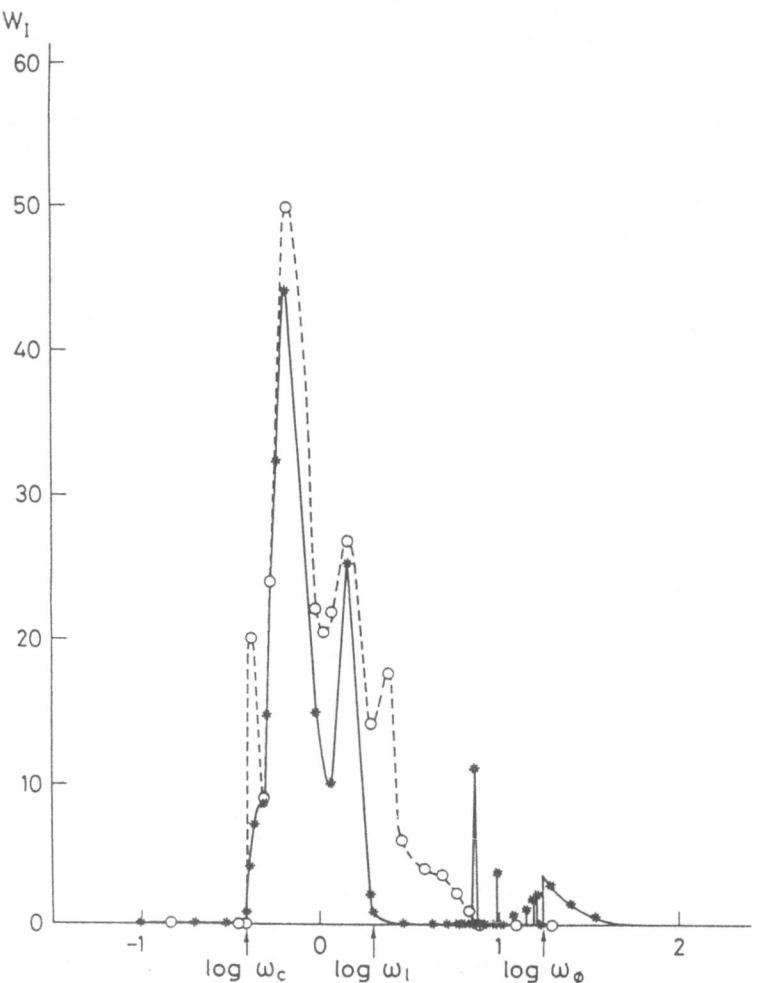

Fig. 3. Classical (o) and quantum (*) ionization probability $W_I = \sum\limits_{n>99} |c_n|^2$ versus field frequency ω_0 after 40 orbital periods corresponding to the initially excited level $n_0=66$. Here $\epsilon_0 = \epsilon n_0^4 = 0.05$. Due to our definition of ionization probability, ω_ϕ is slightly less than $n_0/2$. The values ω_c and ω_I are the critical frequencies (at fixed ϵ_0 and n_0) for transition to classical chaos and to quantum localization, respectively. Notice that the ionization probability reaches its maximum at a frequency $\omega_0 \approx 0.7$ which is less than the frequency for one-photon transition from level 66 to 67.

Notice that the frequency axis is in logarithmic scale. The frequency ω_ϕ is the critical threshold frequency for one-photon ionization: for $\omega > \omega_\phi$ the numerical data are in excellent agreement with the theoretical expression for the ionization rate $\gamma_\phi \sim 1.67\ \epsilon_0^2 n_0^2 / \omega_0^{13/3}$ (ionization probability in one period of the external field). For $\omega < \omega_\phi$ the ionization probability drops to negligible values since the field strength is small. As expected there are two or three photon resonant narrow peaks for some particular frequency value. What it is completely unexpected is the large peak at much lower frequencies in the range $\omega_c < \omega_0 < \omega_1$.

As a matter of fact the presence of this peak is a consequence of the two phenomena described above: the transition to chaotic motion in the classical system and the quantum localization of classical chaos. Indeed ω_c is given by (5) with ϵ_0 the applied field, $\epsilon_0 = 0.05$ while ω_1 is given by (10) with $\epsilon_0 = 0.05$ and $n_0 = 66$. For $\omega > \omega_c$ the classical motion is chaotic while for $\omega > \omega_1$ quantum localization takes place.

In conclusion, we would like to remark that even though in this paper we have limited our discussion to the interaction of microwave fields with highly excited states, the phenomena we have described are of a quite general nature and should also be observed in the interaction of laser fields with initially excited lower levels. It is our belief that the study of the manifestations of classical chaos in quantum mechanics will open new exciting possibilities for our understanding of the radiation matter interaction.

References

[1] J.E. Bayfield, P.M. Koch, Phys. Rev. Lett. 33, (1973), 258.

[2] N.B. Delone, V.P.Krainov, V.A. Zon, Zh Eksp. Teor. Fiz. 75 (1978) 445.

[3] J.G. Leopold, I.C. Percival, Phys. Rev. Lett. 41, (1978) 944; J. Phys. B 12, (1979) 709.

[4] K.A.H van Leeuwen, G.V. Oppen, J.B. Bowlin, P.M. Koch, R.V. Jensen, O. Rath, D. Richards and J.G. Leopold, Phys. Rev. Lett. 55 (1985) 2231.

[5] G. Casati, B.V. Chirikov, D.L. Shepelyansky, Phys. Rev. Lett. 53, (1984) 2525.

[6] G. Casati, B.V. Chirikov, I Guarneri, D.L. Shepelyansky, Phys. Rev. Lett. 57, (1986) 823.

[7] G. Casati. B.V. Chirikov. I. Guarneri. D.L. Shepelyansky, Phys. Rev. Lett. 56 , (1986) 2437.

[8] G. Casati, B.V. Chirikov, I. Guarneri, D.L. Shepelyansky, Physics Report 154 (1987) 77.

[9] G. Casati, I. Guarneri, D.L. Shepelyansky, Phys. Rev. A, 36 (1987) 3501.

[10] G.Casati, B.Chirikov,I.Guarneri, D.L.Shepelyansky, "Two-Dimensional Localization of Diffusive Excitation in the Hydrogen Atom in a Monochromatic Field". Phys. Rev. Lett. December 1987 (In press).

[11] G. Casati, B.V. Chirikov, I. Guarneri, D.L. Shepelyansky, " Hydrogen Atom in a Monochromatic Field: Chaos and Dynamical Photonic Localization", to appear on a special issue of the IEEE Journal of Quantum Electronics.

[12] D.L. Shepelyansky, Int. Conf. on Quantum Chaos, (Plenum 1985),187.

[13] J.E. Bayfield, L.A. Pinnaduwage, Phys. Rev. Lett. 54, (1985), 313; J. Phys. B, 18 (1985) L49: J.N. Bardsley, J.E. Bayfield, L.A. Pinnaduwage, B. Sundaram. Phys. Rev. Lett. 56, (1986), 1007.

[14] V.M. Alekseev and M.V. Yakobson, Phys. Reports 75 (1981) 287.

[15] B.V. Chirikov, Phys. Report 80 (1981), 157.

[16] G. Casati, B.V. Chirikov, J. Ford, F.M. Izrailev, Lectures Notes in Physics, Springer 93 (1979), 334.

[17] B.V. Chirikov, F.M. Izrailev, D.L. Shepelyansky, Soviet Scientific Reviews 2C (1981), 209.

[18] S. Fishman, D.R. Grempel R.E. Prange, Phys. Rev. A29 (1984), 1639.

[19] B.V. Chirikov. D.L. Shepelyansky Radiofizika 29 (1986), 1041.

[20] D.L. Shepelyansky, Phys. Rev. Lett. 56 (1986), 677.

[21] J.E. Bayfield, in "Fundamental Aspects of Quantum Theory" eds. V.

Gorini and A. Frigerio (Plenum Press, New York, 1986).

[22] J.E. Bayfield and P.M. Koch, private communication.

[23] Proceedings of the II International Conference on Quantum Chaos; Ed. T. Seligman and H.Nishioka; Lectures Notes in Physics, vol.263, Springer-Verlag 1986.

QUANTUM LIMITATION OF CHAOTIC DIFFUSION AND UNDERTHRESHOLD IONIZATION

Italo Guarneri
Dipartimento di Fisica Nucleare e Teorica dell'Universita'
and Istituto Nazionale di Fisica Nucleare, Sezione di Pavia
Via Bassi, 6
I-27100 Pavia, Italy

ABSTRACT. The Kicked Rotator is a prototype of the Quantum Localization phenomenon that strongly inhibits the possibility of chaotic excitation in systems subjected to time periodic perturbations. Its dynamics yields a key to the analysis of a simple model for microwave ionization of highly excited Hydrogen atoms.

1. This talk and Casati's are meant to report about extensive investigations by B.V. Chirikov, D.L. Shepelyansky and ourselves on the problem of microwave ionization of highly excited hydrogen atoms.
Leaving to Casati the presentation of specific results, I shall here illustrate those aspects of our work that have a direct bearing on the general problem of Quantum Chaos.
The ionization of highly excited H-atoms by microwave fields can be studied both in a quantum and in a classical framework; the latter approach has a definite interest because of the large quantum numbers involved. The subject of this talk will be the most important conclusion that can be drawn from the comparison of classical and quantum predictions, namely:
- Chaos plays an essential role in the classical ionization process;
- Quantization has an inhibitory effect on classical chaotic motion, that must be overcome in order to get strong ionization.

2. The simplest mathematical model for a H-atom in a microwave field is that of a particle moving along a line, subject to a Coulomb potential and to a perturbing monochromatic electric field /1/.
Since we are here interested more in methods than in specific results there will be no need to describe this model in full detail; a discussion of its general qualitative features should be sufficient to illustrate the basic ideas that, in our opinion, have a potentially wider applicability.
The model is described by an Hamiltonian of the type:

189

A. Amann et al. (eds.),
Fractals, Quasicrystals, Chaos, Knots and Algebraic Quantum Mechanics, 189–194.
© 1988 by Kluwer Academic Publishers.

$$H(x,p,t) = H_0(x,p) + \varepsilon V(x)\cos \omega t \qquad (1)$$

where H_0 is the Hamiltonian of the unperturbed system. The phase plane (x,p) can be divided into two regions according to the nature of the unperturbed motion defined by H_0. Whereas in the region $H_0>0$ the motion is unbounded, the region $H_0<0$ corresponds to bound, strictly periodic orbits. Here the motion is integrable, and action-angle variables (I,θ) can be introduced, so that $H_0(x(I,\theta),p(I,\theta))$ is a function of I alone. Then the angular frequency of bound motion is $\Omega(I)=dH_0/dI$. The two regions are divided by a curve ("separatrix"), defined by $H_0=0$, which is an orbit itself. As the separatrix is approached from within the bound state region, $\Omega(I)\to 0$.

When the monochromatic perturbation $\varepsilon V(x)\cos \omega t$ is added, chaos is generated in the proximity of the separatrix. Specifically, a "stochastic layer" appears near the separatrix; orbits which leave from states within the layer wander erratically and eventually diffuse into the region $H_0>0$, i.e., ionize. The width of the layer depends on the perturbation parameters ε, ω. For a given ω, the value of ε which is required in order to bring the border of the layer down to include a given initial state defines the threshold field ε_c for stochastic ionization in that state; ε_c can be quantitatively estimated /1/.

3. The complexity of the motion in the chaotic regime demands a statistical description, that is usually obtained via some Markovian approximation. Since the onset of chaos is determined by resonances between the external frequency and high harmonics of the unperturbed motion, the 'diffusive time scale' - i.e., the scale on which the loss of memory due to chaos justifies a Markovian approximation - is of the order of the unperturbed orbital period. Because the latter increases as the orbits move out towards continuum, the diffusive scale is not uniform. This difficulty can be circumvented, and a simple statistical description of the chaotic motion near the separatrix can be obtained, by constructing an appropriate Poincare's section, as follows.

First, we cast the time-dependent problem into Hamiltonian conservative form, by introducing a new pair of conjugate variables (N,ϕ) and then considering the Floquet Hamiltonian /4/:

$$H_F(I,\theta,N,\phi) = H_0(I) + \varepsilon V(I,\theta)\cos\phi + \omega N \qquad (2)$$

The motion described by (2) develops over hypersurfaces $H_F=const.$. On such a surface $H_F=\lambda$, points with $\theta=0$ ('perihelion states') define a 2-dim. variety π, that can be parametrized by just N and ϕ. The orbit leaving from any such point (N,ϕ) at a time t_0 will return to π, i.e., to perihelion (unless ionization occurs meanwhile) at some other point $(\bar N,\bar\phi)$. The map $(N,\phi)\to(\bar N,\bar\phi)$ is canonical, and in the proximity of the separatrix it can be easily given an approximate explicit form. Indeed, from (2) we see that $\delta N=\bar N-N$ is the change in energy divided by ω, and $\delta\phi=\bar\phi-\phi$ is ω times the time between two subsequent passages at perihelion. Then, at 1st order in ε

$$\delta N \cong -\omega^{-1} \; \varepsilon \int\limits_{t_0}^{t_0+2\pi\Omega^{-1}} V'_x(x(t))\dot{x}(t)\cos\omega t \; dt \qquad (3)$$

which is just the work done by the perturbation along the unperturbed orbit x(t) (of period $2\pi\Omega^{-1}$) divided by ω. Notice that, since Ω is a function of the unperturbed energy H_0, at the lowest perturbative order it is a function of $\lambda-\omega N$, that will be denoted by $\Omega(N)$.

In the proximity of the separatrix, for x(t) we can take the unperturbed separatrix motion itself, for which $\Omega=0$. Then, after some rearrangement of variables we get

$$\bar{N} \cong N - \varepsilon\omega^{-1} \int\limits_{-\infty}^{+\infty} V'_x(x(t))\cos(\omega t+\phi)\dot{x}(t)dt$$

and, taking into account a simmetry of the separatrix,

$$\bar{N} = N + k \sin\phi \qquad k = \varepsilon\omega^{-1} \int\limits_{-\infty}^{+\infty} V'_x(x(t))\dot{x}(t)\sin \omega t \; dt \qquad (4)$$

where the "Arnol'd-Melnikov integral" k can be explicitly evaluated as a function of ε, ω /2/. It follows that the sought for canonical map is, at 1st order in ε,

$$\bar{N} = N + k \sin\phi$$
$$\bar{\phi} = \phi + 2\pi\omega\Omega^{-1}(\bar{N}) \qquad (5)$$

Notice that this map is defined for all bound states ($\lambda-\omega N<0$) but carries some of them to $\lambda-\omega N>0$, and thus it explicitly accounts for ionization.

4. Upon introducing a fictitious Hamiltonian K(N) in such a way that $K'(N)=2\pi\omega\Omega^{-1}(N)$, (5) is recognized to describe the evolution over one period of a system with the time-periodic Hamiltonian:

$$K(N) + k \cos\phi \sum_{n=-\infty}^{+\infty} \delta(t-n) \qquad (6)$$

which can be depicted as an integrable system (K(N)) periodically subjected to a kick of strength k cosϕ. Here, K(N) is uniquely defined by the form of $H_0(I)$ and by the value λ. The class of periodically kicked systems that are obtained from (6) under different choices of K(N) has a distinguished member: the kicked rotator, defined by $K(N)=N^2/2$. The map (5) associated with the kicked rotator is the Standard (or Chirikov's) Map, and provides a useful local approximation also for generic maps of the type (6) /2/.

The essentials of the dynamics (6) are that, if $K''(N)>0$, then for k above some (computable) chaotic threshold ϕ changes at random and N grows diffusively in time:

$$\overline{\delta N^2} \cong D\tau \qquad\qquad D \cong \frac{k^2}{2} \qquad\qquad (7)$$

(τ = number of iterations). The classical picture emerging from the above analysis is therefore the following: close to the separatrix, the electron moves along an almost unperturbed orbit; the effect of the perturbation is concentrated near the perihelion, and can be approximately depicted as a 'kick' which throws the electron on a new orbit, or directly into continuum. In the chaotic regime, these kicks come at random, in the sense that the sequence of jumps from one orbit to another is similar to a random walk, and this leads to ionization.

5. Now we come to the quantum part, and specifically ask what the quantum version of the map (5) may be. In that the discrete-time dynamical system (5) is a subsystem of te Hamiltonian system (2), its quantum counterpart must in principle be deduced from the quantization of (2). Nevertheless, a valuable insight on the effect of quantization can be obtained from a direct quantization of (6). According to a general prescription of the quantum Floquet theory, we should represent N by $\hat{N} = -i\, \partial/\partial\phi$, $0 \leq \phi \leq 2\pi$ ($\hbar$=1) with periodic boundary conditions /4/. Then we easily get the following "quantum map" that defines the evolution in one period of a quantum system with the Hamiltonian (6):

$$\overline{\psi} = S\psi = e^{-i\hat{K}}\, \hat{P}\, e^{-ik\,\cos\phi}\, \psi \qquad\qquad (8)$$

where $\hat{K} = K(\hat{N})$ is defined only in the 'bound space' $\lambda - \omega\hat{N} < 0$, and $\hat{P}$ is the projector onto this bound space.

Our present understanding of the nature of the quantum dynamics (7) is essentially based on our knowledge of the properties of the quantum kicked rotator that would be obtained from (8) by taking $K = 1/2\, \hat{N}^2$ and dropping the projector $\hat{P}$.

The kicked rotator has been attracting a great interest since the very beginning of investigations on Quantum Chaos. Even though its mathematical theory is still wanting, its essential properties are by now fairly well understood, and will be summarized below.

6. Let us consider the quantum and the classical dynamics of the kicked rotator. On the classical side, we choose an ensemble of initial data with a given $N = N_0$ and randomly distributed phases ϕ. We shall compare the evolution of this ensemble with the quantum evolution of a wave packet, initially concentrated on the N_0-th imperturbed eigenstate (of $\hat{K}$). The evolution of the packet in N-space can be studied by expanding the wave function in eigenvectors of $\hat{N}$, i.e., by a Fourier expansion of $\psi(\phi)$, in $(0,2\pi)$. The integer eigenvalues of $\hat{N}$ then label sites of a 1-d lattice and we shall study how the wave packet spreads on this lattice.

Suppose that the perturbation parameter k is so chosen, that the classical motion is fully chaotic. Then, following the evolution of the classical ensemble, the spread $\delta N^2 = (N - N_0)^2$ grows linearly in time, according to (7).

Instead, the quantum packet spreads on the N-lattice only up to a "break-time" τ^*, after which a steady-state oscillatory regime is entered. The different dynamical possibilities represented in the classical microcanonical ensemble, that in the classical case develop independently along wildly different chaotic orbits, in the quantum case strongly interfere and cause the diffusion to stop. Exceptions to this typical behaviour occur, when the frequency of the kicks is resonant with the internal frequencies of the rotator, or when it is very close to resonant values; there are, however, strong indications that the set of all resonant and quasi-resonant frequencies has zero measure /3/.

This phenomenon, that the quantum wave packet remains localized in N-space in spite of the classical unbounded diffusion is known as "Quantum Localization of Classical Chaos". This denomination is deliberately reminiscent of the Anderson Localization in solid state physics; indeed, a formal connection between the rotator and a "disordered-lattice" problem can be established /5/. In both cases the localization appears to be due to a complicated interference effect, and the connection can be carried so far as to predict that the average steady state distribution is exponential (though, of course, in the Anderson case it would be exponentially localized in space and not in momentum, as in the rotator case).

In the semiclassical regime, the width l of this distribution, i.e., the maximum spread attainable by the wave packet, can be estimated by a deep argument devised by the Novosibirsk group. According to that argument, $l \simeq D$, i.e., the localization length is approximately given by the classical diffusion coefficient (7) /6/.

7. Both the classical map (5) and its quantum version (8) have a definite similarity with the kicked rotator problem. Again, the quantum motion is conveniently studied on the $\tilde{N}$ lattice; moreover, since changes in N are just changes in energy, divided by ω, a jump from the N-th to the N+k-th site will correspond to the emission of k photons. However, (8) differs from the kicked rotator, due to the form of $\tilde{K}(\tilde{N})$ and to the projector $\tilde{P}$. The first difference should not modify the above described localization picture, even though there is no room here for an explicit justification. On the other hand, $\tilde{P}$ should not affect the evolution of wave packets as long as these do not significantly invade the 'continuum'.

Suppose that we start with a packet concentrated on some site N_0 in the bound state region ($\lambda - \omega N_0 < 0$), and with a value of the perturbation parameter k well above the classical chaotic threshold. In the classical case, as we know, all orbits would eventually ionize: but in the quantum case a rotator-like picture would apply, at least in the early stages of the evolution, when the packet does not yet appreciably leak into continuum-i.e., when its spread on the N-lattice is still small compared to the number of photons needed to reach the continuum from N_0. But then, if the localization length is small enough, the packet will never have a chance to reach the continuum, and the only route to ionization will be the small tail in the final (quasi) stationary

distribution, that will be exponential in the number of absorbed photons.
This picture was fully confirmed by numerical simulation of the quantum model (1) /1/2/.

In this situation, the strong ionization produced in the classical case by the onset of chaos would be suppressed by quantum localization.

In other words, the classical condition for chaotic ionization is not sufficient in order to get strong ionization in the quantum system. Some additional condition should be met, in order that the quantum localization becomes ineffective. This condition can be obtained, by estimating the localization length via the classical diffusion coefficient and by requiring that it becomes comparable with the number of photons for ionization. The mechanism of quantum 'diffusive' ionization that will be activated under this condition, and its physical implications, will be illustrated in Casati's talk.

REFERENCES

1. G. Casati, B.V. Chirikov, I. Guarneri, D.L. Shepelyansky 'Relevance of Classical Chaos in Quantum Mechanics: the Hydrogen Atom in a monochromatic Field', Phys. Reports 154, 2, 1987 and references therein.
2. G. Casati, I. Guarneri, D.L. Shepelyansky, Phys. Rev. A
3. G. Casati, J. Ford, I. Guarneri, F. Vivaldi, Phys. Rev. A 34, 1413 (1986) and references therein
4. J. Bellissard, in Trends and Developments in the Eighties, edited by S. Albeverio and Ph. Blanchard (World Scientific, Singapore 1985)
5. S. Fishman, D.R. Grempel and R.E. Prange, Phys. Rev. A 29, 1639 (1984); D.L. Shepelyansky, Phys. Rev. Lett. 56, 677 (1986).
6. B.V. Chirikov, F.M. Izrailev and D.L. Shepelyansky, Sov. Sci. Rev. Sec. C2, 209 (1981).

A RO-VIBRATIONAL STUDY OF REGULAR/IRREGULAR BEHAVIOUR OF THE CO-Ar SYSTEM

Stavros C. Farantos
Department of Chemistry, University of Crete,
and Institute of Electronic Structure and Laser,
Research Center of Crete, P.O. Box 1527
711 10 Iraklion, Crete, Greece.

and

Jonathan Tennyson
Department of Physics and Astronomy,
University College London,
Gower street, London WC1E 6BT, England.

ABSTRACT. Classical and quantum mechanical rotational – vibrational calculations have been performed for the bound states of the van der Waals system CO-Ar. From the nodal patterns of the eigenfunctions and the predominant coefficients in the basis set expansion, it is shown that the regular/irregular classical and quantum behaviour are in qualitative agreement. Regular/localized states are associated with those states which have one, two and three quanta in the bending mode but no excitation in the stretching mode. For the present system it is found that the projection of the total angular momentum on the body fixed z-axis is an almost good quantum number. Thus mixing among the zero order basis functions occur through the potential part of the hamiltonian.

1. Introduction

Non-linear mechanics has made considerable advances, during the last twenty years in understanding the regular/chaotic motion of dynamical systems [1]. This has influenced chemical physics to a great extent [2]. Although most of the work in this area is theoretical a few intriguing experiments have also been performed [3,4]. The basic problem is one of understanding the quantum behaviour of a molecule when its classical dynamics show a transition from regular to chaotic motion. Regular excited states are important in developing a state specific chemistry [5].

A. Amann et al. (eds.),
Fractals, Quasicrystals, Chaos, Knots and Algebraic Quantum Mechanics, 195–206.
© 1988 by Kluwer Academic Publishers.

An explicit comparison of classical to quantum dynamics would require the solution of the time dependent Schrodinger equation. However, for the conservative hamiltonians the wavefunctions, at any instant of time, can be written as a linear combination of the eigenfunctions. Therefore conclusions can be drawn by studying the behaviour of the time-independent eigenvalues and eigenfunctions of the molecule. Thus it has been found that the nodal structure of the eigenfunctions has a regular pattern at energies where the phase space is occupied by quasiperiodic trajectories and irregular patterns for the chaotic regions of phase space [2]. Hose and Taylor [6] demonstrated that if an eigenstate $|\psi_m>$ has a projection on a basis function $|\varphi_i>$ greater than 0.707 i.e.

$$| < \varphi_i |\psi_m > |^2 > 0.5 \qquad (1)$$

then the quantum numbers which are used to assign φ_i are almost good quantum numbers for ψ_m. Such a state can be characterized as regular. As far as the eigenvalues are concerned it has been shown that the distribution of the spacing of neighboring levels is Poisson-like for regular states [7] and Wigner-like [8] for chaotic states.

The above criteria, and a few others proposed in the literature [2] are not exact definitions of quantum chaos since there are limitations in their application. However, after the plethora of studies on model [9] and realistic potentials [10], a general conclusion can be drawn: when the regular/irregular regions of phase space are large enough compared to $\hbar$, then the corresponding eigenstates also show regular/irregular behaviour. It turns out that generally there is a "slaggishness" in the transition to molecular quantum chaos. Nevertheless it is not an overstatement to say that the phase space structure, particularly at high excitation energies where quantum calculations are difficult, is a good diagnostic for tracing regular localized or resonant states. These regular regions, which are usually embedded in the chaotic sea, may have a significant effect in dynamical processes. An example is a recent study on the isomerization process of HCN → HNC [11]. It has been shown that a multiple resonance among the vibrational modes, located at the top of the barrier to isomerization, inhibits the reaction.

Van der Waals (vdW) complexes seem most appropriate to further investigate the regular/chaotic correspondence between classical and quantum molecular mechanics. This is because of the relatively small number of bound states, which vdW species can support, and therefore small density of states which implies that the differences between classical and quantum mechanics will be more pronounced. A recent study on the rotionless HCl-Ar [12] has shown that classical chaos appears at very low energies, lower than the zero point energy. In quantum mechanics chaos is reflected in the complexity of the nodal patterns found in the eigenfunctions of the complex.

In this article we report results on the CO–Ar complex. We study the role of rotational excitation in the quantum chaotic behaviour of the system. There are, compared to vibrational studies, only a few investigations of irregular behaviour induced by rotational excitation [13].

Experimentally the breakdown of regular rotational structure through Coriolis interactions is a promising area of study. For example stimulation emission pumping experiments on formaldehyde have demonstrated the importance of Coriolis coupling in mixing vibrational states [4]. CO–Ar is also amenable to experimental observations [14].

2. Computational methods

A pairwise additive interaction potential was used [15]. This consists of two 6-exp Buckingham functions which describe the C–Ar and O–Ar interactions and an extended Rydberg type function for the diatomic CO [16]. The potential has a triangular minimum below the Ar–CO dissociation limit of 105 cm^{-1}. The potential barriers to the linear configurations Ar–O–C and Ar–C–O are 30 cm^{-1} and 50 cm^{-1} respectively. The potential is steeper for Ar approaching C than O which, as we shall see, has an influence on the first excited bending states.

The classical trajectories were integrated in a six dimensional configuration space described by $\underline{r}$, the distance of O from C, and $\underline{R}$, the distance of Ar from the center of mass of CO. The details are given in ref. [17]. The aim of this study was to investigate the effect of the chaotic motion of Ar on the CO oscillator and the flow of rotational energy initially put on CO to the other degrees of freedom. The characterization of the trajectories as regular/chaotic was made by studying the autocorrelation function of the diatomic frequency fluctuations;

$$C_\omega (\tau) = \lim_{T \to \infty} \frac{1}{T} \int_0^T \Delta\omega \, (t+\tau) \, \Delta\omega \, (t) \, dt \Big/ \int_0^T \Delta\omega^2(t)dt \qquad (2)$$

Regular motion of Ar relative to CO results in an oscillatory autocorrelation function whereas chaotic motion gives a decaying, usually multiexponential, curve.

The quantum calculations involve the solution of the time independent Schrodinger equation in a body fixed coordinate system (R,r,θ). Where cosθ = $\underline{r}.\underline{R}/|\underline{r}.\underline{R}|$. The z-axis was embedded along the R coordinate [19]. In a given body fixed system the vibrational and rotational coordinates can be identified along with the Coriolis coupling terms. The ro-vibrational hamiltonian is generally written [20];

$$H = K_V + K_{VR} + V(r,R,\theta) \tag{3}$$

K_V is the vibrational hamiltonian

$$K_V = - \frac{\hbar^2}{2\mu R} \frac{\partial^2}{\partial R^2} R - \frac{\hbar^2}{2\mu_d r} \frac{\partial^2}{\partial r^2} r - \frac{\hbar^2}{2} \left(\frac{1}{\mu R^2} + \frac{1}{\mu_d r^2} \right) \frac{1}{\sin\theta} \frac{\partial}{\partial\theta} \left(\sin\theta \frac{\partial}{\partial\theta}\right) \tag{4}$$

where μ and μ_d are the CO-Ar and CO reduced masses respectively. K_{VR} describes the rotation of the system together with the coupling of rotation to vibration. It is a null operator for $J=0$ states. Its explicit form is given in ref. [20].

The calculations, which are variational in nature, used the angular basis functions which are products of spherical harmonics and the rotation matrix;

$$\Phi_{j,k}^{J,M} = Y_{j,k} (\theta,\gamma) \, D_{M,k}^{J*} (\alpha,\beta,0) \tag{5}$$

where α, β and γ are the three Euler angles for the rotation of the body fixed to space fixed system. These basis functions ensure the correct quantum number of the total angular momentem J, and its projection M, on the z-spaced fixed axis. k is the projection of the total angular momentum on the body fixed z-axis and j the CO angular momentum. These functions are symmetrised to the total parity of the system [20].

Assuming a polynomial Legendre expansion for the potential

$$V(r,R,\theta) = \sum_\lambda V_\lambda(r,R) \, P_\lambda (\cos\theta) \tag{6}$$

it can be shown that coupling of different j states occurs only through the potential function;

$$< j,k \mid V \mid j',k' > = \delta_{kk'} \sum_{\lambda} g_\lambda(j,j',\lambda) \, V_\lambda \, (r,R) \tag{7}$$

where g_λ are the Gaunt coefficients [19]. On the other hand coupling between different k states is obtained only through the kinetic operator K_{VR}.

$$<\Phi_{j,k}^{J,M} \mid K_{VR} \mid \Phi_{j',k'}^{J,M} > = [\frac{J(J+1) - 2k^2}{2\mu R^2}] \, \delta_{kk'} \delta_{jj'} -$$

$$\delta_{jj'} \frac{\hbar^2}{2\mu R^2} [C_{j,k}^{+} \, C_{j,k}^{+} \, \delta_{k'(k+1)} + C_{j,k}^{-} \, C_{j,k}^{-} \, \delta_{k'(k-1)}] \tag{8}$$

where

$$C_{jk}^{\pm} = [\, j(j+1) - k \, (k \pm 1)]^{1/2} \tag{9}$$

The maximum coupling is for $k=0$. For van der Waals systems this coupling is often neglected meaning that k is a good quantum number.

To identify the regular character of the eigenstates we examine the magnitude of the expansion coefficients according to Hose-Taylor criterion and we inspect the nodal structure of the projections of wavefunctions on (R,θ) plane. In these calculations r is frozen to its equilibrium value. Nevertheless we have another three variables of the wavefunction to define. These are the three Euler angles which are taken equal to zero. In this case the body fixed and space fixed axes coincide. Plots of the (R,θ) coordinate with $\alpha=\beta=\gamma=0$ have the effect of projecting out the $M=k=0$ components of the wavefunctions.

3. Results and Discussion

The classical dynamics of the rotationless CO-Ar shows an early transition to chaotic behaviour as was found for HCl-Ar [12]. Chaos is associated with the R-stretching and θ-bending modes whereas the diatomic vibrational action variable is adiabatically conserved. The chaotic motion of Ar results in a randomization of the phase of CO oscillator as can be deduced from the decaying autocorrelation function of frequency fluctuations in fig.1. Only for energies as low as 0.001 eV above the minimum of the CO-Ar interaction potential, does C_ω shows an oscillatory behaviour. Further excitation of the R-stretching mode results in strong mixing with the bending mode [17].

On rotationally exciting the CO the opposite trends are observed. For low j excitation there is strong mixing of j with the bending mode and finally with the R-stretching mode as well. At appropriate energies rotational predissociation can be seen. For higher j values $(j>10)$ large variations of j occur only during a hard collision of Ar with CO. As j increases further it gradually becomes an adiabatic invariant. At this situation we have decoupling of the R-coordinate from the bending mode and the motion of the system becomes regular. The dynamics is mainly governed by the repulsive part of the potential and the changes of j can be described by a kicked rotor model [17].

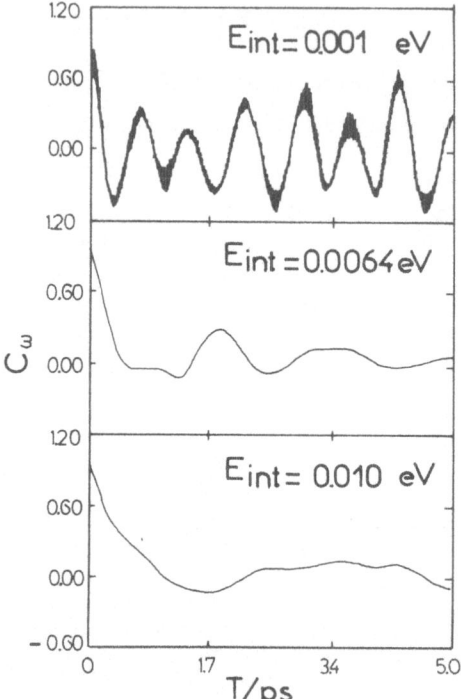

Figure 1

The vibrational CO frequency auto-correlation function at different interaction energies and for initial j=v=0.

Fig.2 shows the time variation of j along with other dynamical variables for a trajectory at 0.001 eV and initially j=10. It can be seen that significant variations of j happen each time R reaches its smallest turning point. That the trajectory behaves regularly can be deduced from the regular oscillations of C_ω.

How does the system behaves quantum mechanically? For J=0 we obtained 21 bound states. These states were computed by performing variational calculations using a basis set comprising all Legendre polynomials with j<10 for the θ coordinate and all Morse oscillator-like functions ($R_e=8a_o$, $D_e=0.005$ E_h, $\omega_e=6x10^{-5}$ E_h [19]) with n<16 [18]. This basis is sufficient to converge the lower states to very high accuracy. Although with a variational calculation we cannot dismiss the possibility of there being further bound states, these are likely to be very diffuse and weakly bound if they exist at all.

Apart from the ground state the only states which show regularity in their nodal patterns and localization in the configuration space are those shown in figures 3a, 3b and 3c. They have no R-stretching excitation and can be assigned 1,2 and 3 quanta in the bending mode respectively. States with stretching excitation, apart from the first one, show relatively complex nodal patterns and they tend to occupy the whole available configuration space (fig.4).

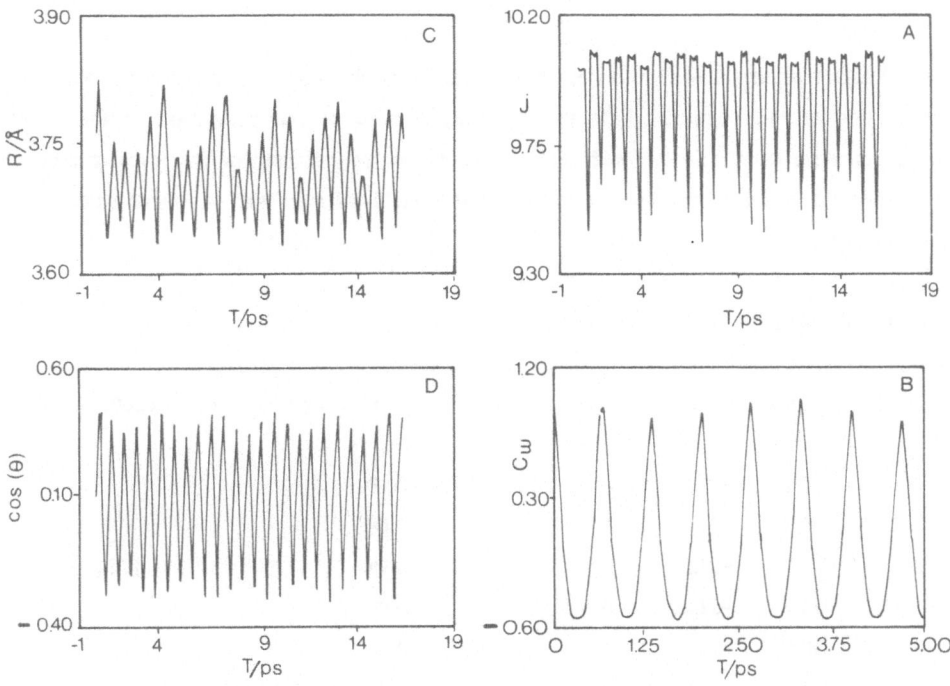

Figure 2

The time variation of some dynamical variables for a trajectory at the interaction energy of 0.001 eV and initially v=0 and j=10.

J>0 calculations were performed using a two-step variational procedure [20]. The 80 lowest solutions of each problem for which k was assumed to be a good quantum number, were used to solve the full problem. This number of functions is much larger than is required for convergence, but was chosen to insure that we obtained most of the bound states of the system.

Results were obtained with k treated as a good quantum number (no Coriolis coupling) together with the fully coupled calculations. Results for the first twenty even parity states for J=10 are given in table 1. Figures 3d, 3e and 3f are the results of calculations with no Coriolis coupling and 3g, 3h and 3i results from fully coupled calculations projected along $\alpha=\beta=\gamma=0$. As can be seen these two sets of figures show only minor changes compared to those of J=0.

Table   1

Lowest 20 levels of the Ar–CO van der Waals complex with J=10 . Frequencies are relative to dissociation of the complex. k is the quantum number of the projection of the total angular momentum on the body fixed z-axis. i describes the ordering of the levels within a given k manifold, which is obtained during the first variational step with k treated as a good quantum number.

Level	Frequencies /cm^{-1}		Assignment					
	No Coriolis	Full	k	i	coeff.	k	i	coeff.
1	−77.0	−77.1	0	1	0.99			
2	−74.3	−74.5	1	1	0.98			
3	−67.3	−67.5	2	1	0.98			
4	−63.4	−63.4	0	2	0.98			
5	−56.6	−57.0	1	2	0.90			
6	−56.3	−56.6	3	1	0.97			
7	−54.0	−53.8	0	3	0.95			
8	−50.5	−51.0	1	3	0.96			
9	−46.8	−46.8	0	4	0.95			
10	−46.2	−46.3	2	2	0.94			
11	−44.1	−43.9	0	5	0.96			
12	−42.9	−43.3	2	3	0.96			
13	−41.5	−41.8	4	0	0.98			
14	−38.1	−39.3	0	6	0.77	1	4	0.54
15	−37.8	−37.0	1	4	0.65	1	5	0.56
16	−36.2	−35.7	0	6	−0.51	1	5	0.73
17	−32.1	−32.7	1	6	−0.59	3	2	0.67
18	−31.6	−32.3	1	6	0.67	3	2	0.64
19	−31.5	−31.8	3	3	0.96			
20	−30.3	−29.8	0	7	0.95			

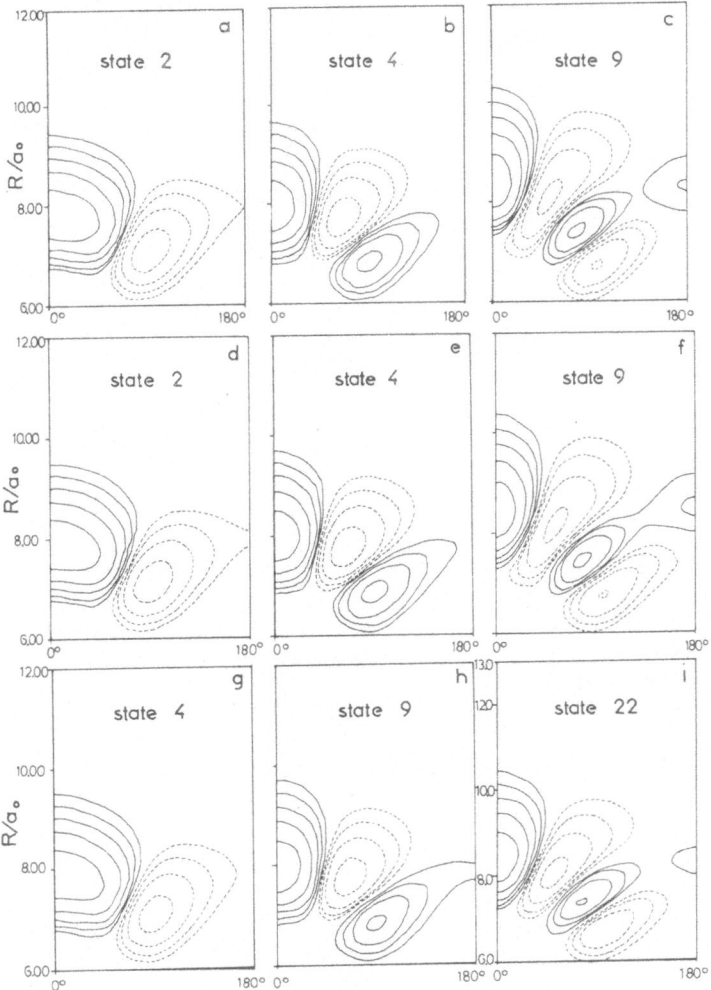

Figure 3

Nodal structure for bending excited states. Contours link points where the wavefunction has 4%, 8%, 16%, 32% and 64% of the maximum amplitude. Solid (Dashed) curves are for positive (negative) amplitude. 3a, 3b and 3c are for J=0 states. 3d, 3e and 3f are for J=10 but k=0 treated as a good quantum number. 3g, 3h and 3i are for fully coupled J=10 states (see text).

The fact that the projections of the eigenfunctions on the (R,θ) plane are independent of the rotational motion, is considered an indication of the conservation of k. Similar results were obtained in the 3-d vibrational calculation of HCl-Ar [12]. In that case, the projections of the wavefunctions on the (R,θ) plane did not change when the vibration of HCl was included in the calculations.

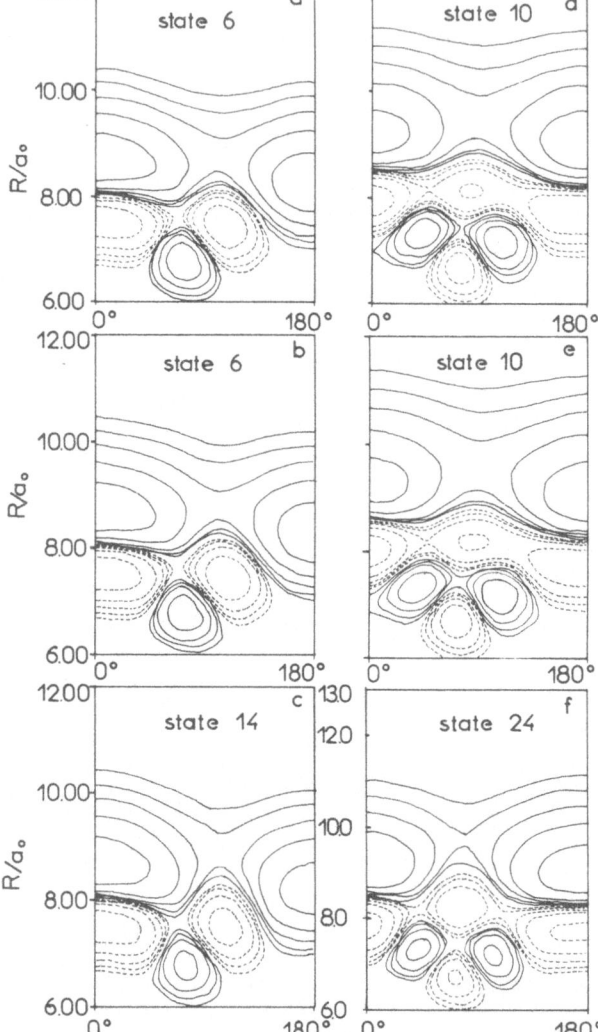

Figure 4

As figure 3 but for states with stretching excitation a) J=0, b) J=10, k=0, and c) J=10, fully coupled calculations.

Inspection of the coefficients in the basis set expansion of the second variational step indeed show that most of the eigenfunctions are characterized by a predominant coefficient (with a particular value of k) which satisfies Hose–Taylor criterion (see table 1). However the highest levels in the fully coupled calculations show differences compared to the highest states of J=0 and J=10 with the Coriolis coupling turned off. Figure 5 shows states close to the dissociation limit. The state from the fully coupled calculations (fig.5c) show smaller complexity in its nodal structure.

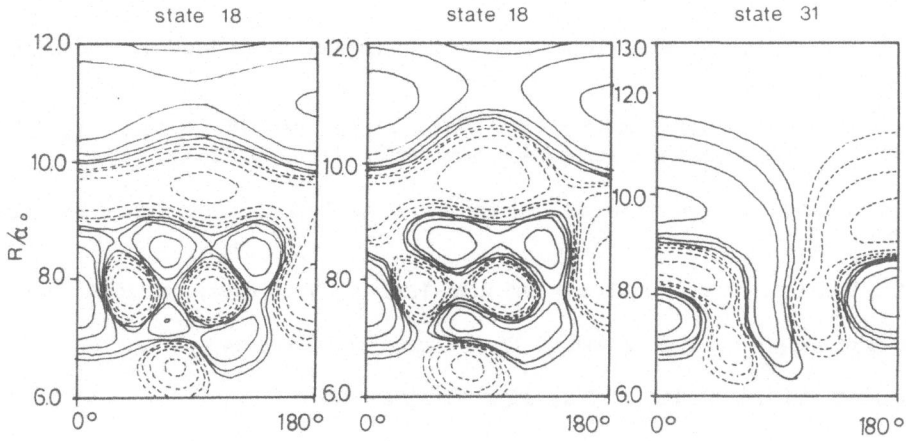

Figure 5

As figure 3 but for states close to the dissociation limit.
a) J=0, b) J=10, k=0 and c) J=10 fully coupled states.

From the above results the following conclusions can be drawn:

1.   The vibrational states with one, two, and three quanta in the bending mode but none in the stretching are regular and localized. R-stretching excitations produce irregular eigenstates. These trends are in qualitative agreement with the classical dynamics. Regularity in the trajectories is observed only at low interaction energies and by increasing the rotation of CO (fig.2).

2. k is an almost good quantum number for CO-Ar. Since coupling among zero order states of different k occurs through the Coriolis term (Eq.(8)), it is concluded that this part of the hamiltonian does not contribute to the mixing of the basis functions.

3. The irregular character of wavefunctions should be attributed to the potential coupling of states with different j (Eq.(7)). The classical dynamics has shown that by increasing j the system becomes more regular. This is reflected in the highest quantum states of CO-Ar for which terms with large j values in equation (7) make a significant contribution.

4. In the classical calculations it was shown that the fluctuation in j are mainly due to the hard collisions of Ar to CO. Therefore, we consider the repulsive part of the potential as the case for erratic behaviour of the wavefunctions. It is the portion of the potential which has the greatest anisotropy in a Legendre expansion, Eq.(7), and hence causes the greatest mixing between j's.

References

1. A.J. Lichtenberg and M.A. Lieberman, "Regular and Stochastic Motion", (*Springer-Verlag*, N.Y. 1983)

2. D.W. Noid, M.L. Koszykowski and R.A. Marcus, *Ann. Rev. Phys. Chem.*, 32, 267, 1981; E.B. Stechel and E.J. Heller, *Ann. Rev. Phys. Chem.*, 35, 563, 1984

3. R.L. Sundberg, E. Abramson, J.L. Kinsey and R.W. Field, *J. Chem. Phys.*, 83, 466, 1985

4. H.L. Dai, C.L. Korpa, J.L. Kinsey and R.W. Field, *J. Chem. Phys.*, 82, 1608, 1985

5. N. Bloemberger and A.H. Zewail, *J. Phys. Chem.*, 88, 5459, 1984

6. G. Hose and H.S. Taylor, *J. Chem. Phys.* 76, 5356, 1982

7. M.V. Berry and M. Tabor, *Proc. Royl. Soc.* London, A356, 375, 1977

8. O. Bohigas, M.J. Giannoni and C. Schmidt, *Phys. Rev. Lett.* 52, 1, 1984

9. E.J. Heller and R.L. Sundberg in "Chaotic Behaviour in Quantum Systems", edited by G. Casati (*Plenum Publishing Co*, 1985)

10. S.C. Farantos and J. Tennyson in "Stochasticity and Intramolecular Redistribution of Energy", edited by R. Lefevbre and S. Muckamel (*Reidel Pub. Co.*, Orsay, 1987)

11. M. Founargiotakis, S.C. Farantos and J. Tennyson, *J. Chem. Phys.*, submitted

12. J. Tennyson, Mol. Phys., 55, 463 (1985); S.C. Farantos and J. Tennyson, *J. Chem. Phys.*, 85, 641, 1986

13. T. Uzer, C.A. Natanson and J.T. Hynes, *Chem. Phys. Lett.*, 122, 12, 1985

14. D.J. Diestler, *J. Chem. Phys.*, 60, 2692, 1974

15. K. Mirsky, *Chem. Phys.*, 40, 445, 1980; J. Manz and K. Mirsky, *Chem. Phys.* 40, 457, 1980

16. J.N. Murrell, S. Carter, S.C. Farantos, A.J. Varandas and P. Huxley, "The Molecular Potential Energy Function", *J. Wiley*, 1984)

17. S.C. Farantos and N. Flytzanis, *J. Chem. Phys.*, in press

18. J. Tennyson, J. Miller and B.T. Sutcliff, *Faraday Discussions*, in press

19. J. Tennyson, *Computer Phys. Comms.*, 42, 157, 1986; *Computer Phys. Report*, 4, 1, 1986

20. J. Tennyson and B.T. Sutcliffe, *Mol. Phys.*, 58, 1067, 1986

Analysis of DNA knots and catenanes allows to deduce the mechanism of
action of enzymes which cut and join DNA strands

Andrzej Stasiak and Theodor Koller
Swiss Federal Institute of Technology
Institut for Cell Biology, ETH-Hoenggerberg
CH-8093 Zurich, Switzerland

ABSTRACT. The topological analysis of DNA knots and catenanes produced
by different enzymes acting on DNA molecules becomes possible by an
electron microscopical technique in which the DNA molecules are covered
with recA protein (1,2). The recA covering of the DNA increases the
thickness of the filaments from 2 nm to 10 nm and by this allows to
distinguish between underlying and overlying segments in knoted or
catenated DNA molecules adsorbed to support films used for electron
microscopy.
     Different enzymes acting on DNA produce specific sets of knots
and catenanes. Analysis of such molecules with the help of recA coating
allowed to conclude about the mechanism of action of two enzymes
studied and about the possible three dimensional shape of DNA molecules
in solution.

INTRODUCTION

DNA molecules are thin and flexible linear heteropolymers composed of
two intertwined strands forming a right handed helix. DNA molecules
have 2 nm in diameter and a length which is proportional to the amount
of the stored genetic information ranging from 1.5 um in simple viruses
up to about 2 meters in each human cell. In a number of processes in
living cells segments from the same or different DNA molecules pass
through each other, get exchanged, inverted, excised or joined. When
these processes mediated by specialized enzymes occur in vitro on
circular DNA molecules like bacterial plasmids, knots and catenanes can
be formed. In living cells, however, DNA knots and catenanes are rather
rare and short lived. DNA knoting and multipe catenation would
interfere with such vital processes like DNA replication or
transcription. Therefore, there is a general tendency of enzymes in the
cell to avoid DNA knoting and catenation or to unknot and decatenate
entangled DNA molecules. However, under in vitro conditions working
with just one kind of purified enzyme and starting with unknoted
circular DNA molecules, it is possible to accumulate knoted or
catenated DNA molecules (2-8).

A. Amann et al. (eds.),
Fractals, Quasicrystals, Chaos, Knots and Algebraic Quantum Mechanics, 207–219.
© 1988 by Kluwer Academic Publishers.

By analysing DNA knots and catenanes one can deduce the way they were formed. This can provide important information about the mechanism of enzyme action. Analysis of formed DNA knots and catenanes was hampered in the past by the methodological inability to determine the type of knots produced. Although naked DNA molecules are well visible in the electron microscope after standard low angle heavy metal shadowing, their small thickness does not allow trustworthy assignment of underlying and overlying DNA segments on crossover points of knoted DNA molecules brought to two dimensions by adsorption to the supporting film. The recA coating technique developed by us is based on the property of recA protein monomers to bind to DNA in a cooperative way forming a 10 nm thick complex with a very regular right-handed helical structure (1). DNA molecules, knoted or catenated, are completely covered with recA and subsequently are adsorbed to supporting films used for electron microscopy (2). The increased thickness of the complexes compared to naked DNA allows on clean metal shadowed preparations to indicate underlying and overlying DNA segments on crossover points and thus to assign a certain type of knot or catenane. In addition the right-handed helical structure of recA-DNA complexes serves as an internal standard of the handedness and allows to avoid the wrong assignement of the handedness of knots and catenanes which can arise by inversion of negatives for photographic enlargements. Also other proteins than recA with similar properties were used successfully for vizualization of knoted DNA molecules (3).

In this paper we review some of the results obtained with DNA knots and catenanes produced by bacteriophage λ integrase (Int) and transposon 3 resolvase (Tn3 res).

MATRIALS AND METHODS

Purified recA protein was kindly provided by Dr.E.DiCapua (this laboratory). Knoted and catenated DNA molecules were prepared in Dr.N.Cozzarelli's laboratory (University of California, Berkeley) in the context of a collaborative project (2, 4-8).

RESULTS

I. The recA coating technique

Figure 1 demonstrates the advantages of the recA coating of DNA as opposed to the spreading of naked DNA. The micrograph shows three initially identical and unlinked with each other circular DNA molecules. Two of them are completely covered with recA forming two thick circular filaments partially intertwined with each other while the third DNA molecule remained uncovered and is visible as a thin circular filament with two cross-over points. It is clear from this micrograph that the covered DNA filament is much thicker and longer than the naked DNA molecule. On shadowed preparations showing only the upper side of adsorbed filaments where the heavy metal is deposited,

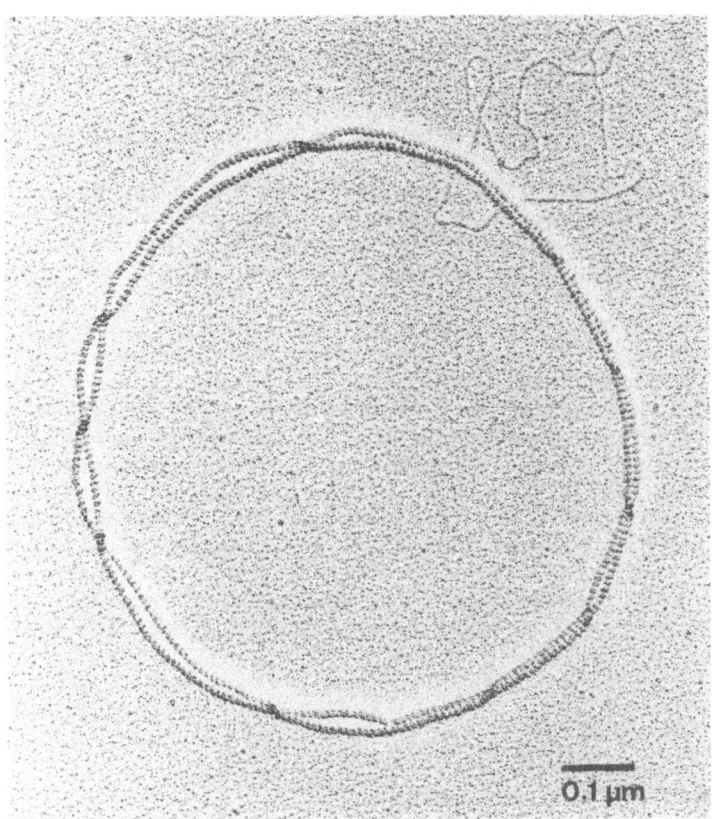

Figure 1: Electron micrograph showing difference of appearance of cross-over points in two intertwined recA covered circular DNA molecules and in naked circular DNA molecule.

the right-handed helical structure of recA-DNA complexes is visible as parallel striations along the filament running from left-down to right-up. RecA coated molecules formed under high magnesium conditions (5 mM magnesium acetate) tend to align side by side (9). In case of recA covered linear DNA molecules such an alignment leads to a regular left-handed wrapping of two filaments with one turn per about 30 striations (9). In case of two unlinked circular filaments an unperturbed left-handed wrapping cannot occur along the entire length of the aligned molecules. Compensation must occur in form of an equal number of right-handed turns. Indeed a close inspection of the intertwined circles in Figure 1 reveales that there is the same number of left- and right-handed crossings of the two circular filaments indicating that the molecules are topologically unlinked. All these crossings are in fact reducible crossings, i.e. which can be removed without breaking of the recA-DNA filaments, in this case simply by unwrapping of the two molecules. Figure 1 shows also crossover points in a naked DNA molecule where the underlying and overlying segments cannot be distinguished.

II. Knots and catenanes formed by bacteriophage λ integrase (Int)

The biological function of λ integrase (Int) is to catalyze the integration of one circular DNA molecule (DNA of bacteriophage) into a second circular DNA molecule (DNA molecule forming bacterial chromosome). For this process to occur Int requires a specific recognition site to be present in one DNA molecule and a second similar but significantly different site to be the present in the other DNA molecule. Specific sites are determined by specific sequences of the nucleotides which are the building blocks of DNA strands. In a simplified description of the integration reaction integrase brings two recognition sites (schematically depicted as arrows in Fig. 2) in close proximity and after cleaving the DNA within each recognition site, joins the front half of one site to the rear half of the other site. As a result two DNA circles are joined. Since the original sequence of nucleotides in both recognition sites is different, after joining the front half of one site to the rear half of the other, the newly formed sites are so much changed that integrase does not recognize them and therefore cannot use them again for a second round of the reaction.

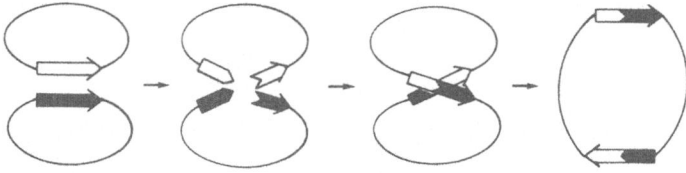

Figure 2:   Schematic representation of integration reaction.

In the natural in vivo reaction of integrase only the joining of two circles like shown in Fig. 2 occurs. However, by means of genetic engineering techniques it is possible to place both Int recognition sites within the same circular DNA molecule and to provide an artificial substrate for the integrase reaction. There are two possible relative orientations of the Int recognition sites within a circular DNA molecule: in the head to tail orientation the two recognition sites (depicted as arrows) are placed in the same direction along the DNA molecule and in the head to head orientation the sites are placed in opposite directions (Fig.3). Depending on the relative orientation of the Int sites, the integrase reaction could lead to relative inversion of the DNA segments flanked by the Int sites (Fig. 3a) or to resolution of the original circle into two smaller circles (Fig. 3b).

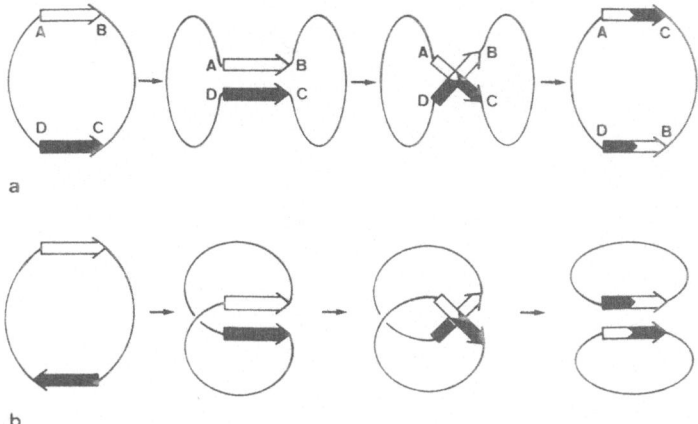

Figure 3: Schematic representation of intramolecular integrase
reaction.
a. Integrase recognition sites in head to head orientation lead to
   reaction resulting in relative inversion of segments flanked by Int
   recognition sites.
b. Integrase recognition sites in head to tail orientation lead to
   resolution of one circular DNA molecule in two smaller DNA circles.

     In order to occur, the Int reaction requires that the DNA is
negatively supercoiled as it is the case in naturally existing DNA
within or isolated from bacterial cells. The notion of supercoiling can
be conveniently explained by comparing the mechanical properties of DNA
to those of a rubber tubing. A linear long piece of elastic tubing can
be brought into the form of a perfect circle just by uniform bending of
the tube in one plane. However, if the tubing in addition to bending is
also twisted by some external force before sealing its ends and
subsequently the twisting force is released, then the tubing which
tends to return to its untwisted form brings the whole circle into a
supercoiled nonplanar shape. Negative supercoiling is defined as
arising due to left-handed twisting before sealing the ends.
     The three dimensional shape of negatively supercoiled DNA
molecules in solution is not yet well characterized since it depends on
the torsional and axial flexibility of the DNA which can vary
significantly under different conditions. Two different shapes have
been proposed so far for supercoiled DNA: namely the plectonemic and
the solenoidal shape (Fig. 4a,b). Using rubber tubing it is easy to
demonstrate that solenoidal supercoiling and plectonemic supercoiling
are interconvertible and that negatively supercoiled molecules show a
left-handed direction of solenoidal supercoils and a right-handed
direction of plectonemic supercoils. Theoretically the integrase
reaction occuring between the recognition sites placed on the same
molecule of solenoidal or plectonemic supercoiled DNA could produce
quite different products (Fig.4). However the knoted and catenated DNA

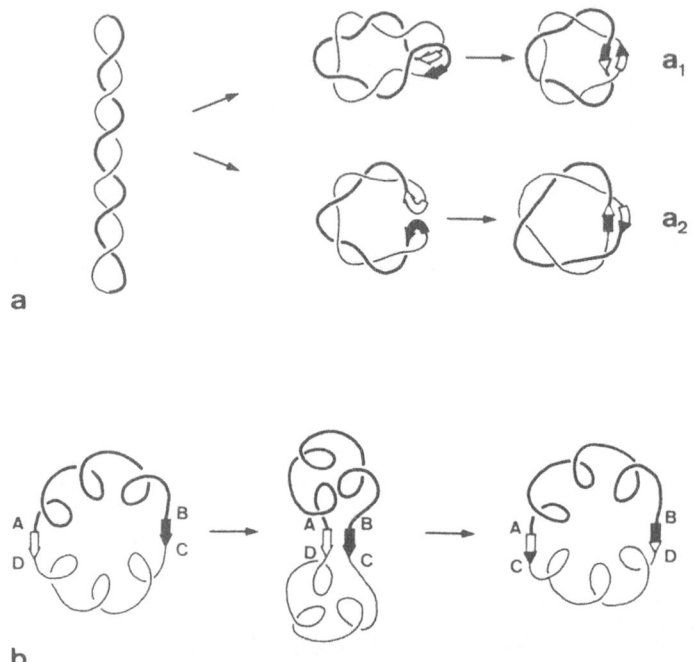

Figure 4: Schematic representation of intramolecular integrase
reaction occurring with supercoiled DNA (Adapted from ref. 5).
a. Plectonemically supercoiled DNA molecule can lead to multiply
   catenated circles when reaction occurs between head to tail oriented
   Int recognition sites (a$_1$) and to torus type of knot when Int
   recognition sites are oriented head to head (a$_2$).
b. Solenoidally supercoiled DNA molecule can lead to unknoted
   solenoidally supercoiled circles just with inverted segments when
   reaction occurs between head to head oriented Int recognition sites.
   (Reaction between head to tail oriented Int sites (not shown here)
   could lead to formation of two uncatenated solenoidally supercoiled
   smaller circles.)

molecules observed in vitro experiments are consistent only with the
possibility that supercoiled DNA in solution exists in the plectonemic
form (6,7). The product molecules obtained from a reaction in which the
Int recognition sequences were placed in head to head orientation on
the same supercoiled DNA molecule belonged to the torus type of knots
with right-handed unreducible crossings (Fig.5a,b). This is exactly the
type of knots one would expect starting from a right-handed
plectonemically coiled molecule (see Fig.4a$_2$). Also the catenanes
arising from supercoiled DNA molecules having head to tail oriented Int

sites were torus type of catenanes with right-handed unreducible
crossing (Fig.5 c,d) precizely as expected for plectonemically coiled
molecules (Fig.4a$_1$).

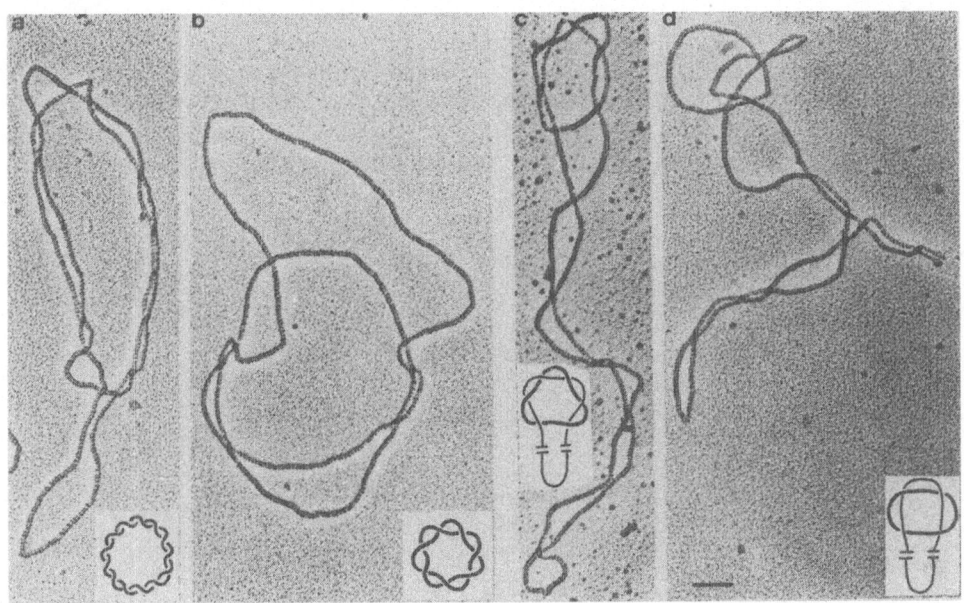

<u>Figure 5</u>: Knots and catenanes obtained after intramolecular Int
reaction occurring with supercoiled DNA. Note that originally
supercoiled DNA appears relaxed after recA-coating (for details see
ref. 2).
a,b. Torus type of knots resulting from head to head orientation of Int
recognition sites. a. Knot with 13 right-handed irreducible cross-over
points (nodes). b. Knot with 7 right-handed nodes.
c,d. Torus type of catenanes resulting from head to tail orientation of
Int recognition sites. In this case, the Int recognition sites were
located asymmatrically on the circle, whereby the two spacings between
the two recognition sites differed by a factor of 5. Therefore the
resulting circles differ by a factor of 5 in length. c. Catenane with 6
right-handed nodes. d. Catenane with 4 right-handed nodes. The bar
indicates 0,1 um.

It is characteristic for such Int reactions that although all
substrate DNA molecules were supercoiled to the same extent the
produced knots and catenanes varied considerably in the number of
nonreducible crossings ranging from knots with just 3 crossings
(trefoils) up to knots with 23 crossings (7). Figure 6 explains how
molecules with the same number of plectonemic supercoils and with the
same separation distance of the Int recognition sites along the DNA
molecule can give rise to catenanes with different numbers of

214

nonreducible crossings. In plectonemically supercoiled DNA random
movements of DNA can lead to "slithering" of facing segments by a
conveyor-belt type of movement of the whole molecule. As a result
sequences separated by a long distance along the DNA molecule can get
very close within the 3-dimensional structure (as measured along the
axis of plectonemic supercoiling). Integrase mediated reactions
occuring between such differently "slithered" recognition sites will
produce catenanes with different numbers of nonreducible crossings.

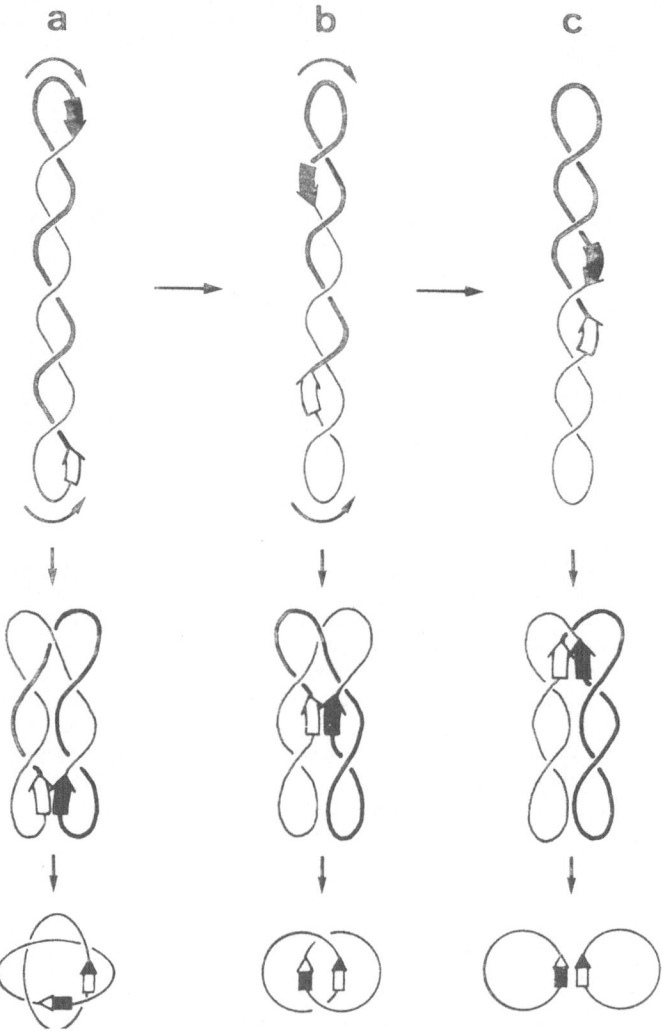

Figure 6: Conveyor-belt type of movement of plectonemically
supercoiled DNA molecules. When differently "slithered" DNA molecules
(a,b,c) get bent to allow reaction to occur then products differ in the
number of nonreducible crossings.

Interestingly there is even the possibility that reaction within plectonemically supercoiled DNA molecules will lead to unknoted or unlinked molecules (Fig. 6c). On the other hand (see below for transposon 3 resolvase) if an enzyme could only work when the two recognition sites are in a defined, fixed position within the 3-dimensional shape of the molecule (fixed distance along the plectonemically coiled molecule), then only one type of product molecules are to be expected.

As already mentioned, the natural function of integrase is to join two circular DNA molecules. The joining reaction results in the formation of an unknoted circle irrespective whether the two circles are plectonemically or solenoidally supercoiled or even if they are not supercoiled. Only when the enzyme acts on two sites within the same DNA molecule then the three dimensional shape of DNA is important. As already discussed in the Introduction, enzymes acting on DNA tend to keep the DNA in an unknoted form, but since under natural conditions integrase never works on two sites within the same circular molecule, its action does not have to be associated with a safeguarding mechanism preventing DNA from getting knoted or multiply catenated.

III. Knots and catenanes formed by transposon 3 resolvase

The biological function of transposon 3 resolvase (Tn3 res) is the resolution of one circular supercoiled DNA molecule into two circular molecules. To a certain extent the Tn3 resolvase action is opposite to that of λ integrase. One significant difference is that the two Tn3 res recognition sequences are identical, thus after joining of the front half of one site to the rear half of the other site, functional recognition sequences are restored and this can lead to an iteration of the reaction.

When Tn3 res acts on negatively supercoiled DNA molecules containing two repeated Tn3 res recognition sites then in 95% of the cases singly interlinked catenanes are produced (5,8). Coming back to the mechanism proposed in the previous Section, the fact that in the case of Tn3 res mostly only one type of catenane is produced suggests that the enzyme has to stabilize the recognition sites in close proximity prior to cutting, end switching and rejoining. Fig. 6 and 7 show how starting with plectonemically supercoiled DNA molecules it is possible to obtain singly linked catenanes by at least two ways. In one way the position of the recognition sites will be such that the enzyme stabilizes one right-handed supercoil between the recognition sites. Cutting, end switching and rejoining reactions in such molecules can lead to complete separation of circles or to singly interlinked catenanes depending on the way the switched ends will cross with each other (Fig.7a,b). In the second way the recognition sites will be brought to a distance allowing stabilization of three plectonemic supercoils between recognition sites. Cutting, end switching and rejoining reactions in such molecules can lead to singly interlinked or doubly interlinked catenanes depending on the way the exchanged ends will cross with each other (Fig.7c,d,). Therefore knowing that the resolvase reaction produces singly linked catenanes does not allow to

216

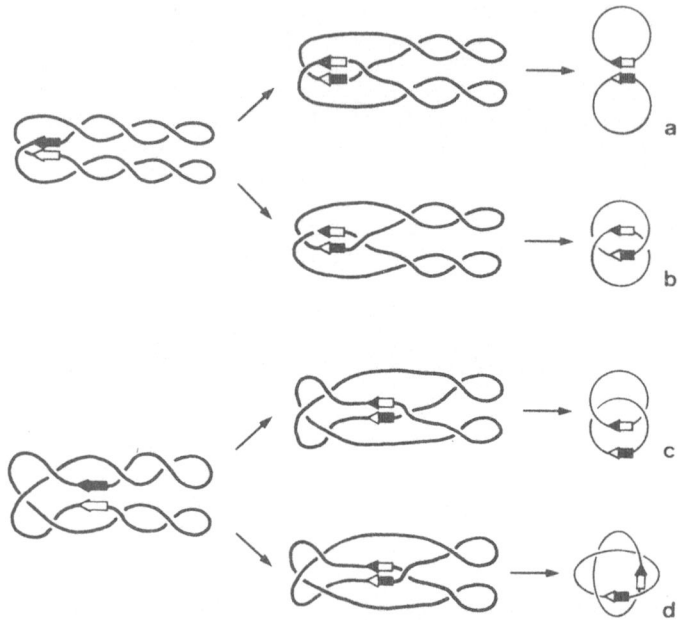

Figure 7: Tn3 resolvase reaction can generate singly interlinked
catenanes by at least two different ways. Three-dimensional shape of
DNA molecules during reaction and direction of crossing by exchanged
ends determine the type of product.

deduce the molecular mechanism of action of this enzyme. To answer
question how many supercoils are trapped by the enzyme between the
recognition sites and how the switched ends cross, it is necessary to
analyze further products of resolvase reaction. As already mentioned
Tn3 resolvase can mediate an iterative reaction. Under in vitro
reaction conditions in about 5% of the cases the enzyme is able to
proceed to further rounds of reaction. As a result 4-noded knots and
figure-8-catenanes are formed (Fig.8).
    The product molecules observed after the iterative reaction are
consistent with a model in which three negative plectonemic supercoils
are stabilized between the two Tn3 res recognition sites and the
exchanged ends cross in the way percepted as right handed (Fig.9).
Iteration of the first reaction would lead to formation of a 4-noded
knot and subsequently to a figure-8-catenane (Fig.8). One could even
predict what type of 6-noded knot would be the product of another round
of reaction (Fig.9). In fact such knots as a result of resolvase action
have been observed (8).

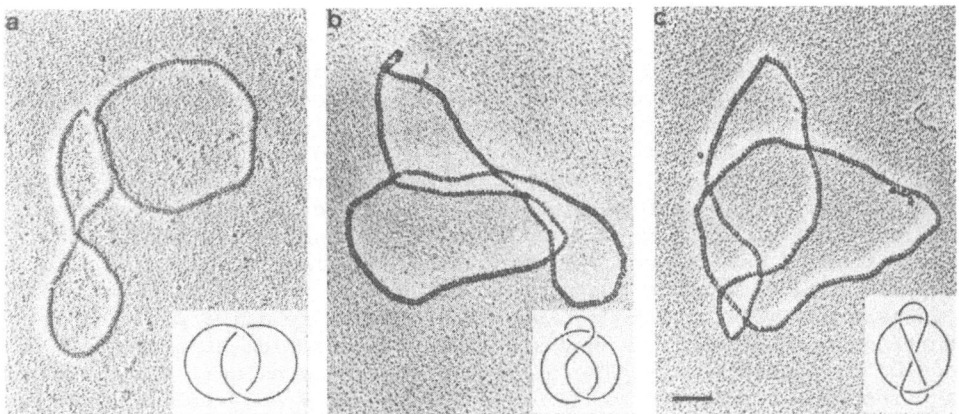

Figure 8: Catenanes and knots obtained during three sequential Tn3 resolvase reactions occurring intramolecularly with a supercoiled DNA molecule. a. Singly interlinked catenane. b. 4-noded knot. c. Figure-8-catenane.
The bar indicates 0.1 um.

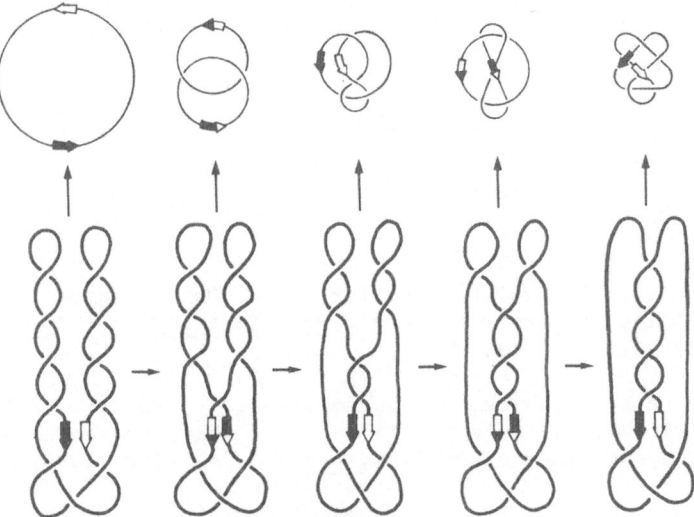

Figure 9: Representation of four sequential Tn3 resolvase reactions and their product knots and catenanes (adapted from ref. 8).

CONCLUSIONS

The analysis of knots and catenanes produced by different enzymes
acting on DNA allowed to determine the three dimensional shape of
superooiled DNA in solution and helped also to elucidate mechanisms
which allow resolving enzymes to avoid multiple catenation of the
produced circular DNA molecules. The mechanism proposed is based on the
stabilization of the recognition sites in a strictly defined relative
position to each other. This stabilization is a necessary preliminary
stage for the cleavage, end switching and rejoining reactions to occur.
There are good reasons to assume that enzymes do not actively "slither"
or reshape DNA molecules to achieve the required relative positions of
the recognition sites, but only stabilize a particular arrangement
produced by random motion. Therefore the three-dimensional shape of the
DNA molecules in solution will influence the speed and efficiency of
such reactions. In case of Tn3 resolvase for example, stabilization of
three negative supercoils between the recognition sites is facilitated
when the molecule is already negatively supercoiled, but also
non-supercoiled DNA can get locally interwound leading to the proper
arrangement of the sites. Since the enzyme only selects one of many
possible DNA arrangements, this mechanism in which the specific shape
of DNA molecules (given by supercoiling or other topological
constraint) interplays with the specific binding geometry of an acting
enzyme, was called a topological filter (10,11).

ACKNOWLEDGEMENTS

We thank Drs. S.J. Spengler and M.A. Krasnow for DNA knots and
catenanes produced by λ integrase and Tn3 resolvase

REFERENCES

1. Stasiak, A., DiCapua, E. & Koller, Th.: 'Unwinding of duplex DNA
   in complexes with recA protein'. Cold Spring Harbor Symp. Quant.
   Biol. 47: 811-820 (1983)

2. Krasnow, M.A., Stasiak, A., Spengler, S.J., Dean, F., Koller, Th. &
   Cozzarelli, N.R.: 'Determination of the absolute handedness of
   knots and catenanes of DNA'. Nature 304: 559-560 (1983)

3. Griffith, J.D. & Nash, H.A.: 'Genetic rearrangement of DNA induces
   knots with unique topology: implications for the mechanism of
   synapsis and crossing-over'. Proc. Natl. Acad. Sci. USA 82:
   3124-3228 (1985)

4. Dean, F.B., Stasiak, A., Koller, Th. & Cozzarelli, N.R.: 'Duplex
   DNA knots produced by Escherichia coli topoisomerase I, structure
   and requirements for formation'. J. Biol. Chem. 260: 4975-4983
   (1985)

5. Krasnow, M.A., Matzuk, M.M., Dungan, J.M., Benjamin, H.W. & Cozzarelli, N.R.: 'Site specific recombination sites'. in: Mechanisms of DNA replication and recombination, Alan R. Liss, Inc., New York pp. 637-659 (1983)

6. Spengler, S.J., Stasiak, A., Stasiak, A.Z. & Cozzarelli, N.R.: 'Quantitative analyzis of the contributions of enzyme and DNA to the structure of integrative recombinants'. Cold Spring Harbor Symp. Quant Biol. 49: 745-749 (1984)

7. Spengler, S.J., Stasiak, A. & Cozzarelli, N.R.: 'The stereostructure of knots and cateananes produced by phage λ integrative recombination: implications for mechanism and DNA structure'. Cell 42: 325-334 (1985)

8. Wasserman, S.A., Dungan, J.M. & Cozzarelli, N.R.: 'Discovery of a predicted DNA knot substantiates a model for site-specific recombination'. Science 229: 171-174 (1985)

9. DiCapua, E., Engel, A., Stasiak, A. & Koller, Th.: 'Characterization of complexes between recA protein and duplex DNA by electron microscopy'. J. Mol. Biol. 157: 87-103 (1981)

10. Boocock, M.R., Brown, J.L. & Sherratt, D.J.: 'Topological specificity in Tn3 resolvase catalysis'. in: DNA replication and recombination, Alan R. Liss, Inc., New York pp. 703-718 (1987)

11. Gellert, M. & Nash, H.A.: 'Communication between segments of DNA during site specific recombination'. Nature 325: 401-404 (1987)

USING KNOT THEORY TO ANALYZE DNA EXPERIMENTS

D.W. Sumners
Department of Mathematics
Florida State University
Tallahassee, Fla. 32306

ABSTRACT. There exist naturally occurring enzymes(*topoisomerases and recombinases*), which, in order to mediate the vital life processes of replication, transcription and recombination, manipulate cellular DNA in topologically interesting and nontrivial ways. These enzyme actions include promoting writhing(coiling up) of the DNA molecules, passing one strand of DNA through another via and enzyme-bridged transient break in one of the strands, and breaking a pair of strands and recombining to different ends. If one regards DNA as very thin string, these enzyme activities are the stuff of which modern knot theory is made! In order to describe and understand these enzyme mechanisms, knot theory is an essential tool. When reacted with unknotted closed circular DNA, each enzyme produces a characteristic family of knots and catenanes. A new experimental technique(*rec A* coating) produces high-resolution electron micrographs of these reaction products. By analyzing these reaction products, one can discover facts about enzyme mechanism. A new topological model(the *tangle* model) for enzyme mechanism will be discussed, and used to analyze experiments on the recombinant enzyme *Tn3 resolvase*.

1. INTRODUCTION

Within the last 4 years[1], a new experimental technique(*rec A* coating of DNA molecules) has been perfected which makes possible the unambiguous determination of crossovers in electron micrographs of DNA. This coating thickens the DNA strands from about 10 angstroms to about 100 angstroms, making it much easier to decide which strand is uppermost in a crossing of two strands in an electron micrograph. This new resolution ability can be exploited in a topological approach to enzymology--one performs experiments in which *topoisomerases*(enzymes producing topological isomers) are reacted with(usually unknotted) closed circular DNA[2,3]. The reaction products are families of DNA knots and catenanes, each enzyme producing its own characteristic family of knotted and catenated DNA circles. The reason that circular DNA is chosen as the reaction substrate is that the circular form of the DNA substrate traps some of the complicated topological changes caused by enzyme action. One wishes ultimately to understand the action of these enzymes on DNA in the cell(*in vivo*). One begins this process by analyzing enzyme action on circular DNA in the laboratory(*in vitro*). The use of circular DNA as a reaction substrate can be regarded as an amplifier--the sub-microscopic action of an enzyme is amplified until it is experimentally observable--in the form of an electron micrograph of a DNA knot or catenane!

A. Amann et al. (eds.),
Fractals, Quasicrystals, Chaos, Knots and Algebraic Quantum Mechanics, 221–232.
© *1988 by Kluwer Academic Publishers.*

The application of knot theory to molecular biology occurs at two levels. Descriptive: Resolution and enumeration of reaction products. The product of an enzyme reaction is an enzyme-specific family of DNA knots and catenanes. The classification results of knot theory must be used to resolve these products into their various knot(catenane) types. Given electron micrographs of DNA, one compares these pictures to tables of knots and catenanes. Predictive:Building mathematical models to calculate and predict enzyme action. It is this latter interaction of mathematics and molecular biology which is the subject of this paper.

This paper deals with a topological model for enzyme mechanism[4,5,6]. The action of many enzymes involve local(near the enzyme) interactions of a pair of DNA strands. The mathematics which can be used to model this local 2-strand interaction is that of the *2-string tangle*. Enzyme action on circular DNA can be viewed as tangle surgery. At *synapsis*, the enzyme naturally separates the circular DNA molecule into two complimentary tangles. The enzyme action is to delete one of the tangles, replacing it by another(*tangle surgery*). One regards these tangles as enzyme *mechanism variables*, and experimental results pose equations relating these variables. One wishes to solve these experimentally posed equations. In full generality, this can be a difficult task. The job is made easier by the realization that most DNA reaction products lie in a completely understood class of knots and 2-component catenanes, the class of *4-plats(2-bridge* knots and catenanes)[7]. Moreover, a great deal can be said about the factorization of 4-plats into tangle summands. The summands of interest are *rational tangles*[8]. Rational tangles are closely related to 4-plats, and likewise form a completely understood class of topological objects. Moreover, they are formed by twisting pairs of strands about each other, and look like electron micrographs! One uses the theory of 3-manifolds[9,10] to prove that most of the tangles of interest which occur in the analysis of a particular DNA enzyme experiment are rational. Once the summands are known to be rational tangles, the analysis becomes a matter of solving tangle equations posed by experiment. This is where the *rational tangle calculus*[11] is employed.

2. KNOTS, CATENANES AND TANGLES

Topology is that branch of mathematics which studies those properties of objects which do not change when the object is elastically deformed. Topological considerations are important in the analysis of macromolecular configuration. A macromolecule does not usually maintain a fixed 3-dimensional configuration. Such a molecule can assume a variety of configurations, driven from one to another by thermal motion, solvent effects, experimental manipulation, etc. *Knot theory* is the branch of topology concerned with the properties of flexible graphs embedded in 3-space.

We will be concerned with the configurations of flexible circles in 3-space. Two topological spaces are *homeomorphic* if there exists a function $f:X \longrightarrow Y$ such that f is 1-1 and onto, and both f and f^{-1} are continuous. We avoid all local pathology by insisting that all functions are *smooth*(infinitely differentiable). A *knot* (K) is a placement of the circle in 3-space; it is a subspace of 3-dimensional Euclidean space(R^3) which is homeomorphic to the unit circle (S^1) in 2-space (R^2). At an intuitive level, we wish to regard two knots {K,K'} as *equivalent* if it is possible to elastically deform one(without breaking it or passing one strand through another) until it can be superimposed upon the other. A homeomorphism $H:R^3 \rightarrow R^3$ is *orientation preserving* if the determinant of its Jacobian matrix(evaluated at the origin) is positive, and *orientation reversing* if this determinant is negative. Unless otherwise specified, all homeomorphisms will be orientation preserving.

Two knots K, K' are *equivalent* if there is a homeomorphism(H) of R^3 to itself which takes K homeomorphically to K'; that is, H is a *homeomorphism of pairs* $H:(R^3,K) \rightarrow (R^3,K')$. This mathematical definition of equivalence of knots agrees with the above intuitive idea of elastic conversion of one configuration to another. A knot, then, is an equivalence class(*knot type*) of placements of the circle in 3-space. The symbol K will denote either a specific representative of the equivalence class, or the equivalence class itself, depending upon context. A *catenane*(of 2 circular components) is a pair of circles in R^3 with the property that they cannot be separated(elastically deformed so that one component lies inside the unit 2-sphere in 3-space, and the other component lies outside the 2-sphere). Chemically, a catenane represents circular molecules bound together by topological bonds instead of chemical bonds. A configuration of 2 or more circles in R^3 is usually called a *link* in mathematics.

The knot K is usually represented by drawing a *projection* of one of its representatives. A projection is a shadow of the knot cast upon a plane in 3-space. By a small rotation of the knot in 3-space, it can be arranged so that no more than 2 strings cross at any point in the projection. At such a 2-string*crossover*, the strand on the bottom is drawn with a break in it. Figure 1 shows some knot projections. Figure 1(a) shows the *unknot*, the unit circle in 2-space(XY space), sitting in 3-space(XYZ space). Figure 1(b) shows the "+" trefoil, 1(c) the "-" trefoil, and 1(d) the "figure 8" knot. The two trefoils of Fig. 1 are mirror images of each other, and are inequivalent knots[7]. Such a pair of inequivalent mirror images forms a *chiral pair* of knots.

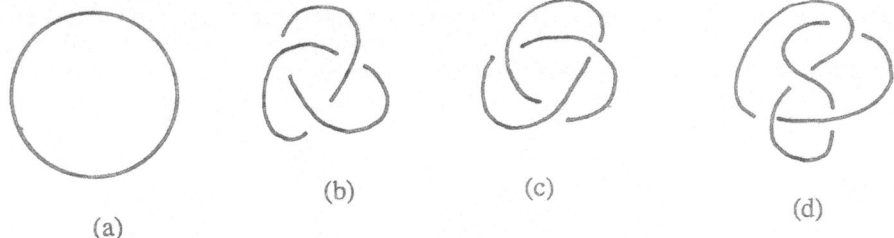

(a)   (b)   (c)   (d)

FIG. 1   Knot Projections

Any given knot type clearly admits infinitely many "different" projections, and recognizing that two completely different projections in fact represent the same knot type can be exceedingly difficult. For any knot type, one usually wishes a projection of minimal complexity--that is, with the minimum number of crossovers. For any given knot type, the number of crossings found in a minimal projection of that knot is the *crossing(crossover) number* for that knot. The knot projections of Fig. 1 are minimal. The problem of resolving knot types is one of the central problems in knot theory[7]. In order to distinguish knots, one devises *topological invariants* for knot types. These invariants are groups, polynomials, numbers, etc. which can unambiguously be attached to a knot type. If K and K' are of the same knot type, then all their topological invariants must be identical. If a given invariant differs for K and K', we can then be sure that they are inequivalent knots. In general, if all known invariants are identical for K and K', we cannot be sure about their resolution. We must either devise a new invariant which distinguishes them, or prove that they are the same knot by direct geometric manipulation. Sometimes, however, it is possible to completely classify specific subfamilies of knots. That is, we can devise a system of topological invariants such that two knots in the subfamily are equivalent if and only if their invariant(s) are identical. We now consider such a subfamily, the family of *4-plats(viergeflechte)*.

(a) {1,3,2}   (b) {2,3,2}

Fig.2   4-Plats

Fig. 2(a) shows the 4-plat knot {1,3,2}, and Fig. 2(b) shows the 4-plat catenane {2,3,2}. These configurations of circles, manufactured by platting 4 strands as shown, are classified by their *Conway* symbols, the vectors {1,3,2} and {2,3,2}. In the Conway symbol {1,3,2}, the integer entries code for the number of positive(right-handed) half-twists between strands as shown. Note that if one rotates one of the projections of Fig. 2 through 180 degrees about an axis in the plane of the paper and perpendicular to the plat, one obtains an equivalent 4-plat whose Conway symbol is the reverse of the one we started with. It turns out that this is the only ambiguity--two 4-plats are equivalent if and only if their Conway symbols are either identical or become identical if one is reversed[7].

4-plats have other nice properties. A knot is *composite* if it admits a projection in which a circle can be drawn on the plane of projection which intersects the knot projection in two points, and such that the drawn circle separates the projection into two knotted arcs. Fig. 3 shows a composite knot. Intuitively, a composite knot is obtained by tying one knot in a string, then another, then glueing the ends of the string together. A knot is *prime* iff it is <u>neither</u> unknotted nor composite. 4-plats are prime[12], and projections like those of Fig. 2 are minimal[13].

Fig. 3   A Composite Knot

Consider now the unit ball B^3 in R^3--the set of all vectors of length ≤ 1 in R^3. The boundary of B^3 is S^2, the unit 2-sphere--the set of all vectors of length 1 in R^3. Thinking of R^3 as XYZ-space, the equator of S^2 is the intersection of S^2 with the XY-plane in R^3. Orient the equator of S^2(put an arrow on it). Select 4 points on the equator(called NW, SW, SE, NE), cyclically arranged so that one encounters them in the order named upon traversing the equator in the chosen direction. This copy of S^2 with 4 distinguished equatorial points will be called the *standard tangle boundary*. A 2-*string tangle*, or just *tangle* for short, will denote any configuration of 2 arcs in a 3-ball, satisfying the following conditions: (i) the arcs meet the boundary of the 3-ball in endpoints only, and all 4 endpoints are in the boundary, and (ii) there is a fixed homeomorphism from the 3-ball to the unit 3-ball B^3 which takes the 4 arc endpoints to the 4 distinguished points {NW,NE,SE,NE}. This fixed homeomorphism is called a *boundary parametrization*[13,14]. By means of this boundary parametrization, we can regard any two tangles as lying inside B^3, and having identical boundaries. Two tangles are *isomorphic* if it is possible to superimpose the arcs of one upon the arcs of the other by means of moving the elastic arcs around in the interior of B^3, leaving the common boundary pointwise fixed. Mathematically, there is a well-understood class of tangles which look like DNA micrographs, and which(like 4-plats) are created by twisting pairs of strands about each other. These tangles are called *rational* tangles, and have been completely classified up to tangle isomorphism by Conway[8]. Like 4-plats, there is a canonical form for tangles, and when written in canonical form, these tangles are classified by a vector with integer entries, each entry corresponding to a number of half-twists. The entries of the classifying vector determine via a continued fraction calculation a rational number which itself classifies the tangle(hence the terminology). Fig. 4 shows some rational tangles, and their classifying vectors and rational numbers. Topologically, a tangle is rational iff it is homeomorphic to the trivial tangle(the tangle {0} of Fig. 4)[8,15]--the homeomorphism to the trivial tangle will in general move the boundary points. A tangle is rational iff it can be "undone" by rotating pairs of boundary points about each other.

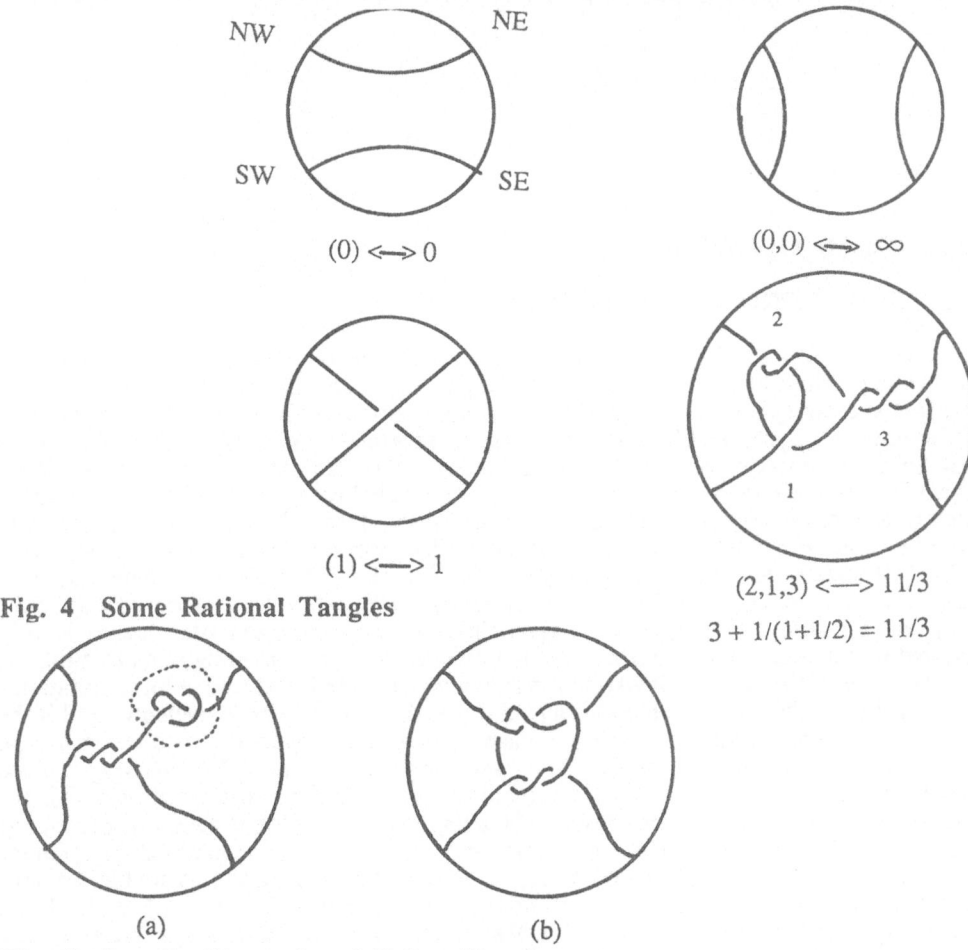

$(0) \longleftrightarrow 0$

$(0,0) \longleftrightarrow \infty$

$(1) \longleftrightarrow 1$

Fig. 4   Some Rational Tangles

$(2,1,3) \longleftrightarrow 11/3$

$3 + 1/(1+1/2) = 11/3$

(a)   (b)

Fig. 5   Locally Knotted and Prime Tangles

Not all tangles are rational. Analogous to the terminology for knots, we say a tangle is *locally knotted* iff it has a local knot in either of its strands(Fig. 5(a)). A tangle is *prime*[15] iff it is <u>neither</u> rational nor locally knotted(Fig. 5(b)).

Given a pair of tangles(A and B), there are two constructions which can be performed on them. One is *tangle addition*, in which the NE point of A is jointd to the NW point of B, and the SE point of A is joined to the SW point of B, forming A#B(Fig. 6(a)). In general, the operation of tangle addition is not commutative, and, even if A and B are rational tangles, A#B may not be a rational tangle. The other operation is the *numerator* construction, in which the NW point of tangle A is joined to the NE point of A, and the SW point of A is joined to the SE point of A, forming N(A), which is either a knot or a link(catenane) of two circular components(Fig. 6(b)).

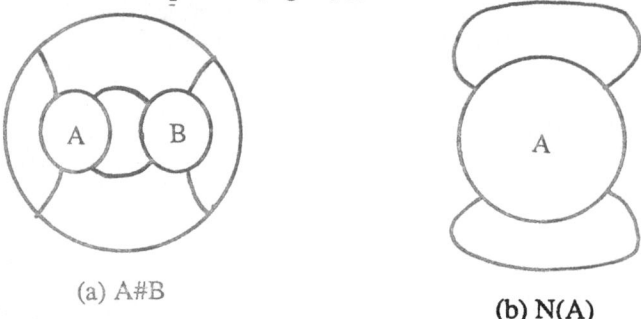

(a) A#B

(b) N(A)

Fig.6   Tangle Operations

3. SITE-SPECIFIC RECOMBINATION

We will now consider the situation of site-specific recombination enzymes operating on covalently closed circular duplex DNA. Duplex DNA consists of two linear backbones of sugar and phosphorus. Attached to each sugar is one of the four bases:A=Adenine, T=Thymine, C=Cytosine, G=Guanine. A ladder is formed by hydrogen bonding between base pairs, where A binds with T, and C binds with G. In the classical Crick-Watson model for DNA, the ladder is twisted in a right-handed helical fashion, with a relaxed-state pitch of approximately 10.5 base pairs per full helical twist. Duplex DNA can exist in closed circular form, where the rungs of the ladder form a twisted cylinder(instead of a twisted Mobius band). In certain closed circular duplex DNA molecules, there exist two short identical sequences of base pairs, called *recombination sites* for a recombinant enzyme. Because of the base pair sequencing, the recombination sites can be locally oriented (reading the sequence from right to left is different from reading it left to right). If one then orients the circular DNA(puts an arrow on it), there is induced a local orientation on each site. If the local orientations agree, this is the case of *direct repeats* , and if the local orientations disagree, this is the case of *inverted repeats* . The recombinase nonspecifically attaches to the molecule, and then the sites are aligned(brought close together), either through enzyme manipulation or random thermal motion(or both), and both sites are then bound by the enzyme. This stage of the reaction is called *synapsis* , and the complex formed by the substrate together with the bound enzyme is called the *synaptic complex.* In a single recombination event, the enzyme then performs two double-stranded breaks at the sites, and recombines the ends in an enzyme-specific manner(Fig. 7). In the following figures, double-stranded DNA is represented by a single strand, and supercoiling is omitted.

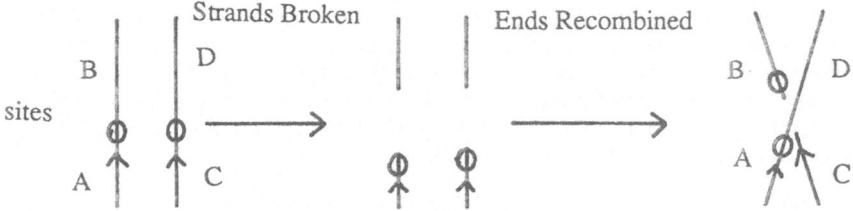

Fig. 7 A SINGLE RECOMBINATION EVENT

We call the unbound DNA molecule before recombination takes place the *substrate*, and after recombination takes place, the *product*. If the substrate is a single circle with direct repeats, the product is a pair of circles, and can form a DNA catenane(Fig. 8). If the substrate is a pair of circles with one site each, the product is a single circle (Fig. 8 read in reverse). If the substrate is a single circle with inverted repeats, the product is a single circle, and can form a DNA knot(Fig. 9).

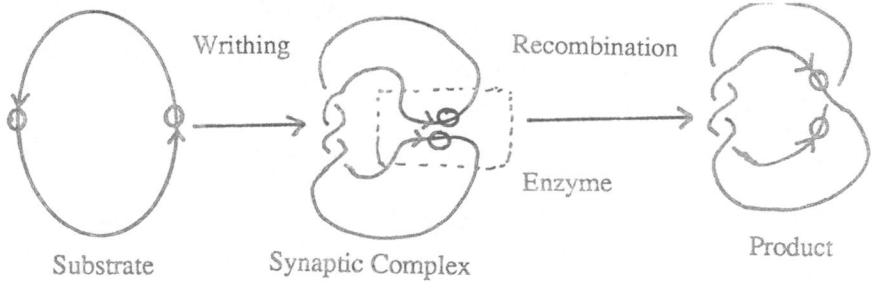

Fig. 8 Hypothetical Recombination Catenane Synthesis

4. THE TANGLE MODEL

In site-specific recombination, two kinds of geometric manipulation of the DNA occur. The first is a global move, in which the sites are juxtaposed, either through enzyme action or random collision(or a combination of these two processes). After synapsis is achieved,the next move is local, and entirely due to enzyme action. Within the region controlled(bound) by the enzyme, the molecule is broken in two places, and the ends recombined. We will model this local move. We model the enzyme itself(or, if necessary, its sphere of influence) as being homeomorphic to the unit ball in 3-space(B^3). The two recombination sites(and some contiguous DNA) form a tangle of two arcs in the enzyme ball. During the local phase of recombination, we assume that the action takes place entirely within the interior of the enzyme ball, and that the substrate configuration outside the ball remains fixed while the strands are being broken and recombined.

For symmetry of mathematical exposition, we take the point of view that the reaction is taking place in the 3-sphere S^3(the unit sphere in 4-space), because the boundary of the recombination ball is homeomorphic to S^2, and this enzyme S^2 functions as an equator in S^3, dividing S^3 into two complimentary 3-balls, glued together along their common boundary. If two tangles(A and B) are identified along their common boundaries, the resulting knot(catenane) can be thought of as N(A#B). In Fig. 9, the dotted circle represents an equatorial circle on the enzyme S^2. The enzyme S^2 in fact divides the substrate into two complementary tangles, the *substrate tangle S* , and the *site tangle T* . The local effect of recombination is to perform tangle surgery, that is, to delete tangle T from the synaptic complex, and replace it with the *recombinant tangle R* . As in Fig. 9, the knot type of the substrate and product each yield an equation in the variables S, T and R. Specifically, an enzyme reaction produces two equations:

SUBSTRATE EQUATION: $\qquad$ N(S#T) = SUBSTRATE

PRODUCT EQUATION: $\qquad$ N(S#R) = PRODUCT

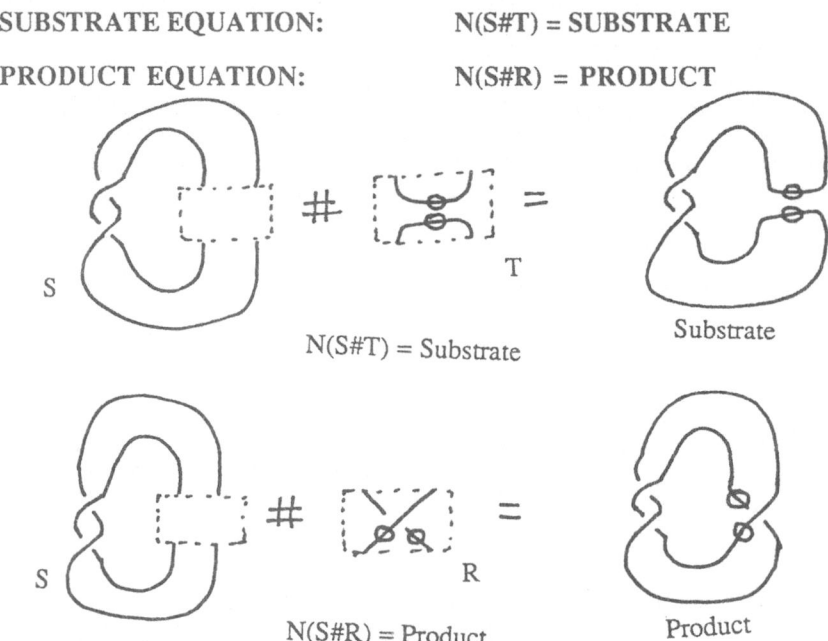

Fig. 9 The Substrate and Product Equations

Ideally, we would like to treat each of S,T and R as *recombination variables*, and to solve the equations posed by experiment for these unknowns. Since a single recombinant event yields only two equations involving three unknowns, the best we can hope for, given only this information, is to solve for any two in terms of a third. The analysis is greatly simplified at this point by making the following biologically reasonable assumption:

BIOLOGICAL ASSUMPTION: *T and R are enzyme-determined constants, independent of the variable geometry of the substrate.*

In some experiments, the substrate may be a large number of circular molecules, all the same knot type, but equipped with different amounts of supercoiling(writhing). Recombination can trap some of this "trivial" geometry, producing a distribution of product knot(catenane) types from a single substrate knot(catenane) type. In such an experiment, then, the substrate tangle S can vary over a number of configurations, but the tangles T and R are enzyme-specific constants, and appear in a number of of pairs of equations, one for each different product. In the experiment discussed in this paper, iterated recombination occurs. This means that one can often obtain enough information about S, T and R to determine them uniquely.

5. Tn3 RESOLVASE

Tn3 Resolvase is a site-specific recombinase which reacts with certain closed circular duplex DNA substrate with directly repeated recombination sites[16,17]. One begins with unknotted DNA substrate, and treats it with resolvase. The principal product[16] of this reaction is known to be the simply-linked catenane of Fig. 8, the 4-plat [2](called the *Hopf Link* in mathematics). Moreover, when endowed with orientation inherited from the parent unknotted substrate, this recombination product has linking number -1. Resolvase is known to act *dispersively* in this situation--to bind to the circular DNA, to mediate a single recombination event, and then to release the linked product. It is also known that resolvase and free(unbound) DNA catenanes do not react. However, in one in 20 encounters, resolvase acts *processively*--additional recombinant strand exchanges are promoted prior to the release of the product, with yield decreasing exponentially with increasing number of strand exchanges at a single binding encounter with the enzyme. Two succesive rounds of recombination produces the figure 8 knot(the 4-plat [2,1,1]); three succesive rounds of recombination produces the figure 8 catenane(the 4-plat [2,1,2]); four successive rounds of recombination produces a 6-crossing knot(the 4-plat [2,1,3]), called 6_2 in the knot tables[7]. The discovery of the DNA 4-plat [2,1,3] substantiated a model for Tn3 resolvase mechanism[17]. Figure 10 shows the DNA 4-plat [2,1,3](from[17]).

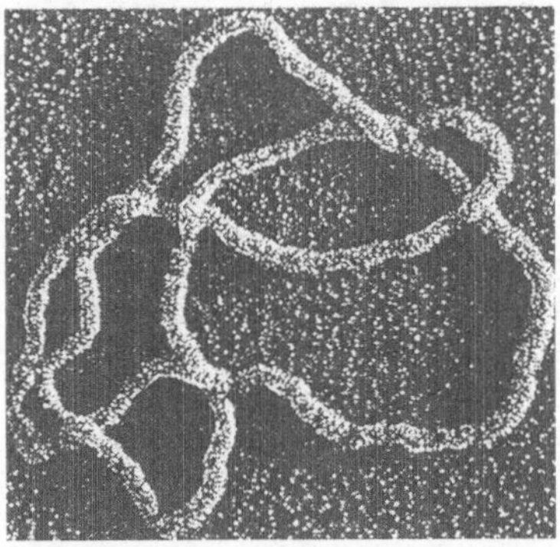

Fig. 10   The DNA 4-plat [2,1,3]

Using the theory of 3-manifolds and the rational tangle calculus, we prove that the experimental results of the first two rounds of recombination uniquely determine the tangles S and R, hence the result of three or more successive rounds of recombination. This theorem is a mathematical proof that the model of [17] is the only explanation for the observed products of Tn3 recombination.

THEOREM : *Suppose tangles S , T and R satisfy the following equations:*

$$N(S\#T) = [1]$$
$$N(S\# R) = [2] \text{ (with linking number = -1)}$$
$$N(S\#R\#R) = [2,1,1]$$

Then   S = {3,0}, R = {-1}, N(S#R#R#R) = [2,1,2] and N(S#R#R#R#R) = [2,1,3].

230

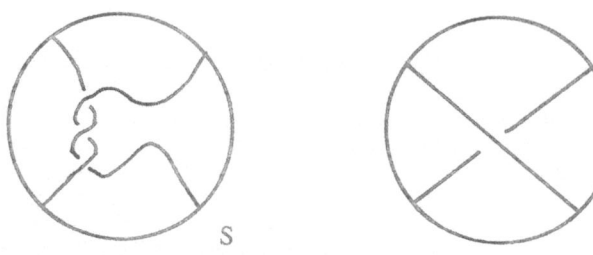

Fig. 11   S = {3,0} and R = {-1}

Proof: The first step in the proof is to argue that R must be a rational tangle. Now R is locally unknotted, because N(S#R) is the 4-plat [2], a catenane with two unknotted components. Any local knot in R would have to turn up in one of the two components of N(S#R). Likewise, The tangles S and T are locally unknotted, because N(S#T) is the unknot.We will use the following tangle facts from [15]: (i) If A is a prime tangle, then the 2-fold branched cyclic cover(A') of A, branched along the two arcs of the tangle, is an irreducible, boundary-irreducible 3-manifold. This means that the inclusion-induced homomorphism on the fundamental group of the torus $(S^1 x S^1)$ boundary of A' is injective. (ii) The tangle A is rational iff its 2-fold branched cyclic cover A' is a solid torus$(S^1 x B^2)$. (iii) If both A and B are prime tangles, then N(A#B) is a prime knot(catenane), and the 2-fold branched cyclic cover N(A#B)' contains an incompressible torus--the common boundary of A and B. In this case the fundamental group of the torus injects into the fundamental group of N(A#B)', and hence the fundamental group of N(A#B)' must be infinite. However, the 2-fold branched cover of a 4-plat is a lens space[7], with finite cyclic fundamental group. This means that if N(A#B) is a 4-plat, then at most one of A and B is a prime tangle. Moreover, if A is a prime tangle, and B is <u>any</u> prime or rational tangle, then (A#B) is likewise a prime tangle[15].   So if R is a prime tangle, then so is (S#R). But N((S#R)#R) is the 4-plat [2,1,1], so R cannot be a prime tangle. This means that R is a rational tangle.

The next step is to show that S is a rational tangle. Suppose that S is a prime tangle. Then T must be a rational tangle, because N(S#T) is the unknot. Taking 2-fold branched cyclic coverings, we have that N(S#T)' = S^3, and so it is possible to obtain the 3-sphere from S' by adding on a solid torus along the boundary of S'. This means that S' is a knot complement. Now R is a rational tangle, and one can show that R#R is locally unknotted[11]. Since N(S#(R#R)) = [2,1,1], we conclude that (R#R) is a rational tangle. Since both R and R#R are rational tangles, this means that R must be an integral tangle(a single horizontal row of half-twists). Taking the 2-fold branched cyclic cover, we have that N(S#R)' = L(2,1), and N(S#R#R) = L(5,3). Since both R and R#R are rational tangles, this means that we can add a solid torus to S' in two ways to obtain the Lens spaces shown. This operation of adding a solid torus is known mathematically as *Dehn Surgery*. The *Cyclic Surgery Theorem*[9], however, asserts that the only way this can happen is for S'to be a Seifert Fiber Space(*SFS*)[10]. The results of surgery on Seifert Fiber Spaces[18] can then be applied to show that S' must be a solid torus, and hence that S must be a rational tangle.

Once we know that both S and R are rational tangles, we can use the rational tangle calculus[11] to solve the iterated product tangle equations of the hypothesis for the answers shown in the statement of the theorem. Since the unoriented 4-plats [2] and [2,1,1] are achiral, the linking number information is crucial in ruling out S = {-3,0} and R = {1} as solutions. The mathematical[11] and biological[19] details of this(and similar arguments for other enzymes) will appear elsewhere.

This work was partially supported by the United States Office of Naval Research.

REFERENCES

1. M.A. Krasnow, A. Stasiak, S.J. Spengler, F. Dean, T. Koller, N.R. Cozzarelli, 'Determination of the absolute handedness of knots and catenanes of DNA',Nature **304**(1983), 559-560.

2. S.A. Wasserman, N.R. Cozzarelli, 'Biochemical topology:applications to DNA recombination and replication', Science **232**(1986), 951-960.

3. D.W. Sumners, 'The role of knot theory in DNA research', in Geometry and Topology, Marcel Dekker(1987), 297-318.

4. D.W. Sumners, 'Rational tangles and recombinant DNA', Am. Math. Soc. Abstracts **47**(1986), 420-421.

5. D.W. Sumners, N.R. Cozzarelli, S.J. Spengler, 'A topological model for site-specific recombination', Abstracts, Cold Spring Harbor 1986 Meeting on Biological Effects of DNA Topology, 72.

6. D.W. Sumners, 'Knots, macromolecules and chemical dynamics', in Graph Theory and Topology in Chemistry, Elsevier(1987), (in press).

7. G. Burde, H. Zieschang, Knots, de Gruyter(1985).

8. J. Conway, 'On enumeration of knots and links and some of their related properties', Computational Problems in Abstract Algebra. Proc. Conf. Oxford 1967, Pergamon(1967), 329-358.

9. M.C. Culler, C.M. Gordon, J. Leucke, P.B. Shalen, 'Dehn surgery on knots', Ann. of Math. **125**(1987), 237-300.

10. H. Seifert, 'Topologie dreidimensionaler gefaserter raume', Acta. Math. **60**(1933), 147-238.

11. C. Ernst, D.W. Sumners, 'A calculus for rational tangles with applications to DNA', (in preparation).

12. H. Schubert, 'Knoten mit zwei brucken, Math. Z. **65**(1956), 133-170.

13. C. Ernst, D.W. Sumners, 'The growth of the number of prime knots', Math. Proc. Camb. Phil. Soc.(1987), (in press).

14. F. Bonahon, L.C. Siebenmann, New Geometric Splittings of Classical Knots, LMS Monograph, (to appear).

15. W.B.R. Lickorish, 'Prime Knots and Tangles', Trans. A.M.S. **267**(1981), 321-332.

232

16. S.A. Wasserman, N.R. Cozzarelli, 'Determination of the stereostructure of of the product of Tn3 resolvase by a general method', Proc. N.A.S.(USA) 82(1985), 1079-1083.

17. S.A. Wasserman, J.M. Dungan, N.R. Cozzarelli, 'Discovery of a predicted DNA knot substantiates a model for site-specific recombination', Science 229(1985), 171-174.

18. W. Heil,'Elementary surgery on Seifert fiber spaces, Yokohama Math. J. 22(1974), 135-139.

19. D.W. Sumners, C. Ernst, N.R. Cozzarelli, S.J. Spengler, 'A topological model for site-specific recombination', (in preparation).

INTRODUCTION TO KNOT AND LINK POLYNOMIALS

P. de la Harpe
Section de Mathématiques
Université de Genève
C.P. 240
1211 Genève 24
Switzerland

ABSTRACT. The mathematical study of knots and links began around 1870.
Now it is a chapter of topology having connections with several other
domains including singularities of functions, dynamical systems, and
the study of various enzymes acting on DNA molecules. Basic definitions
are given, which are in particular intended to make existing knot tables
intelligible, and various examples are described.

In 1928, J.W. Alexander first introduced a polynomial invariant for
knots which now bears his name. In 1985, V.F.R. Jones defined another
one variable polynomial. This has inspired a two variable link polynomial and, later, several other polynomial invariants. The notes offer
an introduction to some of these recent ideas.

1. KNOTS AND LINKS FROM A TOPOLOGICAL POINT OF VIEW

Knots have long been used by sailors or drawn for decorative purposes.
For example, Figure 1 is the picture of a roman bas relief from about
the third century A.D., copied from the cover of the issue of the journal *Science* where the report [WDC] appeared.

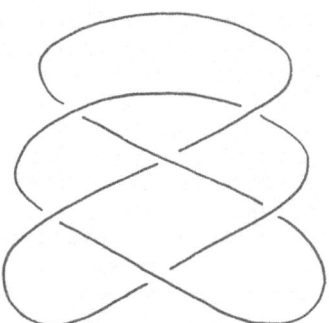

Figure 1. A roman diagram of the knot 6_2.

233

A. Amann et al. (eds.),
Fractals, Quasicrystals, Chaos, Knots and Algebraic Quantum Mechanics, 233–263.
© *1988 by Kluwer Academic Publishers.*

C.F. Gauss and J.B. Listing had some thoughts on knots and links in the first half of the last century, and so did J.C. Maxwell in his study of electricity and magnetism. But the first mathematical paper on knots is one by the physicist P.G. Tait [Ta] in 1877.

To put a long story short, Tait came to knots as follows (we quote from [Kn]). He was greatly impressed with Helmholtz's famous paper on vortex motion (1858). Early in 1867 he devised a simple but effective method of producing smoke rings. At this time, Sir William Thomson (later Lord Kelvin) was paying frequent visits to Tait in Edinburgh. The displays of smoke rings that Thomson witnessed in Tait's lecture room had an influence on Thomson's theory of vortex atoms [Tho]. In turn Thomson's theory was Tait's motivation to understand the structure of knots, as shown by the following quotation from the introduction of [Ta].

"I was led to the consideration of the forms of knots by Sir W. Thomson's Theory of Vortex Atoms, and consequently the point of view which, at least at first, I adopted was that of classifying knots by the number of their crossings (...). The enormous numbers of lines in the spectra of certain elementary substances show that, if Thomson's suggestion be correct, the form of the corresponding atoms cannot be regarded as very simple. For though there is, of course, an infinite number of possible modes of vibration for every vortex, the number of modes whose period is within a few octaves of the fundamental mode is small unless the form of the atom be very complex. (...) Are there after all very many different forms of knots with any given small number of crossings? This is the main question treated in the following paper, and it seems, so far as I can ascertain, to be an entirely novel one."

Today — or rather between 1900 and 1985 — knot theory is essentially a chapter of topology. There are many introductions to or reviews of knots, among which we wish to quote [Go], [MW2] and [Th1].

Let us review some of the basic definitions. A knot is a simple closed curve in the usual space $\mathbb{R}^3$ or in the 3-sphere $S^3 = \mathbb{R}^3 \cup \{\infty\}$. We understand here that curves are smooth, namely that knots are tame, not wild. (It would be equivalent for what follows to ask that curves are polygonal.) A knot is often viewed as being oriented.

Two knots K and K' are ambient isotopic if there exists an orientation preserving homeomorphism h of $\mathbb{R}^3$ (or of S^3) such that h(K) = K'. (By a theorem of Moise [Moi], the same notion is obtained if one requires more regularity for h. For example, one may ask h to be a diffeomorphism.) If K and K' are oriented, it is moreover required that h is compatible with these orientations.

For example, the basic theorems of Jordan and Schönflies (or easier particular cases of them for smooth curves, see Theorems 9.2.1 and 9.4.8 in [BG]) imply that two plane knots are ambient isotopic. A knot is trivial if it is equivalent to a plane knot. The (equivalence class of a) plane knot is also called the unknot.

A plane curve is generic if its only singular points are double points with transverse tangents. (See [DT] for a discussion of these curves.) A regular projection of a knot K is an orthogonal projection p of $\mathbb{R}^3$ onto some plane E such that the knot projection p(K) is generic. Suppose E is oriented, so that it makes sense to define "over" and "under" with respect to E (because $\mathbb{R}^3$ is oriented). The corresponding knot

diagram is the knot projection together with an indication of which
branch is over and which branch is under at each crossing. Figure 2 represents one knot projection and two of the eight associated knot diagrams (see which one represents the unknot).

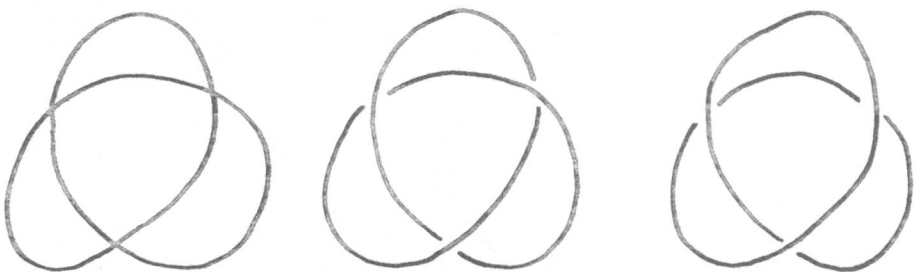

Figure 2. A knot projection and two knot diagrams.

Given a knot K in $\mathbb{R}^3$, it is a fact that "almost all" projections of
$\mathbb{R}^3$ onto planes are regular with respect to K. Given a knot diagram D in
some oriented plane E, it is obvious that D represents some knot K, well
defined up to ambient isotopy - for example K can be viewed as being
"almost" in E, but for points near the crossings of D.

In general, it is quite hard to recognize whether two diagrams define ambient isotopic knots. For example H. Tietze [Ti] has constructed
the two diagrams in Figure 4 and Figure 5. One represents a trefoil knot
(equivalent to the second picture of Figure 2) and the other the unknot,
but it could take some time to the reader to discover which is which. An
easier exercise is to check that the two diagrams of Figure 3 represent
the same knot.

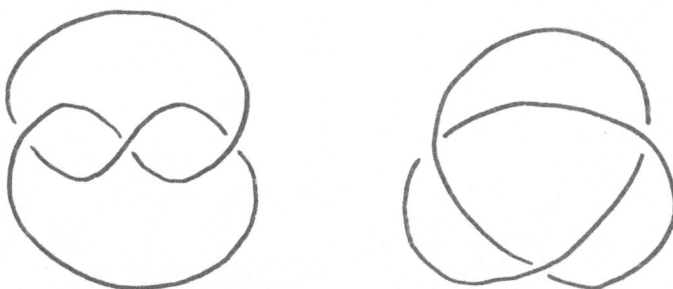

Figure 3. Two diagrams for the same knot.

The <u>crossing number</u> of a knot K is the minimum number of double
points of a projection of a knot ambient isotopic to K. The diagram of
Figure 6, copied from the cover of [Ro], is one of the eight glorious
emblems of Tibetan Buddhism. This diagram has 9 crossings, but it is obvious that the crossing number of the corresponding knot is not more
than 7. In fact, it is precisely 7.

Figure 4.   One of the Tietze's diagram.

Figure 5.   The other Tietze's diagram.

Figure 6. A Tibetan diagram of the knot 7_4.

Let K be a knot with crossing number n. If $n \leq 2$, then $n = 0$ and K
is the unknot. If $n = 3$ then K is one of the two trefoil knots of Figure
7. If $n = 4$ then K is the so-called bretzel knot, or figure eight knot,
of Figure 8. If $n = 5$ then K is either one of the two knots of Figure 9
or one of their mirror images (with diagrams obtained by changing all
crossings).

The number $a(n)$ of ambient isotopy classes of nonoriented knots
with crossing number n seems to be given for small n by the following
table.

n =	0	1	2	3	4	5	6	7	8	9
a(n) =	1	0	0	2	1	4	8	16	51	116

Little is known about the asymptotic behaviour of $a(n)$; see [ES]. However, usual tables of knots up to 9 crossings do not show these 199
knots, but have 84 entries only. See [Rol] and [BZ], as well as [Th1]
for an excellent guide to older tables. The reading of tables requires
a few more notions.

We define first the connected sum $K_1 \# K_2$ of two oriented knots K_1
and K_2 which is well defined on oriented ambient isotopy classes.
Choose a plane E dividing $\mathbb{R}^3$ in two closed halfspaces $\mathbb{R}^3_+$ and $\mathbb{R}^3_-$. Up to
ambient isotopy, one may assume that K_1 is in $\mathbb{R}^3_+$, with one small straight
arc l_1 in E and with $K_1 - l_1$ in $\mathbb{R}^3_+ - E$. Similarly one may assume that K_2 is
in $\mathbb{R}^3_-$ with one small straight arc l_2 in E and with $K_2 - l_2$ in $\mathbb{R}^3_-$. One may
moreover assume that l_1 and l_2 are the same arc with its two opposite
orientations. Then $(K_1 - l_1) \cup (K_2 - l_2)$ is again an oriented knot, which is
$K_1 \# K_2$. It is true and important that $K_1 \# K_2$ does not depend on any of the
choices described above. The operation # is commutative, associative,
and has a neutral element which is the unknot. Figure 10 shows a diagram
of a connected sum of two knots with 3 crossings (one of the two so-
-called granny knots).

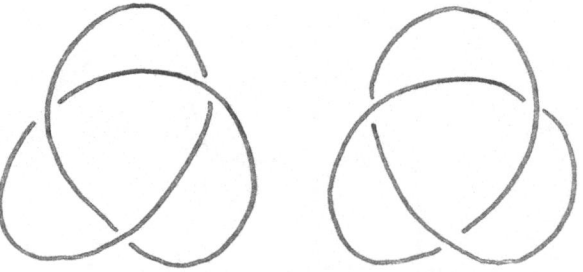

Figure 7.   The trefoil knots.

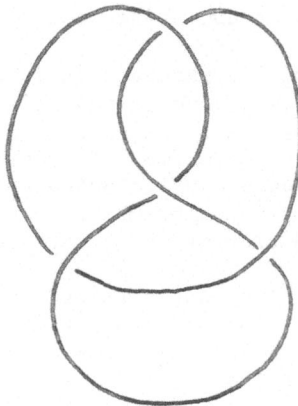

Figure 8.   The bretzel knot.

Figure 9.   The knots 5_1 and 5_2.

240

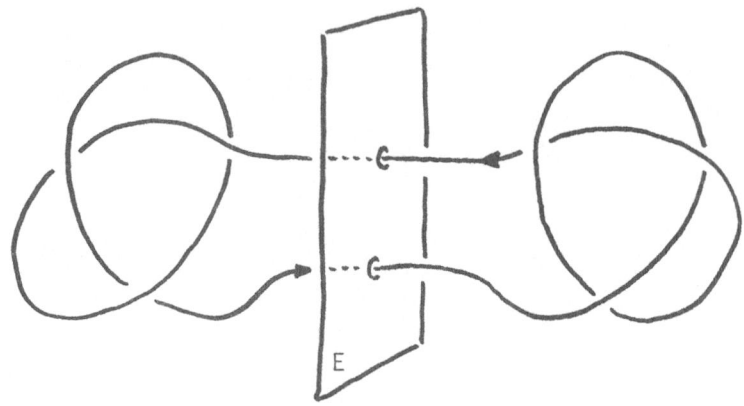

Figure 10. A composite knot.

A non trivial knot is prime if, whenever it is oriented and if it is written as a connected sum $K_1 \# K_2$, one of the K_j's is the unknot. It is composite otherwise. By convention, the unknot is not prime. (Recall that, similarly, the number 1 is not prime.) A theorem of Schubert (1949) shows unique factorization of knots : any oriented knot can be written as a connected sum $K_1 \# \ldots \# K_n$ of oriented prime knots, with the K_j's well defined up to a permutation. In view of this, most tables list prime knots only (unfortunately!).

Let K be an oriented knot. The knot obtained by inverting the orientation is the reverse knot −K. The knot obtained by reflection in a plane is the mirrored knot (or obverse) K*. The knot $(-K)^* = -(K^*)$ is the inverse knot. A knot K is said to be

reversible if   K = −K
amphicheiral if   K = K*
involutive if   K = −K*.

One says also

chiral for "not amphicheiral" (K ≠ K*)   .

Chirality seems to be a crucial property of knots from (among others) a chemist's point of view [Wa].

Caution : When discussing nonoriented knots, some writers use "amphicheiral" to mean "amphicheiral or involutive" as defined above. For these authors, the knot 8_{17} is amphicheiral, but it is not below, where we follow Conway [Co].

Usual tables list only one of the four knots K, −K, K*, −K*, whether they are ambient isotopic or not. The five logical possibilities are illustrated as follows.

(i)   The trefoil knot 3_1 is reversible (easy) and chiral (Dehn,

1914) - hence it is not involutive.

(ii)   The bretzel knot 4_1 is reversible and amphicheiral - hence also involutive.

(iii) The knot 8_{17} is chiral and involutive - hence it is not reversible.

(iv)   The knot 9_{32} is chiral and not involutive and not reversible.

(v)   If K denotes one of the 4 oriented 9_{32}'s, the connected sum K#K* is amphicheiral and not involutive - hence it is not reversible.

Following Thistlethwaite, we say that two unoriented knots K, K' are <u>equivalent</u> if K' is ambient isotopic to one of K, K*. The number b(n) of equivalence classes of unoriented prime knots with crossing number n is given by the following table ([Th1]), whose author expresses in [Th2] a wish for independent verification, which seems to be missing for n = 12 and n = 13).

n =	3	4	5	6	7	8	9	10	11	12	13
b(n) =	1	1	2	3	7	21	49	165	552	2176	9988

<u>Links</u> are a natural generalization of knots. A link with r components (r = 1,2,3,...) is a subset of $\mathbb{R}^3$ (or S^3) consisting of r disjoint simple closed curves, called the <u>components</u> of the link. In particular, a knot is a link with one component. One defines as above ambient isotopy for (oriented) links, link projections, crossing numbers, link diagrams, ... The unique factorization theorem has been extended from knots to links by Hashizame (1958). Figure 11 shows five examples of link diagrams.

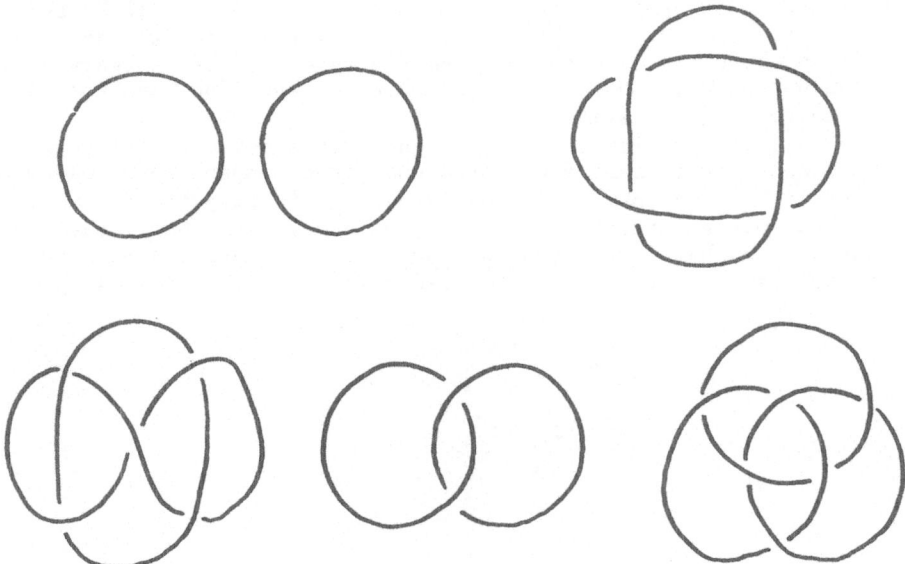

Figure 11.   Five link diagrams with 2 and 3 components.

242

There are many more diagrams in Rolfsen's book [Rol].

Note about the vocabulary in chemistry: A link is called a <u>catenane</u>. A chiral pair (K,K*) with K ≠ K* is a pair of <u>enantiomers</u> (it seems that K ≠ -K* is also required, but this is not always clear).

2.   EXAMPLES

Knots and links appear at many places in topology.

As a first example, there is a result going back to Alexander (1920) which has attracted a lot of interest and which can be quoted (imprecisely) as follows. For any orientable connected compact <u>3-dimensional manifold</u> M without boundary, there exists a link $L \subset S^3$ and a smooth map f of M onto S^3 with the following properties:

(i) $f^{-1}(L)$ is a finite set of disjoint simple closed curves in M.
(ii)   The restriction $M - f^{-1}(L) \to S^3 - L$ of f is a finite covering, say with n sheets.
(iii) Over each component L_j of L, the restriction $f^{-1}(L_j) \to L_j$ of f is a finite covering, with $n_j < n$ sheets.

More precisely, f is a ramified covering [Lin], ramified over L and with ramification set $f^{-1}(L)$. One may moreover require <u>one</u> of the following conditions

(iv)   n ≤ 3 and L is a knot (Hilden and Montesinos (1974 - 1976)).
(iv') L is the figure eight knot [HLM].

There are other deep connections between the theory of links and that of 3-manifolds, such as the theorem of Lickorish and Wallace: any 3-manifold as above may be obtained from the sphere S^3 by "surgery on a link" (take out a small tubular neighbourhood of the link, and put it back with appropriate twisting).

Given an oriented link L in $\mathbb{R}^3$, it has proved quite useful to study <u>Seifert surfaces</u> for L, namely oriented connected compact surfaces smoothly embedded in $\mathbb{R}^3$ with (oriented) boundary L. It is easy to show that each link has such a Seifert surface (indeed several of them), the minimum of their genus being the <u>genus of the link</u>. We shall restrict ourselves on this subject to showing two pictures in Figures 12 and 13.

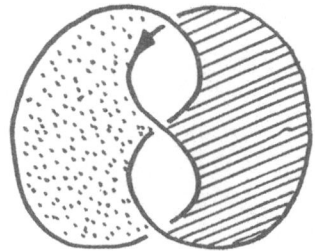

Figure 12.   A Seifert surface of genus 1 for the trefoil knot.

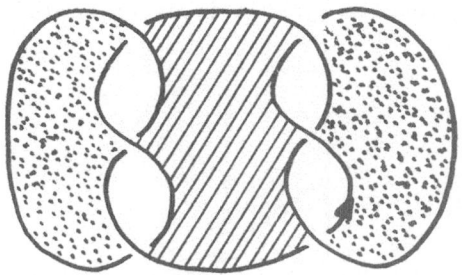

Figure 13. A Seifert surface of genus 2 for the granny knot.

But we would like now to mention sources of examples which relate knot theory to problems outside topology.

2.1. Links and singularities

Last century mathematicians met great successes in studying complex curves (which have in particular two real dimensions), and considered then as one of their main challenges the understanding of complex surfaces (four real dimensions). Now surfaces may be studied via ramified coverings of the complex plane C^2, and this point of view draws attention to singular plane curves (the discriminants). Thus much attention has been given to singular points of analytic functions $f(z_1,z_2)$ describing plane curves. Then knot theory entered the scene via a construction described below, used by K. Brauner (1928), K. Kähler (1929), W. Bureau (1934) and many others later. Historically, this was of decisive importance for the mathematical growth of knot theory.

Let $f(z_1,z_2)$ be an <u>analytic function</u> of two complex variables defined in a neighbourhood of the origin $0 \in C^2$. Assume that $f \neq 0$ and that $f(0) = 0$. For technical reasons, assume also that f is <u>reduced</u>, namely that one cannot find analytic functions g,h with $g(0) = \overline{0 = h}(0)$ such that $f = gh^2$ or $f = g^2$ near the origin. Then $V = f^{-1}(0)$ is a (possibly singular) complex curve in a neighbourhood of the origin of C^2, and V contains the origin.

If the origin is a <u>regular point</u> for f, namely if

$$\left(\frac{\partial f}{\partial z_1}(0) \quad , \quad \frac{\partial f}{\partial z_2}(0) \right) \neq (0 \ , \ 0)$$

then V is smooth, and there are indeed holomorphic local coordinates (w_1,w_2) around the origin such that V is locally defined by the equation $w_2 = 0$. Otherwise the origin is a <u>singular point</u> for f and for V, and the first question is that of understanding the local topology.

For $r > 0$, let L_r be the intersection of V with the sphere S_r of equation $|z_1|^2 + |z_2|^2 = r^2$. For r small enough, L_r is a link whose ambient isotopy class does not depend on r, and which is simply denoted by L. For example the link associated to $f(z_1,z_2) = z_1^2 + z_2^3$ is the trefoil knot of Figure 3, and that associated to $f(z_1,z_2) = z_1^3 + z_2^6$ is the link of Figure 14.

The relationship between f and L is quite interesting, as partly

244

shown by the following results (listed in order of increasing difficulty).

Figure 14.   The link associated to the singular
surface of equation $z_1{}^3 + z_2{}^6 = 0$.

(i)   The link L is a knot if and only if f is irreducible, namely
        if and only if one cannot find analytic functions g,h with
        g(0) = h(0) = 0 and f = gh near the origin.
(ii)   Let f_1, f_2 be two functions as above. Let V_1, V_2 be the corre-
        sponding surfaces and let L_1, L_2 be the corresponding links.
        Then L_1 and L_2 are equivalent if and only if there exists a
        local homeomorphism h (not a diffeomorphism in general!) such
        that $h(V_1) = V_2$.
(iii)   The links one may obtain by the construction above are known:
        they are some of the so-called iterated torus links [MW2]
(iv)   The link L is the trivial knot if and only if the origin is
        a regular point for f.

More about this in [Mi], [BK], [MW2]. See also [BW].

2.2. Trajectories of vector fields

Let (p,q) be a pair of coprime integers. Consider the autonomous system
of ordinary differential equations (a vector field) on $\mathbb{R}^4$

$$\dot{x}_1 = px_2 \qquad \dot{x}_3 = qx_4$$
$$\dot{x}_2 = -px_1 \qquad \dot{x}_4 = -qx_3 \ .$$

The corresponding flow is found by a straightforward integration:

$$\begin{array}{ll}
x_1(t) = x_1\cos(pt) + x_2\sin(pt) & x_1(0) = x_1 \\
x_2(t) = -x_1\sin(pt) + x_2\cos(pt) & x_2(0) = x_2 \\
x_3(t) = x_3\cos(qt) + x_4\sin(qt) & x_3(0) = x_3 \\
x_4(t) = -x_3\sin(qt) + x_4\cos(qt) & x_4(0) = x_4
\end{array}$$

We are interested here in the vector field induced by the one a-
bove on the sphere S^3 of equation $x_1^2 + x_2^2 + x_3^2 + x_4^2 = 1$. In S^3, each
orbit is periodic and defines a knot. The orbit with $x_1 = x_2 = 0$ and the
orbit with $x_3 = x_4 = 0$ are both unknots, and constitute together a link
equivalent to the fourth link of Figure 11. Any other orbit is a so-
called <u>torus knot</u> of type (p,q), and is in particular a trefoil knot
if (p,q) is $(2,3)$ or $(3,2)$.

A study of knots and links appearing as orbits of vector fields
has been made by R.F. Williams and co-workers [BW1], [BW2], [FW]. Given
a flow defined by a vector field in $\mathbb{R}^3$ or S^3, the following questions
are natural :
Is there any closed orbit?
If yes, is there any knotted closed orbit?
If yes, are there infinitely many knot types appearing?
In 1972, P. Schweitzer created a surprise when he showed that the first
answer, for S^3, may be no [Ros]. But the three questions have positive
answers for <u>flows with positive entropy</u> [FW]. There are also very precise theorems and (observed but not yet proved (!)) facts for the quite
specific example of Lorenz's equation [BW1].

This direction of research may help to understand <u>chaotic dynamical systems</u>. But a lot of work has yet to be done there.

2.3. Knots in physics?

The movement of a fluid in $\mathbb{R}^3$ is described by a time dependent vector
field, the fluid velocity v. Since Helmholtz (1858), one is also interested in the field curl v, the integral lines of which are called <u>vortex
lines</u> (see e.g. [So], Chapter IV). A smoke ring is a popular materiali-
zation of a closed vortex line. I am not aware of any work trying to
decide whether knotted vortex lines may exist. Has anybody studied the
equations or taken a picture of a <u>knotted smoke ring</u>?

One of the crucial ingredients in the understanding of condensed
matter is that of <u>lines of defect</u>. These lines contribute to make metals
hard, because in a perfect crystal the crystalline planes may softly
slide on each other. Does one know if and how these lines are closed?
knotted? linked? As the theory of defects has already a rich interaction
with algebraic topology [Mer], it would not be surprising if knot theory
was also of interest for the physics of ordered media.

There has been some speculation about molecules in polymers being
knotted. We do not know much on this, except that it has motivated the
following very natural question : how often is a <u>random closed curve</u>
knotted ? Part of the answer would be to make the question precise.
Computer simulations suggest that long curves are almost surely knotted
[MW], [Su1].

Let us conclude this section by qoting a nice result of Conway
and Gordon [CG], even if it has little to do with physics. Choose at
random seven distinct points in $\mathbb{R}^3$, and join each pair of these points
by an arbitrary simple smooth curve, these 21 curves being disjoint
(outside the seven nodes). The resulting figure is an embedding in $\mathbb{R}^3$
of the complete graph K_7. In this graph there are $\frac{1}{2} 6!$ Hamiltonian cy-
cles, namely 360 simple closed curves going through each of the 7 nodes.

Theorem: at least one of these 360 curves is knotted.

2.4. Knots and DNA molecules

Linked molecules seem to have been discussed as early as 1912, and syn-
thesized in the 1960's [Si], [Wa]. The synthesis of molecules which are
actually knotted is discussed by Walba (1985) as an interesting challenge, not yet as an experimental fact.

But we would like to report here on recent investigations on DNA
molecules. See [DSKC], [WDC], [Su4], as well as the advertisement [Ko],
and the reviews [Su2], [Su3], [WC].

Living cells contain chromosomes, and each of these contains a
long double stranded DNA molecule. The simplest cells are bacteria, such
as Escherichia Coli, and are known as prokaryotes. Their DNA molecules
are usually circular (in both chromosomes and plasmids), and may then
be described by knots. (In eukaryote cells, nuclear DNA is often linear,
not closed. But these cells have also mitochondria and chloroplasts,
where genetic material is shaped into circles which may be linked in a
very spectacular way.) The knot model is of course an approximation, as
the Crick-Watson DNA double helix consists of two strands, and constitutes a ribbon rather than a curve - but still, this model fits with
pictures and has proven useful.

During replication and recombination, various enzymes act on the
DNA molecule. Some, during this action, change the topological configuration of the molecule, and are called topoisomerases. Their topological
effect can be observed as electron micrographs have been taken (at least
in vitro), and these show actual knots and links. We have stated above
that there are

$$\sum_{n=3}^{6} a(n) = 15$$

ambient isotopy classes of nontrivial unoriented knots with at most 6
crossings. All have been observed, as well as several knots with $n \geq 7$
crossings.

Moreover, biologists have learned to deduce enzymatic mechanism
from the analysis of which knots are observed, depending on which enzyme
works. (For example, the trefoil knots produced by some enzymes are all
right handed, while other enzymes produce both kinds of trefoil knots.)
More on this in [Su3] and [Su4].

3. POLYNOMIAL INVARIANTS

We shall denote by L the set of ambient isotopy classes of oriented
links in the 3-sphere S^3. A link invariant is a map I defined on L such
that if L_1 and L_2 are two links with $I(L_1) \neq I(L_2)$ then L_1 and L_2 cannot
be ambient isotopic. (Of course, $I(L_1) = I(L_2)$ does not show that L_1 and
L_2 are ambient isotopic !!!)

The most important link invariant is the group of a link : I(L) is
the fundamental group (or first homotopy group) of the link complement
S^3-L. Given a diagram of L, there is a standard procedure (Wirtinger
1905) to write down a presentation of the group of L. The group of a

link L is the same as that of -L, L* or -L*. But otherwise the group is a very strong invariant. For example :

(i)   A knot is trivial if and only if its group is infinite cyclic (Dehn 1910 and Papakyriakopoulos 1957).

(ii) Given a group G, there are at most two unoriented equivalence classes of prime knots with group G [CGLS], [Wh].

(iii) We do not know any example of a pair K, K' of unoriented prime knots which are inequivalent and have isomorphic groups.

However, there are known examples of unequivalent composite knots which have the same group (the square knot and the granny knot).

There are excellent introductions to link groups, and we shall not duplicate them. See among many others [Rh1], [Th1], or the usual textbooks [Rol], [BZ].

Other link invariants take their values in integers, such as the crossing number and the genus discussed above. But we shall center what follows on <u>polynomial invariants</u>, for which I(L) is a Laurent polynomial in one or two variables with integral coefficients.

3.1. The Alexander-Conway polynomial

Historically, the first polynomial invariants are the Alexander polynomials, defined for knots by Alexander in 1928. We shall give below a definition of the first of these polynomials which can be taught to a computer, and we shall refer to the literature for the usual Alexander's definition (and mention a variant due to de Rham [Rh2]). To be more precise, Alexander's definition provides a Laurent polynomial in $\mathbb{Z}[t,t^{-1}]$ which is only defined up to a multiplication by some $\pm t^n$. Tables such as those of [BZ] show a polynomial

$$\sum_{n=0}^{N} a_n t^n$$

normalized by $a_N > 0$ and $a_0 \neq 0$. It is much later (1969) that J.H. Conway formulated <u>his</u> coherent choice for the normalizations, so that the Alexander-Conway polynomial is really a polynomial invariant [Co].

The (first) Alexander polynomial is quite powerful to distinguish unoriented equivalence classes of knots :

(i)   A prime knot with Alexander polynomial 1 has at least 11 crossings (see [Co], and the diagram on page 173 of [Rol]).

(ii) Among the 84 unoriented equivalence classes of prime knots with at most 9 crossings, there are only five pairs which are not distinguished by this invariant : $(6_1,9_{46})$, $(7_4,9_2)$, $(8_{14},9_8)$, $(8_{18},9_{24})$ and $(9_{28},9_{29})$. The 12965 classes up to 13 crossings yield 5639 different polynomials.

(iii) Any polynomial $f(t) \in \mathbb{Z}[t,t^{-1}]$ with $f(1) = 1$ and $f(t) = f(t^{-1})$ is the Alexander polynomial of some knot.

However, there are known examples of infinite families of prime knots defining a single Alexander polynomial. See for example [QW].

Now we give Conway's definition of the Alexander polynomial. Say that three oriented links L_+, L_-, L_0 are <u>skein related</u> if they can be represented by three diagrams which are <u>identical outside</u> of a small neighbourhood of one crossing point where they look as in Figure 15 (see also Figure 16).

Figure 15.   Skein related links.

Now the Alexander–Conway polynomial of an oriented link L is the Laurent polynomial $\Delta_L(t)$ in a formal variable $t^{1/2}$ with coefficients in Z defined by

$$\Delta_0(t) = 1$$

if 0 is the unknot and by

$$\Delta_{L_+}(t) - \Delta_{L_-}(t) + (t^{\frac{1}{2}} - t^{-\frac{1}{2}})\Delta_{L_0}(t) = 0$$

whenever L_+, L_-, L_0 are skein related. Of course, one has to prove that this definition makes sense, and this is not quite a trivial step, but we refer to [Ka1] or [LM] for this. Let us illustrate the definition by a computation of Δ_T where T denotes the right–handed trefoil knot. Con-

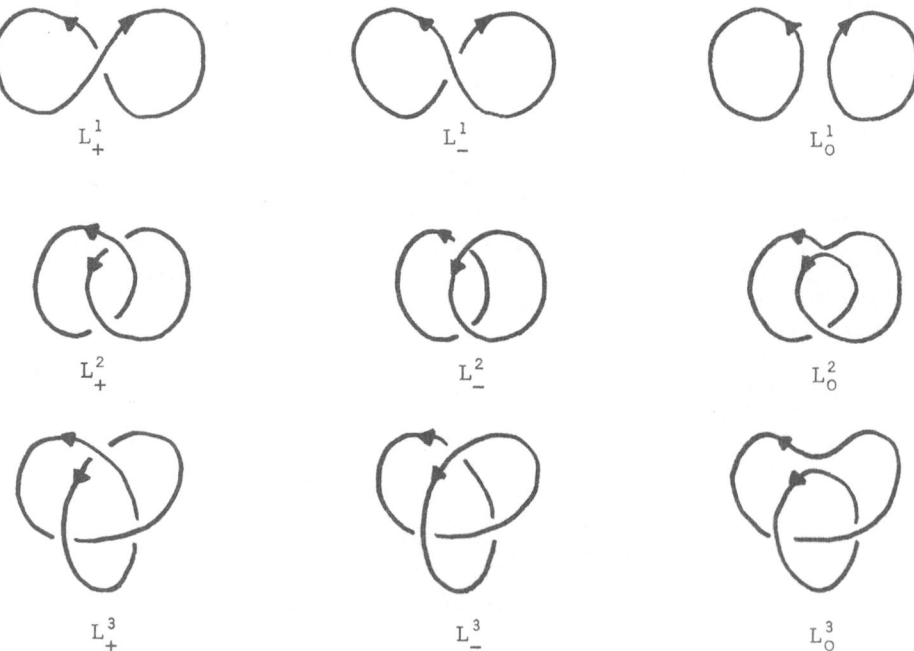

Figure 16.   Three skein related triples.

sider three triples of skein related links as in Figure 16.

Denote by 0 [respectively 00, L, T] the unknot [resp. the trivial link with two components, the linked pair of unknots, the trefoild knot] appearing in Figure 16. From the first triple one has

$$\Delta_0(t) - \Delta_0(t) + (t^{\frac{1}{2}} - t^{-\frac{1}{2}})\Delta_{00}(t) = 0$$

and $\Delta_{00}(t) = 0$. From the second triple

$$\Delta_L(t) - \Delta_{00}(t) + (t^{\frac{1}{2}} - t^{-\frac{1}{2}})\Delta_0(t) = 0$$

and $\Delta_L(t) = -t^{\frac{1}{2}} + t^{-\frac{1}{2}}$ because $\Delta_0(t) = 1$. And finally

$$\Delta_T(t) = 1 + (t^{\frac{1}{2}} - t^{-\frac{1}{2}})^2 = t - 1 + t^{-1}$$

in agreement with tables, which show [BZ]

$$\Delta_T^{BZ}(t) = t\Delta_T(t) = t^2 - t + 1.$$

Let us mention three more properties of the invariant $L \to \Delta_L(t)$.

(i)   If L has an odd number of components, and in particular if L is a knot (one component), then $\Delta_L(t)$ is a Laurent polynomial in t (an even polynomial in $t^{1/2}$!). If L has an even number of components, then $t^{1/2}\Delta_L(t)$ is a Laurent polynomial in t.

(ii)   The invariant is multiplicative under connected sums:
$$\Delta_{K\#L}(t) = \Delta_K(t)\Delta_L(t).$$

(iii)   The invariant does not distinguish reverse links :
$\Delta_L(t) = \Delta_{-L}(t)$. For an actual knot K, the invariant does not distinguish it from its mirror image: $\Delta_{K*}(t) = \Delta_K(t)$.
(For a link L and for L*, see Proposition 5.3 in [Ka2]).

3.2. The Jones polynomial

In the early 1980's, V. Jones was working on finding invariants for sub-factors, a problem in the theory of von Neumann algebras [Jo2]. This research had a spectacular offspring [Jo1] in producing a completely new Laurent polynomial invariant $L \to V_L(t) \in \mathbb{Z}[t^{1/2}, t^{-1/2}]$ for oriented links.

One may now define this invariant in several (equivalent) ways, independently of the theory of operator algebras. Here is one, in the same spirit as that of $\Delta_L(t)$ above:
$$V_0(t) = 1$$
if 0 is the unknot and
$$t^{-1}V_{L_+}(t) - tV_{L_-}(t) - (t^{\frac{1}{2}} - t^{-\frac{1}{2}})V_{L_0}(t) = 0$$

if L_+, L_- and L_0 are skein related. (The sign convention is that of [Jo3], not that of [Jo1] !) The reader should find it easy to compute this invariant for the right-handed trefoil knot T :
$$V_T(t) = t + t^3 - t^4.$$

Properties (i) and (ii) of $t \to \Delta_L(t)$ mentioned at the end of Section 3.1 hold again for $t \to V_L(t)$, but property (iii) has to be changed in a very important way:

$$V_L*(t) = V_L(t^{-1}) \quad .$$

In particular, if L is a link such that $V_L(t^{-1}) \neq V_L(t)$, then L* and L are <u>not</u> ambient isotopic, so that L is chiral. For example, the computation above shows that <u>the trefoil knot is chiral</u>. (But this invariant is not complete: the table of [Jo3] shows in particular chiral knots L with $V_L(t^{-1}) = V_L(t)$.)

For another very nice computational definition of $V_L(t)$ due to L. Kauffmann, see [Ka3] or the Section 8 of [HKW].

To illustrate the use of the Jones' polynomial, let us describe one of the so-called <u>Tait conjectures</u>. Say that a link is <u>alternating</u> if it has a diagram showing, along each of its components, crossings which are alternatively over and under. Of course, an alternating link, which has by definition one alternating diagram, may have another diagram which is not alternating, as Figure 8 and Figure 17 show.

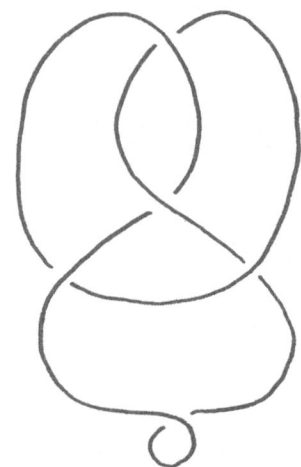

Figure 17. A non alternating diagram of the alternating figure eight knot.

It is known that any knot with crossing number at most 7 is alternating (see Figures 1, 3 and 6 to 10), but it can be shown that the knot 8_{19} is not alternating (see § 10 of [HKW]). Thistlethwaite has computed that, among the 12965 unoriented prime knots of up to 13 crossings, precisely 6236 are nonalternating [Th2].

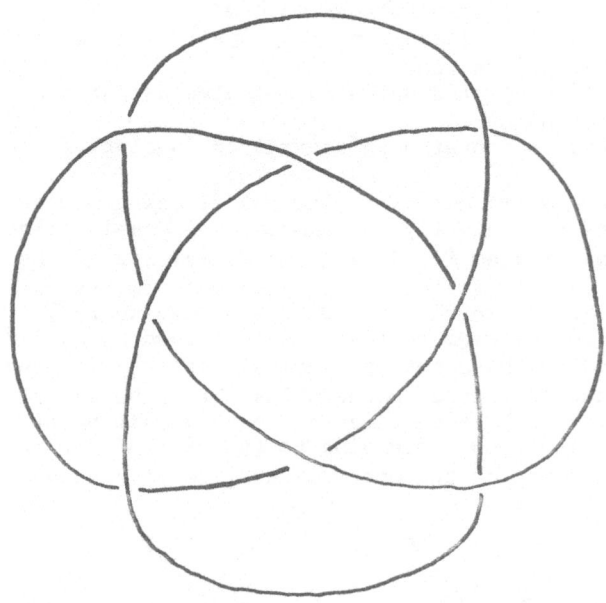

Figure 18. The knot 8_{19} is not alternating.

Say that a link diagram is <u>reduced</u> if "there is no clearly redun-
dant crossing ", or more precisely if there is no simple closed curve in
the plane of the diagram which intersects the diagram in exactly two
points which lie near a crossing point. Diagrams which are not reduced,
as those of Figures 17 and 19, are easy to recognize and easy to simplify, namely to replace by diagrams having less crossings.

The following fact was used by Tait a century ago and has been known
later as one of "Tait conjectures". Since 1986, it is a theorem of Kauffmann and Murasugi. (See [HKW] and [Tu1].) We state only a particular case,
and refer to [Tu1] for more general results.

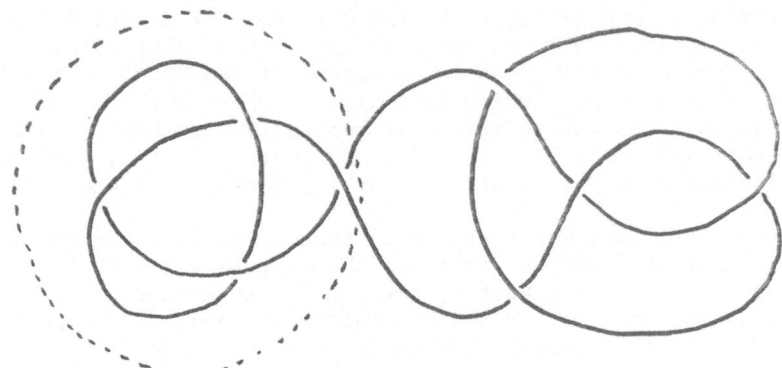

Figure 19. A diagram which is not reduced.

<u>Theorem.</u> Let K be an alternating prime knot.
  (i)   Two reduced alternating diagrams of K have the same number of
         double points, say n.
  (ii)   Any diagram of K which is not alternating has at least n + 1
         double points.
  (iii) If K is moreover amphicheiral or involutive, then n is even.

It follows from (i) and (ii) that the crossing number of an alternating prime knot K is exactly the number n of double points of any reduced alternating diagram for K (see the definition of the crossing number after Figure 5). It follows from (iii) that the trefoil knot or the
knots of Figure 9, for which n = 3 or n = 5, are chiral.
    We shall also describe another result proved by Kauffmann and Mura-
sugi with the aid of the Jones' polynomial, because it provides a nice
case study for the history of mathematics. Let D be an oriented link
diagram. The plane of D being oriented, each double point is locally of
one of two possible types, according to the sign of the rotation which
brings the over branch to the under branch, as in Figure 20.

Figure 20. The two types of crossings.

The <u>Tait number</u> of D is the sum T(D) of these signs over all double
points of D. (It is mistakenly called the writhe number in many papers,
including [HKW], but the writhe number is something else [Po].) The Tait
number does <u>not</u> change if the orientation is reversed on each component
of D. Consequently T(D) is well defined if D is a diagram for an <u>unori-
ented knot.</u>
    During the twenty last years of the XIXth century, C.N. Little has
been a frantic table maker for knots (references in [Th2]). In [Lit], he
states (and proves!) the following as a theorem: Let K be a knot with
crossing number n, and let D_1,D_2 be two diagrams for K with n double
points; then $T(D_1) = T(D_2)$. But this is <u>not</u> true as discovered by K.Per-
ko in 1974. More precisely, the two diagrams listed as 10_{161} and 10_{162}
in Rolfsen's tables [Rol], which have distinct Tait numbers and which
were consequently thought to represent non equivalent knots, do indeed
represent the same knot. This explains why Table I in [BZ] stops with
10_{165}, while Appendix C in [Rol] and earlier tables list knots up to
10_{166}.
    However Little was not that wrong, because it is now a theorem that
he was right for alternating knots, so that the Tait number T(K) is well
defined for an alternating knot K, but <u>not</u> for an arbitrary knot. If K
is an alternating knot, it is obvious that T(K*) = -T(K). In particular,
if T(K) ≠ 0, then K is chiral. The converse does not hold, because the
knot 8_4, which is alternating and has Tait number zero, is chiral. This
should bring some light to the interesting discussion of D.M. Walba

(part V in [Wa]).

(First proof of the claim about 8_4: If Δ is the Alexander polynomial of 8_4, then the "determinant" $\Delta(-1) = 19$ is a prime number of the form $4k+3$, and this implies chirality by the result of § 9 in [Re]. Second proof: If V is the Jones polynomial of 8_4, then

$$V(t) = t^{-3} - t^{-2} + 2t^{-1} - 3 + 3t - 3t^2 + 3t^3 - 2t^4 + t^5 \neq V(t^{-1})$$

so that 8_4 is chiral.)

3.3. Other polynomial invariants

The work of V. Jones got a new subject started. Very soon after [Jo1], at least 5 groups of authors have independently proposed a two variable polynomial invariant for oriented links [FYHLMO], [PT], [LM], [Hos]. One may define this invariant

$$L \longrightarrow P_L(1,m) \in \mathbb{Z}[1,1^{-1},m,m^{-1}]$$

as before by $P_O(1,m) = 1$ and by

$$(*) \quad 1P_{L_+}(1,m) + 1^{-1}P_{L_-}(1,m) + mP_{L_O}(1,m) = 0$$

whenever L_+, L_-, L_O are skein related. This invariant is sometimes called the Fyhlmo polynomial, or the Jones-Conway polynomial; it contains the invariants of Sections 3.1 and 3.2

$$\Delta_L(t) = P_L(i,i(t^{\frac{1}{2}} - t^{-\frac{1}{2}}))$$

$$V_L(t) = P_L(it^{-1},i(t^{-\frac{1}{2}} - t^{\frac{1}{2}}))$$

but is strictly more powerful than these two specializations [LM]. (Caution: one should not believe that any odd relation in the style of (*) will work ! Only very special relations produce link invariants.)

Slightly later, another one variable polynomial

$$L \longrightarrow Q_L(x) \in \mathbb{Z}[x,x^{-1}]$$

for underlined links has been discovered [BLM], [Ho]. This has been generalized by Kauffman as a two variable polynomial

$$L \longrightarrow F_L(a,x) \in \mathbb{Z}[a,a^{-1},x,x^{-1}] \ .$$

See [Ka4] and [Lic]. There are further interesting polynomial invariants studied by Murakami [Mu] and others.

All these invariants – V_L, P_L, Q_L, F_L – are puzzling, in part because they lack good interpretations in terms of algebraic topology. But some coherence could well come back from the theory of quantum groups, itself motivated by quantum statistical mechanics and the "quantum inverse scattering problem". See [Dr], [Man], [Tu2], [Ve]. Though we can hardly enter this circle of ideas here, we want to end this report by discussing one part of the necessary background, namely the connection between braids and links.

4. BRAIDS

4.1. Braid groups

Consider an integer $n \geq 1$ and n points $P_1 , \ldots , P_n$ in the horizontal x-y-plane $\mathbb{R}^2$ of $\mathbb{R}^3$. A braid with n strings is a family of n nonintersecting descending oriented smooth curves with origins $(P_1, z_i) , \ldots , (P_n, z_i)$ in some horizontal plane and with ends $(P_1, z_f) , \ldots , (P_n, z_f)$ in some other parallel plane, with $z_f < z_i$. Two braids are ambient isotopic if one is the image of the other by an homeomorphism of $\mathbb{R}^3$ which is the identity if z is either very large or very small, namely if $z \gg z_i$ or if $z \ll z_f$. In mathematics, braids seem to go back to Gauss.

In 1925, E. Artin realized that (ambient isotopy classes of) braids with n strings can be composed in the evident way, as Figure 21 shows, and constitute the n-string braid group B_n. Of course B_1 has only one element and B_2 is isomorphic to the group $\mathbb{Z}$ of all integers, but the B_n's for $n \geq 3$ are very interesting groups. The inverse of a braid is given by a mirror image and the braid with straight strings is the neutral braid: see Figure 22.

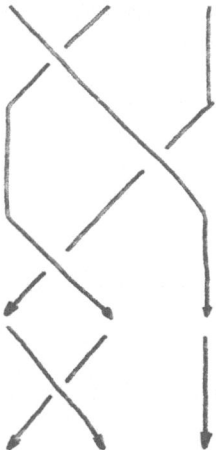

Figure 21. A composition of two braids with three strings.

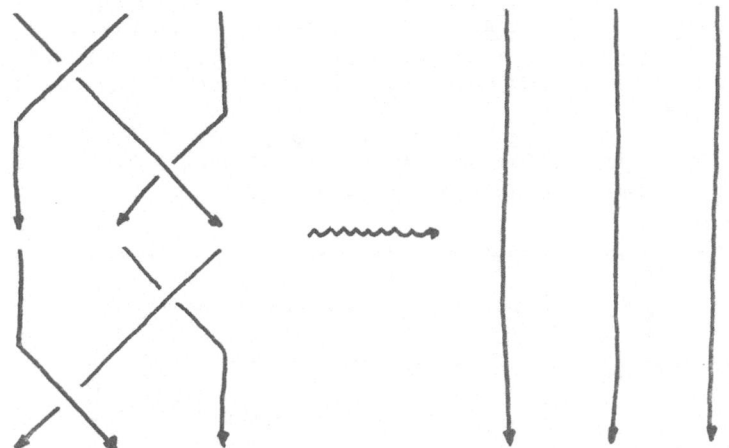

Figure 22. A diagram to illustrate $\beta\beta^{-1} = 1$ in B_3.

For $j \in \{1, \ldots, n-1\}$, we denote by σ_j the braid of Figure 23. It is easy to check that the braids $\sigma_1, \ldots, \sigma_{n-1}$ generate B_n and that they satisfy

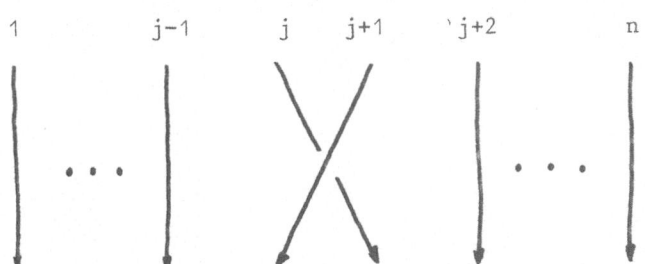

Figure 23. The jth generator of B_n.

$$\sigma_j\sigma_k\sigma_j = \sigma_k\sigma_j\sigma_k \quad \text{if } |j-k| = 1 ,$$

$$\sigma_j\sigma_k = \sigma_k\sigma_j \quad \text{if } |j-k| \geq 2 .$$

It is a result of E. Artin that the relations above constitute a presentation of the group B_n.

Any braid $\beta \in B_n$ can be <u>closed</u> and produces this way a well defined (ambient isotopy class of) oriented link $\hat{\beta}$, as in Figure 24. An early (and now easy) <u>theorem of J.W. Alexander</u> (1923) shows that any oriented link is a closed braid. (In fact, E. Artin [Ar1] quotes much earlier proofs.) But the Figure 25 shows that several braids may close up to the same knot.

A <u>Markov move of type one</u> is an operation on the disjoint union of the braid groups which replaces a braid β in B_n for some n by a conjugate $\gamma\beta\gamma^{-1}$, with $\gamma \in B_n$. A <u>Markov move of type two</u> replaces $\beta \in B_n$ by $\beta\sigma_n \in B_{n+1}$

256

or by $\beta\sigma_n^{-1} \in B_{n+1}$, or replaces one of $\beta\sigma_n$, $\beta\sigma_n^{-1}$ by $\beta \in B_n$.

A <u>theorem of A.A. Markov</u> (1936) shows that two braids $\alpha \in B_m$ and $\beta \in B_n$ close up to the same oriented link $\hat{\alpha} = \hat{\beta}$ if and only if there exists a finite number of Markov moves $\gamma_j \to \gamma_{j+1}$ of one of the two types defined above such that

$$\alpha = \gamma_0 \to \gamma_1 \to \ldots \to \gamma_{k-1} \to \gamma_k = \beta .$$

See [Mor] for a proof. One has to be careful that some γ_j may have to lie in B_p for some p much larger than max(m,n).

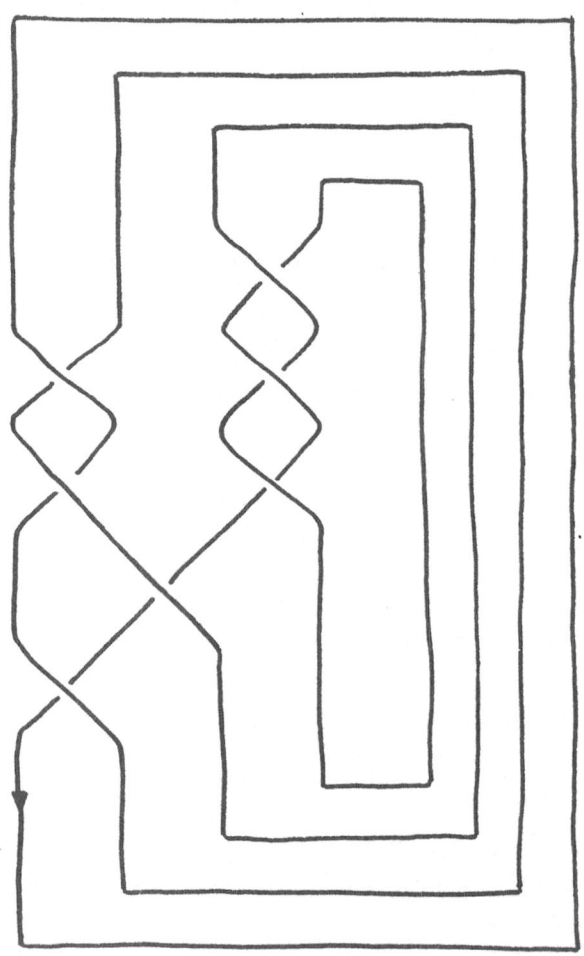

Figure 24. A closed braid is a link.

More about braids in [Bi] and [Mag], and also in Section III.C of [Wa]. See also [Ar2] for an exposition by the Master himself.

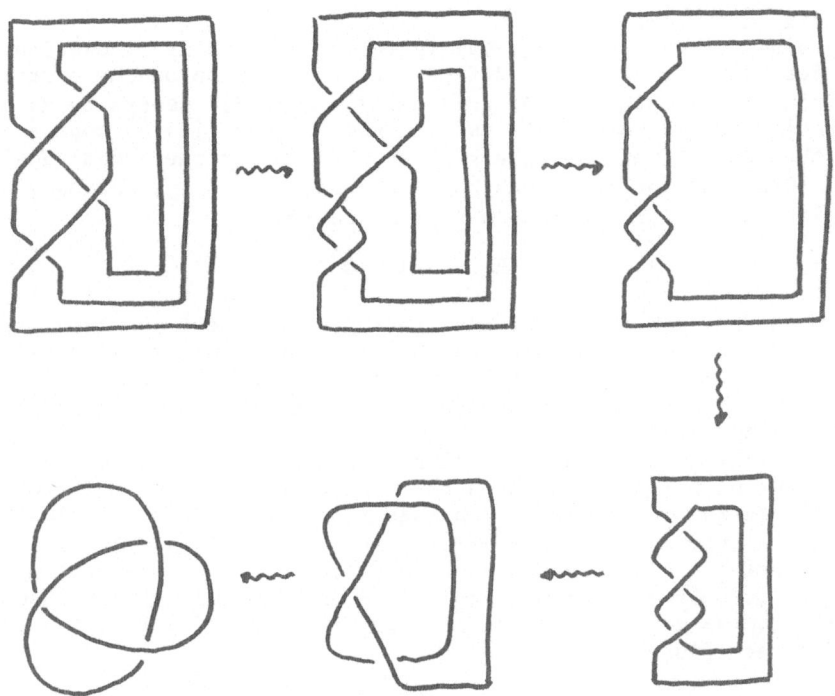

Figure 25.   The two braids $\alpha = \sigma_2\sigma_1\sigma_2\sigma_1 \in B_3$ and
$\beta = \sigma_1^3 \in B_2$ close up to a trefoil
knot $\hat{\alpha} = \hat{\beta}$.

4.2. Towards some recent work

Let V be a vector space of dimension k, say with basis $e_1, \ldots, e_k$. The
tensor products $V \otimes V$ and $\otimes^3 V = V \otimes V \otimes V$ have dimensions k^2 and k^3 respectively, and typical basis elements $e_a \otimes e_b$ and $e_a \otimes e_b \otimes e_c$. We denote by
End(V) the algebra of all linear endomorphisms of V (or of all k-by-k
matrices), which is of dimension k^2. Similarly, End($V \otimes V$) and
End($V \otimes V \otimes V$) have dimensions k^4 and k^6.
    Consider $R \in$ End($V \otimes V$). Then R defines naturally an endomorphism
$R_{1,2} = R \otimes \mathrm{id}_V \in$ End($V \otimes V \otimes V$) which acts identically on the third V. Similarly R defines endomorphisms $R_{1,3}$ and $R_{2,3}$ in End ($\otimes^3 V$). Several problems in scattering theory (the Yang side) and in statistical mechanics
(the Baxter side) have focused attention to (one form of) the so-called
quantum Yang-Baxter equation

$$R_{1,2}R_{2,3}R_{1,2} = R_{2,3}R_{1,2}R_{2,3} \quad (QYB)$$

which is an equality in End($\otimes^3 V$) and in which the unknown is R (or more
precisely a one parameter family $R(\theta)$, but we shall ignore this here).
What we shall remark is that several solutions of (QYB) have been worked
out, some using an advanced theory of "deformations" of Lie algebras and

representations [Ji].

Suppose now that $R \in \text{End}(V \otimes V)$ is invertible and is a solution to (QYB). For each $n \geq 1$, we may define a representation of the n-string braid group B_n as follows. For each $j \in \{1, \ldots, n-1\}$, denote by R_j the endomorphism of $\otimes^n V$ which acts as R on the jth and $(j+1)^{th}$ copies of V and as the identity on the others. Then, because of the formal analogy between (QYB) and the relations among the generators $\sigma_1, \ldots, \sigma_{n-1}$ of B_n, the assignment $\sigma j \to R_j$ defines a <u>representation</u>

$$\pi_R : B_n \longrightarrow \text{Aut}(\otimes^n V)$$

of B_n on the space $\otimes^n V$. It follows that the assignment $\beta \to \text{trace}(\pi_R(\beta))$ defines a scalar valued mapping on the disjoint union of all braid groups. If parameters are suitably introduced, the values of this mapping are not scalars any more, but rather functions, and in interesting cases Laurent polynomials.

Variations on this theme provide mappings from the union of the B_n's to Laurent polynomials which take the same values on two braids related by a Markov move, namely mappings from ambient isotopy classes of oriented links to Laurent polynomials. It has been found that all polynomial invariants discussed in Section 3 above fit in this setting. The full story is only currently being worked out. To our present knowledge, the most precise written account of these ideas is, for the time being, that of Turaev [Tu2].

* * * * * * * * * * * * * *

Many thanks to Claude Weber, without the help of whom I wouldn't have written this report. I am also grateful to Bernard Dudez, Dale Husemoller, Vaughan Jones, Jean-David Rochaix, François Rothen and Hans Wenzl for their generous help and sharing of ideas.

REFERENCES

Ar1   E. Artin : 'Theorie der Zöpfe'. *Hamb.Abh.4* (1925) 47 – 72
        = Collected papers 416 – 441.

Ar2   E. Artin : 'The theory of braids'. *American Scientist 38* (1950)
        112 – 119 = Collected papers 491 – 498.

BG   M. Berger and B. Gostiaux : *Géométrie différentielle: variétés,
        courbes et surfaces*. Presses Univ.France 1987.

Bi   J.S. Birman : *Braids, links and mapping class groups*. Princeton
        Univ.Press 1974.

BW1   J.S. Birman and R.F. Williams : 'Knotted periodic orbits in dy-
        namical systems – I : Lorenz's equations'. *Topology 22*
        (1983) 47 – 82.

BW2   J.S. Birman and R.F. Williams : 'Knotted periodic orbits in dy-
        namical systems – II : knot holders for fibered knots'.
        Contemporary Math.20 (1983) 1 – 60.

BW   M. Boileau and C. Weber : 'Le problème de Milnor sur le nombre
        gordien des noeuds algébriques'. *L'Enseignement math.,
        Monographie 31* (1983) 49 – 98.

BLM   R.D. Brandt, W.B.R. Lickorish and K.C. Millett : 'A polynomial
        invariant for unoriented knots and links'. *Invent.math.84*
        (1986) 563 – 573.

BK   E. Brieskorn and H. Knörrer : *Ebene algebraische Kurven*. Birk-
        häuser 1981. (English translation 1986.)

BZ   G. Burde and H. Zieschang : *Knots*. W. de Gruyter 1985.

Co   J.H. Conway : 'An enumeration of knots and links, and some of
        their algebraic properties' in *Computational problems in
        abstract algebra*, edited by J. Leech. Pergamon Press 1969
        (pages 329 – 358).

CG   J.H. Conway and C.McA. Gordon : 'Knots and links in spatial
        graphs'. *J. of Graph Th.7* (1983) 445 – 453.

CGLS   M. Culler, C.McA. Gordon, J. Luecke and P.B. Shalen : 'Dehn
        surgery on knots'. *Ann. of Math.125* (1987) 237 – 300.

DSKC   F.B. Dean, A. Stasiak, T. Koller and N.R. Cozzarelli : 'Duplex
        DNA knots produced by Escherichia Coli Topoisomerase I'.
        J.Biol.Chem.260 (1985) 4795 – 4983.

260

DT   C.H. Dowker and M.B. Thistlethwaite : 'Classification of knot projections'. *Topology and its Appl.* $\underline{16}$ (1983) 19 - 31.

Dr   V.G. Drinfel'd : *Quantum groups*. Intern. Congress of Math., Berkeley 1986.

ES   C. Ernst and D.W. Sumners : 'The growth of the number of prime knots'. To appear in *Math.Proc.Camb.Phil.Soc.*

FW   J. Franks and R.F. Williams : 'Entropy and knots'. *Trans.Amer. Math.Soc.* $\underline{291}$ (1985) 241 - 253.

FYHLMO P. Freyd, D. Yetter, J. Hoste, W.B.R. Lickorish, K.C. Millet and A. Ocneanu : 'A new polynomial invariant of knots and links'. *Bull.Amer.Math.Soc.* $\underline{12}$ (1985) 239 - 246.

Go   C.McA. Gordon : *Some aspects of classical knot theory*. Lecture Notes in Math. $\underline{685}$ (Springer 1978) 1 - 60.

HKW   P. de la Harpe, M. Kervaire and C. Weber : 'On the Jones polynomial'. *L'Enseignement math.* $\underline{32}$ (1986) 271 - 335.

HLM   H.M. Hilden, M.T. Lozano and J.M. Montesinos : *Universal knots*. Lecture Notes in Math. $\underline{1144}$ (Springer 1985) 25 - 59. Also : 'On knots that are universal'. *Topology* $\underline{24}$ (1985) 499 - 504.

Ho   C.F. Ho : 'A new polynomial for knots and links - preliminary report'. *Abstracts Amer.Math.Soc.* $\underline{6} - 4$ (1985), 300, abstract 821-57-16.

Hos   J. Hoste : 'A polynomial invariant of knots and links'. *Pacific J.Math.* $\underline{124}$ (1986) 295 - 320.

Ji   M. Jimbo : 'A q-difference analogue of U(g) and the Yang-Baxter equation'. *Letters in Math.Phys.* $\underline{10}$ (1985) 63 - 69.

Jo1   V.F.R. Jones : 'A polynomial invariant for knots via von Neumann algebras'. *Bull.Amer.Math.Soc.* $\underline{12}$ (1985) 103 - 111.

Jo2   V.F.R. Jones : 'A new knot polynomial and von Neumann algebras'. *Notices Amer.Math.Soc.* $\underline{33}$ (1986) 219 - 225.

Jo3   V.F.R. Jones : 'Hecke algebra representations of braid groups and link polynomials'. To appear in *Ann. of Math.*

Ka1   L.H. Kauffman : 'The Conway polynomial'. *Topology* $\underline{20}$ (1981) 101 - 108.

Ka2   L.H. Kauffman : *Formal knot theory*. Princeton Univ. Press 1983.

Ka3   L.H. Kauffmann : 'State models and the Jones polynomial'. *Topology* $\underline{\underline{26}}$ (1987) 395 – 407.

Ka4   L.H. Kauffmann : 'An invariant of regular isotopy'. Preprint, Chicago 1985.

Kn   C.G. Knott : *Life and scientific work of P.G. Tait.* Cambridge Univ. Press 1911.

Ko   G. Kolata : 'Solving knotty problems in math and biology'. *Science* $\underline{\underline{231}}$ (28 March 1986) 1506 – 1508.

Lic   W.B.R. Lickorish : 'A relationship between link polynomials'. To appear in *Math.Proc.Camb.Philos.Soc.*

LM   W.B.R. Lickorish and K.C. Millett : 'A polynomial invariant of oriented links'. *Topology* $\underline{\underline{26}}$ (1987) 107 – 141.

Lin   D. Lines : 'Revêtements ramifiés'. *L'Enseignement math.* $\underline{\underline{26}}$ (1980) 173 – 182.

Lit   C.N. Little : 'Non-alternate ± knots'. *Trans.Roy.Soc.Edinburgh* $\underline{\underline{39}}$ (1900) 771 – 778.

Mag   W. Magnus : *Braid groups : a survey.* Lecture Notes in Math. $\underline{\underline{372}}$ (Springer 1974) 463 – 487.

Man   Yu.I. Manin : 'Some remarks on Koszul algebras and quantum groups'. Preprint, 1987.

Mer   N.D. Mermin : 'The topological theory of defects in ordered media'. *Rev.Mod.Phys.* $\underline{\underline{51}}$ – 3 (1979) 591 – 648.

MW1   F. Michel and C. Weber : *Topologie des germes de courbes planes à plusiers branches.* Book to appear.

MW2   F. Michel and C. Weber : 'Noeuds et entrelacs'. Preprint, Genève 1987.

MW   J.P.J. Michels and F.W. Wiegel : 'Probability of knots in a polymer ring'. *Physics Letters* $\underline{\underline{90A}}$ – 7 (26 July 1982) 381 – 384.

Mi   J. Milnor : *Singular points of complex hypersurfaces.* Princeton Univ. Press 1968.

Moi   E.E. Moise : 'Affine structures in 3-manifolds VIII. Invariance of the knot types; local tame embedding'. *Ann. of Math.* $\underline{\underline{59}}$ (1954) 159 – 170.

262

Mor   H.R. Morton : 'Threading knot diagrams'. *Math.Proc.Camb.Phil.*
                Soc.99 (1986) 247 – 260.

Mu   J. Murakami : 'The parallel version of link invariants'. Pre-
                print, Osaka University.

Po   W.F. Pohl : 'DNA and differential geometry'. *Math.Intell.3 – 1*
                (1980) 20 – 27.

PT   J.J. Przytycki and P. Traczyk : 'Invariants of links of Conway
                type'. To appear in *Kobe J.Math.*

QW   T.C.V. Quach and C. Weber : 'Une famille infinie de noeuds
                fibrés cobordants à zéro et ayant même polynôme d'Alexan-
                der'. *Comment.Math.Helv.54* (1979) 562 – 566.

Re   K. Reidemeister : *Knotentheorie*. Springer 1932 and Chelsea 1948.

Rh1   G. de Rham : *Lectures on introduction to algebraic topology'*.
                Tata Institute, Bombay 1969.

Rh2   G. de Rham : 'Introduction aux polynômes d'un noeud'. *L'Enseigne-*
                ment Math.13 (1967) 187 – 194 = Oeuvres 588 – 595. Personal
                note : During the academic year 1966/67, G. de Rham was
                giving his "Cours d'analyse supérieure" in Lausanne, and
                the winter term programme was not far from that of [Rh1].
                But on the 30th of January 1967, he said he would digress
                and entertain his audience with some idea he had been
                thinking about the day(s?) before. The lecture was a sketch
                of [Rh2].

Rol   D. Rolfsen : *Knots and links*. Publish or Perish 1976.

Ros   H. Rosenberg : 'Un contre-exemple à la conjecture de Seifert
                (d'après P. Schweitzer)'. *Séminaire Bourbaki 434* (Juin 1973).

Si   J. Simon : 'Topological chirality of certain molecules'.
                Topology 25 (1986) 229 – 235.

So   A. Sommerfeld : *Thermodynamics and statistical mechanics*.
                Academic Press 1964.

Su1   D.W. Sumners : 'Knot theory, statistics and DNA'. *Kem.Ind.35–12*
                (1986) 657 – 661.

Su2   D.W. Sumners : 'The role of knot theory in DNA research' in
                Geometry and topology – manifolds, varieties, and knots,
                edited by C. McCrory and T. Shifrin. M. Dekker 1987.

Su3   D.W. Sumners : 'The knot theory of molecules'. *J.Math.Chem. 1*
                (1987) 1 – 14.

Su4   D.W. Sumners : 'Knots, macromolecules and chemical dynamics'.
To appear in *Graph theory and topology in chemistry*, edited
by R.B. King and D. Rouvray, Elsevier.

Ta   P.G. Tait : 'On knots'. *Trans.Roy.Soc.Edingburgh (1867-7) =*
Scientific Papers Vol.I, Cambridge Univ.Press (1898) 273 -
317. (See also 270 - 272 and 318 - 347).

Th1   M.B. Thistlethwaite : 'Knot tabulations and related topics' in
Aspects of topology. In memory of Hugh Dowker 1912 - 1982,
edited by I.M. James and E.H. Kronheimer. Cambridge Univ.
Press 1985 (pages 1 - 76).

Th2   M.B. Thistlethwaite : 'A spanning tree expansion of the Jones
polynomial'. *Topology* $\underline{26}$ (1987) 297 - 309.

Tho   W. Thomson (Baron Kelvin) : 'On vortex atoms'. *Proc.Roy.Soc.*
Edinburgh $\underline{6}$ (1867) 94 - 105 = *Phil.Mag.$\underline{34}$* (1867) 15 - 24 =
Mathematical and physical papers, vol.$\overline{IV}$, Cambridge Univ.
Press, 1910, pages 1 - 12. (This volume has several pictures
of knots and links on pages 46, 121 and 122.)

Ti   H. Tietze : *Ein Kapitel Topologie. Zur Einführung in die Lehre*
von den verknoteten Linien. Teubner 1942.

Tu1   V.G. Turaev : 'A simple proof of the Murasugi and Kauffman
theorems on alternating links'. To appear in *L'Enseigne-*
ment math..

Tu2   V.G. Turaev : 'The Yang-Baxter equation and invariants of links'.
Preprint, Leningrad 1987.

Ve   J.L. Verdier : 'Groupes quantiques (d'après V.G. Drinfel'd)'.
Séminaire Bourbaki $\underline{685}$ (Juin 1987).

Wa   D.M. Walba : 'Topological stereochemistry'. *Tetrahedron* $\underline{41}$
(1985) 3161 - 3212.

WC   S.A. Wasserman and N.R. Cozzarelli : 'Biochemical topology :
applications to DNA recombination and replication'.
Science (23 May 1986) 951 - 960.

WDC   S.A. Wasserman, J.M. Dungan and N.R. Cozzarelli : 'Discovery of
a predicted DNA knot substantiates a model for site-specific recombination'. *Science* $\underline{229}$ (12 July 1985) 171 - 174.

Wh   W. Whitten : 'Knot complements and groups'. *Topology* $\underline{26}$ (1987)
41 - 44.

LONG RANGE DYNAMICS AND SPONTANEOUS SYMMETRY BREAKING IN MANY-BODY SYSTEMS

F. Strocchi
International School for Advanced Studies,
Strada Costiera, 11
34014 Trieste
Italy

ABSTRACT. The algebraic approach to quantum mechanics is used to describe the quantum properties of many-body systems, in particular when the dynamics involves long range interactions. In contrast with the short range case, in the presence of long range interactions the time evolution of essentially localized observables involves variables at infinity; this means that infinitely delocalized "classical-like" observables have to be introduced for a complete description of the system. They play a crucial rôle in the phenomenon of spontaneous symmetry breaking (generalized Goldstone's theorem). In particular, the occurrence of an energy gap (at zero momentum) for the generalized Goldstone's boson spectrum is related to a non-trivial "classical" motion of the variables at infinity, induced by the effective dynamics (in a factorial representation). As explicit examples we discuss the Heisenberg model in the molecular field approximation, the BCS model for superconductivity, and the electron gas in uniform background as a model of metals.

1.   QUANTUM MECHANICS OF INFINITE SYSTEMS. ALGEBRAIC STRUCTURE AND STATES

Quantum Mechanics (QM) was invented to describe atomic physics i.e. systems with a finite number of degrees of freedom, and this is the field where the theory proved so successful. Later, it turned out that QM could be used also for other branches of physics, like quantum theory of electromagnetic radiation, quantum optics, solid state physics, nuclear matter, phase transitions etc., i.e. for physical systems which are conveniently described by infinite degrees of freedom. The point is that for (macroscopic) systems consisting of a large number N of constituents a simple description of their bulk behaviour suggests to neglect effects of order $1/N$ and $\frac{1}{V} = \frac{n}{N}$ (n being the density), and to consider the so-called "thermodynamical limit" $N \to \infty$, $V \to \infty$, $n = N/V$ fixed. (This means that the atten-

265

A. Amann et al. (eds.),
Fractals, Quasicrystals, Chaos, Knots and Algebraic Quantum Mechanics, 265–285.
© *1988 by Kluwer Academic Publishers.*

tion is focused on "intensive" quantities like mean energy per particle, pressure, density correlations, mean electromagnetic properties etc.). The quantum theory of such systems requires then an extension of the ordinary QM to the case of infinite degrees of freedom ($N \to \infty$) , briefly QM_∞. According to the fields to which it has been applied, such an extension of QM goes under the names of Second Quantization, Quantum Field Theory, Many-Body Theory, Quantum Statistics etc. and its success has proved so remarkable in the last twenty years, much beyond the theoretical expectations that the physical relevance of such theory is not less than that of ordinary QM. Actually, many of the newly emerging theoretical structures associated to the infinite degrees of freedom turned out to be crucial for understanding fundamental phenomena like collective effects, condensation, spontaneous symmetry breaking, phase transitions etc. and they provide a common unifying framework for apparently different physical situations[1].

A fruitful approach to QM_∞, pioneered in particular by Segal[2] and Haag[3], distinguishes two different structures at the basis of QM:

i) the <u>algebra of canonical variables</u>* $\mathcal{O}$ and

ii) their representations as operators in a Hilbert space $\mathcal{H}$ of <u>states of the system</u>.

Such a distinction is not so crucial in the case of finite degrees of freedom, since in this case, by the Von Neumann theorem, all the irreducible representations of the Weyl algebra generated by the canonical variables are unitary equivalent[1][4] and therefore the choice of one instead of another is purely a matter of convenience, devoid of any physical implication. The situation changes drastically in the case of infinite degrees of freedom, where the existence of inequivalent representations is at the basis of relevant physical phenomena like phase transitions, spontaneous symmetry breaking, collective effects etc. and the choice of the representation is strictly related to the physical properties of the system.

The first step is therefore that of identifying the algebra $\mathcal{O}$ of canonical variables or of the observables, if the emphasis is on the observable character of the basic variables. For technical reasons it is convenient to consider $\mathcal{O}$ as a <u>C*-algebra</u> namely i) <u>closed</u> under the * operation (corresponding to taking the adjoint: e.g. $(\exp i\alpha_i q_i)^* = \exp(-i\alpha_i q_i)$, α_i real etc.), and ii) <u>normed</u>, i.e. such that every element

---

* For concreteness one may think of the algebra generated by the q_i, p_i's or better by their Weyl exponentials $\exp i\alpha_i q_i$, $\exp i\beta_i p_i$, which are bounded operators.

$A \in \mathcal{A}$ is provided with a norm $\|A\| \geq 0$, $(\|A\|=0$ iff $A=0)$, such that

$$\|A*A\| = \|A\|^2 ,$$

iii) <u>norm closed</u> (i.e. complete in the norm topology). A <u>representation</u> π of $\mathcal{A}$ is a homomorphic mapping

$$\pi: A \rightarrow \pi(A) ,$$

into the (bounded) operators of a Hilbert space $\mathcal{H}_\pi$. Isomorphic mappings define faithful representations and in this case the Hilbert space norm $\|\pi(A)\|_{\mathcal{H}}$ coincides with $\|A\|$. A vector $\Psi \in \mathcal{H}_\pi$ is <u>cyclic</u> for the representation π if $\mathcal{H}_\pi = \overline{\pi(\mathcal{A})}\Psi$ (the bar denoting the strong closure). Two representations π_1, π_2 are <u>equivalent</u> if there is a unitary mapping from $\mathcal{H}_1$ to $\mathcal{H}_2$ such that $\pi_2(A) = U \pi_1(A)U^{-1}$, $\forall A \in \mathcal{A}$. Clearly, every (normalized) vector $\Psi \in \mathcal{H}$ defines a (normalized) <u>linear functional</u> $\omega_\Psi(A) \equiv (\Psi, A\Psi)$ on $\mathcal{A}$, $\omega_\Psi(1)=1$, which is <u>positive</u>, $(\Psi, A*A\Psi) = \|A\Psi\|^2 \geq 0$. This is also true for expectation values defined by density matrices

$$\omega_\rho(A) \equiv Tr(\rho A), \quad \rho = \Sigma_i \lambda_i P_i, \quad P_i = \text{one-dim. projector}$$

Quite generally a <u>state</u> on $\mathcal{A}$ is a (normalized) positive linear functional ω and it is called <u>pure</u> if it cannot be decomposed as a convex sum $\lambda_1\omega_1+\lambda_2\omega_2$, $\lambda_i \geq 0$, $\lambda_1+\lambda_2=1$ of two different states ω_1, ω_2.

Every state ω defines through the GNS construction a representation π_ω with a cyclic vector $\Omega \in \mathcal{H}_{\pi_\omega}$ such that

$$\omega(A) = (\Omega, A\Omega), \quad \forall A \in \mathcal{A}$$

(for details see Ref. 4, vol. II). Thus, in particular the ground state correlation functions completely determine the representation and therefore the (corresponding) Hilbert space of states.

It is not difficult to recognize in the above definitions the precise algebraic formulation of familiar concepts of QM (with the basic advantage that the so obtained framework is rich enough to allow a smooth transition to QM$_\infty$). Along the same lines, one can define a <u>symmetry transformation</u> β on the canonical variables (or on the observables) as a <u>*-automorphism</u> of $\mathcal{A}$ (i.e. an invertible mapping which preserves the algebraic structure including the *). As an example one may consider the time translations α^t and the space translations $\alpha_{\vec{x}}$, with the simple physical interpretation that $\alpha^t(A)$ describes the time evolution of A and $\alpha_{\vec{x}}(A) \equiv A_{\vec{x}}$, the observable A translated by $\vec{x}$. In the following, we will always assume that the space translations (or at least some discrete subgroup of them) define *-automorphisms of $\mathcal{Q}$. As we will see below the definition of α^t on $\mathcal{Q}$ is a deep dynamical problem.

A <u>symmetry</u> is <u>unbroken</u> in a representation π if there is a (anti) unitary operator on $\mathcal{H}_\pi$ such that

$$\pi(\beta(A)) = U \pi(A) U^{-1} , \quad \forall A \in \mathcal{Q} .$$

Clearly in this case the physical description of the system by the states of $\mathcal{H}_\pi$ is symmetric under β i.e. there exists a mapping of the states which preserves the transition probabilities. Otherwise, the symmetry is said to be <u>broken</u>; this means that the symmetry exists at the algebraic level (in particular at the level of commutation relations and of Heisenberg equations of motion), but it is not a symmetry of the expectation values in the sense that one cannot find a one-to-one mapping $\psi \rightarrow \psi^\beta$ of the states of $\mathcal{H}_\pi$ which preserves the transition probabilities (by Wigner's theorem such correspondence would be described by a unitary or an antiunitary operator).

It is not the point to emphasise here the tremendous impact that the above ideas of spontaneously broken symmetries had on the recent developments in theoretical physics both at the level of Elementary Particle physics as well as for Solid State physics and Statistical Mechanics. (For a brief account see e.g. Ref. 1).

2. LOCAL ALGEBRA. NON—RELATIVISTIC DELOCALIZATION. ESSENTIALLY LOCALIZED OBSERVABLES.

A further important idea for the quantum description of infinitely extended systems is the realization[5] that only measurements which are localized in space can actually be performed and therefore the local properties of an infinitely extended system have a prominent status. It is therefore convenient to consider the algebra $\mathcal{Q}$ as generated by elements which have localization properties. More precisely, to any bounded region V one associates the (Von Neumann) algebra $\mathcal{Q}_V$ of variables localized in V (or of observables which can be measured in the volume V); one then considers the <u>local algebra</u>

$$\mathcal{Q}_o \equiv \bigcup_V \mathcal{Q}_V \tag{2.1}$$

and its norm closure $\mathcal{Q} = \overline{\mathcal{Q}_o}^{\|\ \|}$ (called <u>quasi-local algebra</u>).

A crucial property to be required for an acceptable physical interpretation of the theory is that the measurements of any two localized elements A, B $\in \mathcal{Q}$ do not interfere in the limit of infinite space separation (<u>asymptotic abelianess in space</u>):

$$\text{norm-}\lim_{|\vec{x}| \to \infty} \; [A_{\vec{x}}^t, B] \; = 0 \;. \tag{2.2}$$

This property is obviously satisfied in the case of relativistic systems
as a consequence of Einstein's causality. In the above language the solution of the dynamical problem is the determination of the time evolution
$A \to A_t$ for any element of $\mathcal{O}$. From a constructive point of view the
theory is defined in terms of finite volume or infrared cutoffed
Hamiltonians H_V and the generated finite volume dynamics α_V^t, which define
one parameter groups of automorphisms of $\mathcal{O}_o$ (more generally of $\mathcal{O}$).
One has then to remove the infrared cutoff, $V \to \infty$ to obtain the algebraic dynamics on $\mathcal{O}_o$.

In the case of <u>dynamics with finite propagation speed</u>, for any
$A \in \mathcal{O}_o$, α_V^t becomes independent of V, for V large enough, and therefore
one easily gets the <u>algebraic dynamics</u> $\alpha^t \equiv \lim_V \alpha_V^t$ as an automorphism of
$\mathcal{O}_o$. More generally, in the case of "short range" dynamics, for any
$A \in \mathcal{O}_o$, $\alpha_V^t(A)$ converges in the norm topology to an automorphism group of
$\mathcal{O}$. (This has been proved[6], e.g., for spin systems with two-body in-
teractions decaying faster than $|\vec{x}|^{-3}$). In this case, the dynamical pro-
blem is conceptually under control and the corresponding theory is well
developed[4][7]. In particular, one may select the physically relevant
representations of $\mathcal{O}$ as those characterized by a time translation invariant state (<u>ground state</u>). This guarantees that the spectrum of the
Hamiltonian is well defined.

It is worthwhile to stress that the existence of the dynamics at the
algebraic level, namely as an automorphism group of $\mathcal{O}$, independent of
the choice of the representation, implies that <u>different phases have</u> the
<u>same dynamics</u> and that their <u>different physical behaviour</u> is solely <u>due</u>
<u>to the different properties of the ground states</u>,i.e. of their correlation
functions yielding inequivalent representations of the <u>same</u> algebra.

The local structure plays also a relevant rôle in the breaking of
continuous symmetries (<u>Goldstone's theorem</u>)[8][1]. A one parameter group
of automorphisms β^λ, $\lambda \in \mathbb{R}$, is <u>generated by a local charge</u> Q_R (affiliated
to some $\mathcal{O}_V$) if $\forall A \in \mathcal{O}_o$

$$\beta^\lambda(A) = \text{norm-}\lim_{R \to \infty} e^{iQ_R\lambda} \; A \; e^{-iQ_R\lambda} \;. \tag{2.2}$$

Modulo technical assumptions (like the limit (2.2) being uniform in λ
together with its derivative) the above equation implies

$$\delta A \equiv \frac{d}{d\lambda} \beta^\lambda(A) \Big|_{\lambda=0} = i \lim_{R \to \infty} [Q_R, A] \;. \tag{2.3}$$

In most of the applications Q_R can be written as the integral of a charge density

$$Q_R = \int_{|\vec{x}| \leq R} j_o(\vec{x},t) \, d^3x \qquad (2.4)$$

and the charge density commutations $[j_o(\vec{x},t), A]$ are (absolutely) integrable in $\vec{x}$ (charge integrability property*, see Refs; 9) for more details)

THEOREM (Non-relativistic Goldstone's theorem)[8-10]

Let

1) β^λ be a one par. group of aut. of $\mathcal{Q}_o$, $[\beta^\lambda, \alpha_{\vec{x}}]=0$, generated by a local charge Q_R, on $\mathcal{Q}_o$

$$\beta^\lambda(A) = \underset{R \to \infty}{n-\lim} \, e^{iQ_R\lambda} \, A \, e^{-iQ_R\lambda} \, , \qquad \forall A \in \mathcal{Q}_o \, ,$$

(and limit is uniform in λ together with its derivative).

2) $\beta^\lambda \alpha_V^t = \alpha_V^t \beta^\lambda$ (symmetry of H_V) ,

3) α_V^t converge in norm to α^t ("short range" dynamics).

Then i) β^λ has unique extension to the algebra $\mathcal{Q} = \overline{\mathcal{Q}_o}^{\| \, \|}$, stable under time evolution, generated by a local charge; ii) $\beta^\lambda \alpha^t = \alpha^t \beta^\lambda$. Hence, if Ψ_o is a translationally invariant state and

4) $\underset{R \to \infty}{\lim} <[Q_R, B]>_{\Psi_o} \neq 0 \, , \qquad B \in \mathcal{Q} \, ,$

5) $Q_R = \int_{|x| < R} d^3x \, \rho(x)$ and the charge density commutators $\langle [\rho(x,t), A] \rangle_o$ are absolutely integrable in x,

then: iii) β^λ is spontaneously broken in the representation π defined by Ψ_o; iv) there exist quasi particle excitations (Goldstone bosons) with infinite lifetime as $k \to 0$ and energy $\omega(k) \to 0$ as $k \to 0$ (no energy gap). Property 5) is crucial for the derivation of the point spectrum at $k \to 0$. It generalizes the causality condition of relativistic quantum field theory. When α_V^t does not converge in norm, as in the case of long range interactions, properties i) and ii) become problematic and their failure is at the root of the evasion of Goldstone's theorem. The relevance of

---

* This property plays a crucial rôle for a derivation of the Goldstone boson spectrum[9]. For the applications, it is enough to have eq.(2.3) and the integrability property satisfied as expectation values on the ground state.

the theorem is that from symmetry considerations one may get <u>exact</u>, non-perturbative information on the <u>energy spectrum</u> at k → 0; for significant applications we mention the theory of ferromagnetism (spin waves), super-fluidity (Landau phonons), crystals (phonons) etc.[11].

It is worthwhile to mention that for non-relativistic systems like e.g. the free bose gas (and also for relativistic systems involving unobservable field algebras) the above framework requires some <u>technical</u> modification namely an enlargement of the quasi-local algebra to get a well-defined algebraic dynamics (see e.g. Refs. 12). The reason is that for non-relativistic systems the speed of propagation is infinite and therefore some delocalization is unavoidable in the time evolution; the norm convergence of α_V^t has then to be replaced by the convergence of the expectation values of α_V^t on a family $\mathbb{F}$ of "physically relevant" representations of $\mathcal{A}_o$. More precisely, the algebraic dynamics has to be defined as the weak limit of α_V^t with respect to the weak topology $\tau_{\mathbb{F}}$ of the states of $\mathbb{F}$. Consequently, the algebraic dynamics is naturally defined on the $\tau_{\mathbb{F}}$ closure M of $\mathcal{A}_o$, rather than on $\mathcal{A}$.[9,10,12] (Additional technical conditions are also needed in order that α^t defines a one parameter group of automorphisms of M.[9,10,12]). However, as it appears in several models under control, when the interaction is of short range, the algebraic dynamics leaves stable a weakly dense subalgebra $\mathcal{A}_1$ of M, with trivial center which to all effects can be considered as the algebra of observables for such non-relativistic systems. Typically, (in particular in the simple models), the enlargement from $\mathcal{A}$ to $\mathcal{A}_1$ corresponds to the inclusion of field variables smeared with test functions of non-compact support, suitably decreasing at infinity (for example test functions in $\mathcal{S}(R^3)$ or more generally in $L^2(R^3)$) [12] (nonrelativistic delocalization). The emerging structure is that the dynamics leaves stable an algebra $\mathcal{A}_1$, which still has some kind of localization (<u>algebra of essential localization</u>)[9], so that the modification of the Haag-Kastler local structure appears more technical than substantial. The crucial point, which can be taken as a characterization of these cases, is that the <u>algebraic dynamics leaves stable an algebra</u> $\mathcal{A}_1 \supset \mathcal{A}_o$, with <u>trivial center</u> (<u>essentially local structure</u>)[9].

In this framework, under general conditions (realized, e.g., in simple models), one parameter groups β^λ, $\lambda \in \mathbb{R}$, of $\mathcal{A}_o$, generated by a local charge Q_R on $\mathcal{A}_o$, can be extended to $\mathcal{A}_1$ with the property that they are still generated by Q_R on $\mathcal{A}_1$. The basic ingredient of the Goldstone's theorem, as discussed in the previous section, is then recovered and the conclusions of the theorem hold as in the standard case.

3. LONG RANGE INTERACTIONS AND VARIABLES AT INFINITY. MODIFIED HAAG-KASTLER LOCAL STRUCTURE.

The general algebraic structure of QM_∞ discussed in the previous section requires some modification in the presence of long range interactions, an unavoidable feature of most many-body systems, since the Coulomb force is at the basis of the structure of matter. The crucial point is that in the case of long interaction the commutator $[H_V, A]$, with H_V the finite volume Hamiltonian and A a localized variable, in general involves in a substantial way variables localized on or near the boundary of V, which in the limit $V \to \infty$ become variables localized outside any bounded region, i.e. the so-called <u>variables at infinity</u>[13]. As a consequence of the asymptotic abelianess in space, the variables at infinity commute with all local variables.[13] Typical examples are the <u>ergodic means</u>

$$\lim_{V \to \infty} \frac{1}{V} \int_V d^3x \ A_{\vec{x}} = A_\infty \quad , \tag{3.1}$$

or the $V \to \infty$ limit of averages around the boundary, like

$$\lim_{R \to \infty} \int d^3x \ f_R(\vec{x}) A_{\vec{x}} \ , \tag{3.2}$$

with $f_R(\vec{x})$ a regular function different from zero only in the region $R < |\vec{x}| < R(1+a)$ and suitably normalized ($\int f_R(x) d^3x = 1$).

It is a technical relevant feature of the above limits (and more generally of the convergence to variables at infinity) that they do not exist in the norm topology but only in a weak topology defined by states which must be sufficiently regular at infinity. We meet here what we consider to be a basic feature of long range interactions; the infrared behaviour of the dynamics requires a careful handling of the thermodynamical limit. To get an <u>algebraic</u> definition of the dynamics, starting from an infrared regularized one, one has to make reference to infrared regular states to remove the infrared cutoff[9,12]. We will denote by $\mathbb{F}$ a family of such "physically relevant" states and by $\tau_\mathbb{F}$ the corresponding weak topology. One can specify conditions[9] which guarantee that the $\tau_\mathbb{F}$-weak limit of α_V^t define an algebraic dynamics α^t as an automorphism group of $M \equiv$ the $\tau_\mathbb{F}$-weak closure of $\mathcal{Q}$. The relevant mathematical and physical property, which distinguishes the long range case with respect to the short range case, is that the quasi-local algebra $\mathcal{Q}$ is not stable under α^t <u>and</u> that the <u>time evolution of local variables involves variables at infinity</u>. Thus, the local framework of Sect. 2 appears too narrow (in a much more substantial way than that mentioned at the end of sect. 2) and global infinitely delocalized variables have to be added to $\mathcal{Q}$ (or more generally to $\mathcal{Q}_1$) for a complete algebraic description of systems with long range interactions. (For the mathematical structures involved we refer to Ref. 9). The above phenomena can be explicitly checked to occur in spin models like the BCS model[9], in simple gauge models with Coulomb-like interaction[15] and in realistic many-body models like the Coulomb gas with uniform background[16].

For the discussion of symmetries and their breaking it is convenient (and always possible[9]) to choose the family of states $\mathbb{F}$ stable under the * automorphisms β of $\mathcal{Q}$ and their inverse β^{-1}, so that (the condition is necessary and sufficient[9]) they can be uniquely extended to automorphisms of M . Furthermore one can show[9] that <u>the symmetries of the finite volume dynamics</u> α_V^t <u>are also the symmetries of the algebraic dynamics</u> α^t (more generally the covariance properties of α_V^t are shared by α^t).

On the basis of the above general considerations and of the experience with explicit models one is led to the following:

<u>Modified Haag-Kastler structure</u>: There exists a weakly dense subalgebra $\mathcal{Q}_1$ of M , $\mathcal{Q}_1 \supset \mathcal{Q}_0$ with the properties that

i) $\mathcal{Q}_1$ has a trivial center

ii) $\mathcal{Q}_1$ is faithfully represented in each factorial representation of $\mathbb{F}$

iii) the algebraic dynamics α^t leaves stable the algebra generated by $\mathcal{Q}_1$ and by the algebra at infinity $\mathcal{Q}_\infty$.

<u>Remarks</u>. In relativistic theories based on the observable algebra, $\mathcal{Q}_1$ can be taken as the local observable algebra; clearly the occurrence of variables at infinity in the time evolution of local variables is compatible with Einstein causality. In the case of mean field lattice models, like e.g. the BCS model, $\mathcal{Q}_1$ can be taken as the quasi-local algebra $\mathcal{Q}$. In general a delocalization is needed for non-relativistic systems where the propagation speed is not finite, and in these cases $\mathcal{Q}_1 \supset \mathcal{Q}$.

If follows from the above structure that in each factorial representation π of $\mathcal{Q}_1$, defined by states of $\mathbb{F}$, since the variables at infinity get frozen to their ground state expectation values, the algebraic dynamics α^t defines an <u>effective dynamics</u> α_π^t as an automorphism of $\mathcal{Q}_1$. This can be interpreted as an <u>effective essential localization</u> of the dynamics in each factorial representation π of $\mathbb{F}$. Symbolically, if

$$\alpha^t(A) = F_t(\mathcal{Q}_1, \mathcal{Q}_\infty) ,$$

(3.3)

then

$$\alpha_\pi^t(A) = F_t(\mathcal{Q}_1, <\mathcal{Q}>_{\infty\pi}) \ . \tag{3.4}$$

This can be regarded as the rigorous formulation and clarification of the so-called <u>seizing of the vacuum</u>, euristically advocated for gauge theories on the basis of two dimensional models. For more details on the above structure see Refs. 9.

From a pragmatic point of view one could take the position that $\mathcal{Q}_1$ is the physically relevant algebra and that the algebraic dynamics is given by α_π^t, so that no variable at infinity ever occurs. The drawback of this point of view is however that in so doing one makes reference in an essential way to a given representation π so that a) the algebraic properties of the time evolution are essentially lost, since α_π^t is in general implementable only in the representation π, b) there is no unifying algebraic treatment of the dynamics of different phases, since they are described by different algebraic structures $(\mathcal{Q}_1, \alpha_\pi^t)$ c) the symmetry properties of the dynamics are lost at the algebraic level and in particular phases related by automorphisms of $\mathcal{Q}_1$, e.g. defined by the states ω and $\omega \circ \beta$, are no longer on equal footing as far as the algebraic dynamics is concerned.

In our opinion, the introduction of variables at infinity appears as a non-dramatic modification of the standard structure and a technical price worth paying to keep the formulation as algebraic as possible. The occurrence of variables at infinity is not only compatible with the basic physical principles, but it also has a physical meaning since it allows to establish links between "phases" with different effective dynamics. In some way, the variables at infinity parametrize at the algebraic level the "infrared" structure of the dynamics and in some sense they trivialize the dependence on the boundary conditions (vacuum seizing) in terms of classical-like quantities, in situations in which, as the results of the long range interactions, the coupling with the boundary gives rise to volume effects in the dynamics.

4.   GENERALIZED GOLDSTONE'S THEOREM[9,10] FOR DYNAMICS WITH LONG RANGE
      INTERACTIONS

From the discussion of the previous section, it follows that the symmetries of the finite volume dynamics α_V^t are no longer shared by the effective dynamics α_π^t, (the property of being symmetric is peculiar to α^t and actually one of the main motivations for its definition), whenever the variables at infinity involved in the time evolution of $\mathcal{Q}_1$ are not pointwise invariant (see eq. (3.4)). This provides a <u>new mechanism of symmetry breaking</u>[9,10] substantially different from the standard Goldstone's mechanism, the latter being characterized by symmetric equations of motion and by non-symmetric ground state. In particular, the spectrum of the elementary excitations associated to this new mechanism of symmetry breaking is no longer constrained to be trivial, namely $\omega(k) \to 0$ as $k \to 0$ (no mass gap).

More precisely, the point is that in each factorial representation, defined by a translationally invariant ground state ϕ_0, the energy spectrum associated to the breaking of a one parameter group of symmetries β^λ, $\lambda \in \mathbb{R}$, generated by a local charge Q_R, is related to the Fourier transform of the charge density commutator, (a crucial rôle is played by the charge integrability condition[9]),

$$<[Q_R,A_t]>_{\phi_0} = <[Q_R,\alpha_\pi^t(A)]>_{\phi_0} = \int_{|x| \leq R} d^3x <[j_0(x),\alpha_\pi^t(A)]>_{\phi_0} \quad (4.1)$$

and, since α_π^t is not symmetric, the above expectation value is in general not independent of t, in the limit $R \to \infty$. Actually one has[9]

$$\frac{d}{d\lambda}<\beta^\lambda(A)>_{\phi_0}\bigg|_{\lambda=0} = i \lim_{R\to\infty} <[Q_R,\alpha_\pi^t(A)]>_{\phi_0} =$$

$$= \frac{d}{d\lambda}<\alpha_\pi^t(A)>_{\phi_0^\lambda}\bigg|_{\lambda=0} \quad , \quad \phi_0^\lambda \equiv \phi_0 \circ \beta^\lambda \quad (4.2)$$

and, if β^λ commutes with the space translations (more generally if $[\beta^\lambda, \alpha_{\underset{x}{\to}}]$ is "unbroken"), the right hand side can be written as

$$\frac{d}{d\lambda}<\alpha_\pi^t(\frac{1}{V}\int_V d^3x\, A_{\underset{x}{\to}})>_{\phi_0^\lambda}\bigg|_{\lambda=0} = \frac{d}{d\lambda}<\alpha_\pi^t(A_\infty)>_{\phi_0^\lambda}\bigg|_{\lambda=0} \quad . \quad (4.3)$$

The energy spectrum is therefore the same as that of the motion defined on the variables at infinity A_∞ by the effective dynamics α_π^t, linearized around the stable point

$$<\alpha_\pi^t(A_\infty)>_{\phi_0} = <A_\infty>_{\phi_0} \quad . \quad (4.4)$$

Since the (effective) time evolution of A_∞ takes place in an abelian algebra (identified by a family of states stable under α_π^t and β^λ) one is led to the study of a "classical" dynamical system. If such motion is finite dimensional, as is the case in the applications so far under control[9,15,16] and as can be guaranteed by general conditions, then the spectrum is quasi periodic and its frequencies ω_j give a discrete non-trivial energy spectrum at $k \to 0$, corresponding to excitations with infinite lifetime in the limit $k \to 0$, (<u>quasi particle spectrum of the generalized Goldstone's bosons</u>). For the mathematical details see Ref.9.

5. EXAMPLES

5.1. BCS model of superconductivity

The BCS model of superconductivity in the spin version suggested by Anderson[17] is described by the following finite volume Hamiltonian

$$H_V = \frac{1}{|V|} \sum_{i,j \in V} \sum_{\alpha,\beta} \sigma_\alpha^i \, A_{\alpha\beta} \, \sigma_\beta^j + \sum_{i \in V} \sum_\alpha C_\alpha \, \sigma_\alpha^i \,, \tag{5.1}$$

where i,j are lattice site indices, α,β denote the spin components,

$$A_{\alpha\beta} = -\frac{T_c}{2} \begin{pmatrix} 1 & i & 0 \\ -i & 1 & 0 \\ 0 & 0 & 0 \end{pmatrix} \,, \quad C_\alpha = -\epsilon\delta_{\alpha,3} \,. \tag{5.2}$$

It is mostly this form of the model which has been discussed in the text-books of quantum solid state[18] and in the mathematical physics litera-ture[19]. However, in the past treatments the general algebraic features discussed in the previous sections were not pointed out. As a matter of fact, this model together with other mean spin models like the Heisenberg model (see 5.2 below) provide the most clear and simple realization of the general structures discussed so far[9].

The algebra of quasi local observables $\mathcal{A}$ is the standard spin algebra on a lattice[20]. The finite volume dynamics α_V^t generated by H_V defines a one-parameter group of automorphisms of $\mathcal{A}$ determined by the following equations of motion

$$\frac{d}{dt} \alpha_V^t(\sigma_\delta^i) = -2 \, \epsilon_{\alpha\delta\gamma} [A_{\alpha\beta} \, \alpha_V^t(\sigma_\gamma^i)\alpha_V^t(\sigma_\beta^V) +$$

$$+ A_{\alpha\beta} \, \alpha_V^t(\sigma_\beta^V) \, {}_V^t(\sigma_\gamma^i)] - 2C_\alpha \epsilon_{\alpha\delta\gamma}\alpha_V^t(\sigma_\gamma^i) \,, \tag{5.3}$$

where

$$\sigma_\alpha^V = \frac{1}{|V|} \; \Sigma_{i \in V} \; \sigma_\alpha^i \; . \tag{5.4}$$

As a family $\mathbb{F}$ of physically relevant states one can take the largest set of states on which the averages (5.4) converge ultrastrongly as $V \to \infty$ to the ergodic means

$$\sigma_\alpha^\infty = \lim_{V \to \infty} \sigma_\alpha^V \; . \tag{5.5}$$

This is the condition of sufficient regularity of the states at large distances needed to guarantee the ultrastrong convergence of α_V^t to a one parameter group of automorphisms α^t of the Von Neumann algebra $M \equiv$ weak closure of $\mathcal{Q}$ with respect to $\mathbb{F}$ (see Ref. 9; see also Ref. 21 for a different approach). To be more concrete we further specify $\mathbb{F}$ as the set of product states ϕ with the property

$$\phi(\alpha^t(\sigma_\alpha^\infty)) = \phi(\sigma_\alpha^\infty) . \tag{5.6}$$

The algebraic dynamics α^t then satisfies the equations (5.3) with $\alpha_V^t(\sigma_\beta^V) \to \sigma_\beta^\infty$, i.e. it corresponds to a rotation $R_{\vec{\sigma}\infty}^{\to}(t)$ of the spin $\vec{\sigma}^i$ "around" the vector $\vec{\sigma}^\infty$. It is then obvious that the <u>time evolution of the local variable</u> $\vec{\sigma}^i$ involves the variable at infinity $\vec{\sigma}_\infty$. Moreover, the algebraic dynamics leaves stable the algebra generated by $\mathcal{Q}$ and the algebra at infinity $\mathcal{Q}_\infty$ generated by $\vec{\sigma}^\infty$. It also follows that there is no dense subalgebra of M with trivial center stable under α^t (<u>quantum bifurcation</u>).

Furthermore, in each factorial representation π of $\mathbb{F}$, the algebraic dynamics gets essentially localized with $\mathcal{Q}$ as the <u>algebra of essential localization</u>, so that the <u>generalized Haag-Kastler locality</u> holds with $\mathcal{Q}$ as the algebra of essentially localized variables. Here, due to the lattice formulation and the simplicity of the interaction no delocalization is involved ($\mathcal{Q}_1 = \mathcal{Q}$), as in relativistic local quantum field theory.

Given a pure product state invariant under space translations $\phi_0^{\vec{n}}$, with

$$\phi_0^{\vec{n}}(\sigma_\alpha^i) = n_\alpha \; , \qquad |\vec{n}| = 1 \; , \tag{5.7}$$

the invariance under time translations requires either $n_\alpha = (0,0 \pm 1)$ or $n_\alpha = (n_1, n_2, \; \epsilon/T_c)$. For each factorial representation π_n determined by

such states the effective dynamics $\alpha^t_{\pi_n}$ is obtained by freezing the variables at infinity in α^t to their ground state expectation value $\vec{n}$

$$\alpha^t_{\pi_n} (\sigma^i_\alpha) = (R_{\vec{n}}(t)\sigma^i)_\alpha . \tag{5.8}$$

Clearly, the effective dynamics leaves stable the algebra of essentially
localized observables $\mathcal{Q}_1 = \mathcal{Q}$, i.e. the stability under time evolution
is regained once the boundary conditions are fixed and the effective dynamics depends on the boundary conditions (seizing of the vacuum). Thus,
different phases are described by different effective dynamics in contrast with the standard local case.

The space rotations around the z axis, define a one parameter group
of automorphisms β^λ , $\lambda \in \mathbb{R}$, of $\mathcal{Q}$, which extend to automorphisms of
by weak continuity as a consequence of the stability of $\mathbb{F}$ under
$(\beta^\lambda)^*$. Furthermore, while the algebraic dynamics α^t is β^λ symmetric
(as is the finite volume dynamics), the effective dynamics $\alpha^t_{\pi_n}$ is not
symmetric if $\vec{n}$ is not aligned to the z direction. In fact, variables at
infinity which are not invariant under β^λ are involved in the algebraic
dynamics α^t (non-symmetric quantum bifurcation[9]) and therefore their
getting frozen to c-numbers spoils the symmetry (spontaneous symmetry
breaking induced by non-symmetric bifurcation).

The group G of kinematical symmetries of $\mathcal{Q}$, here the rotations
around the z axis, and the effective dynamics $\alpha^t_{\pi_n}$ generate a larger group
$\mathcal{G}$ of automorphisms of $\mathcal{Q}$ (dynamical group of symmetries), here the
group generated by rotations around the z axis and around $\vec{n}$. The group
$\mathcal{G}$ is not a symmetry of the finite volume dynamics for $\vec{n}$ not aligned to
the z axis. Finally, the effective dynamics commutes with rotations
around $\vec{n}$ (dynamical symmetries).

Under the effective dynamics the variables at infinity are not
pointwise stable and since the algebra at infinity is abelian (as a consequence of asymptotic abelianess) one has a "classical" dynamical system
whose time evolution is given by

$$\overset{\infty}{\sigma}_\alpha(t) \equiv \alpha^t_{\pi_n} (\overset{\infty}{\sigma}_\alpha) = (R_{\vec{n}}(t) \overset{\infty}{\sigma}(0))_\alpha , \tag{5.9}$$

i.e. by rotations around $\vec{n}$ with frequency $\omega = 2T_c$. The generalized
Goldstone theorem predicts the existence of generalized Goldstone excitations with energy spectrum at $k \to 0$, given by the classical motion of
the variables at infinity. Here, one has a periodic motion of frequency
$\omega = 2T_c$ and therefore the Goldstone bosons are quasi particles, i.e.

excitations with infinite life time in the limit $k \to 0$, with <u>energy</u> spectrum characterized by a <u>gap</u> $\omega = 2T_c$, as $k \to 0$.

5.2 Heisenberg model in the molecular field approximation

The Heisenberg model of ferromagnetism is not soluble and as a simplify-ing approximation one may consider its molecular field version. It is defined by replacing the local spin-spin interaction

$$H_V = - \sum_{i,j \in V} J_{ij} \, \vec{\sigma}^i \cdot \vec{\sigma}^j \; , \tag{5.10}$$

with J_{ij} a finite range potential by the (long range) interaction of the spin at the i-th site with the operator average of the spin inside the volume V

$$\sum_{j \in V} J_{ij} \, \vec{\sigma}^j \to \frac{\bar{J}}{|V|} \sum_{j \in V} \vec{\sigma}^j \; ,$$

so that

$$H_V \to \bar{H}_V \equiv - \frac{\bar{J}}{|V|} \sum_{i,j \in V} \vec{\sigma}^i \cdot \vec{\sigma}^j \; . \tag{5.11}$$

From the algebraic point of view such approximation is slightly better than the crude mean field approximation, by which, in eq. (5.10), $\vec{\sigma}^j$ is replaced by its expectation value $\langle \vec{\sigma}^j \rangle$. In fact $\bar{H}_V$ still does not make reference to the choice of a representation and furthermore the space rotational symmetry is preserved.

The model, as well as other simple spin models in the molecular field approximation, can be treated exactly[9] as the BCS model above. The different features concern the symmetry properties of the model (which corresponds to $A_{\alpha\beta} = -\bar{J} \, \delta_{\alpha\beta}$, $C_\alpha = 0$ in eq. (5.1)). In particular there is no restriction on $\vec{n}$ for the time translation invariance of the pure product states ϕ_o^n (eq.(5.7)).

The symmetries of the finite volume dynamics are the space rotations $\beta^{\vec{n}}$; they are also symmetries of α^t but not of $\alpha^t_{\pi_m}$. The dynamical group generated by $\alpha^t_{\pi_m}$ and $\beta^{\vec{n}}$ here coincides with original group rotations ($\mathcal{G}$ = G). The motion of the variables at infinity $\vec{\sigma}^\infty$ is the group of rotations around $\vec{n}$, with frequency 4J. This is the <u>energy gap</u> at $k \to 0$ of the <u>generalized Goldstone bosons</u> associated to the spontaneous break-ing of rotations by the mechanism of <u>non-symmetric bifurcation</u>, in each representation defined by the states $\phi_o^{\vec{n}}$.

6. COULOMB FERMI GAS IN UNIFORM BACKGROUND

From the many-body point of view a relevant question is the possible
relation between the plasmon energy gap and the energy gap in the BCS
theory of superconductivity and more generally the relation between energy
spectrum at k → 0 and spontaneous symmetry breaking. This is also one of
the deep questions in the gauge theory of elementary particles, in particular at the grand unified scale and for the so-called U(1) problem. As
we shall see, the framework discussed so far provides a general understanding[16] of the mechanism which underlies the above phenomenon of mass/
energy gap generation in the electron gas. In particular, we will show
that the dynamics of the electron gas is characterized by the occurrence
of variables at infinity.

To clearly display these features we will consider the Coulomb Fermi
gas in uniform background commonly regarded as a model of the theory of
metals. The model is obtained by neglecting the ion dynamics and by
approximating the ions with a uniform background of charge density ρ_B.
Roughly the model can be regarded as the zero order expansion of the full
theory in the ratio m/M between the electron and the ion mass.

6.1 The infrared regularization and the removal of the infrared cutoff

The model is defined by the formal Hamiltonian

$$H = \frac{1}{2} \int \nabla\psi^* \; \nabla\psi \, d^3x - \int \psi^*(x)\psi(y)V(x-y)\rho_B \; d^3x d^3y$$

$$+ \frac{1}{2} \int \psi^*(x)\psi^*(y)V(x-y)\psi(y)\psi(x)d^3x d^3y \quad , \qquad (6.1)$$

where $V(x-y) = e^2/|x-y|$ and ψ^*, ψ are the electron creation and annihil-
ation operators. Clearly, an infrared regularization is needed to give
a meaning to the above Hamiltonian and we will choose[16] to cut the
Coulomb potential

$$V(x) \to V(x) \; f_L(x) \quad ,$$

with $f_L(x) = f(|\vec{x}|/L)$ and f a regular function which is one inside a
sphere of radius one and vanishes outside a sphere of radius 1+ε; (as
will be clear in the following the results are independent of the particular form of f).

The so-obtained infrared cutoffed Hamiltonian H_L generates an infrared cutoffed dynamics α_L^t on the (quasi) local gauge invariant algebra
$\mathcal{Q}$ generated by ψ, ψ^* at t = 0 and the problem is the removal of such

cutoff. As a family of physically relevant states $\mathbb{F}$ we consider those states ϕ which are sufficiently regular at infinity such that

A)   The correlation functions

$$|\vec{x}|^{-1} \ \phi(A(\rho(x)-\rho_B)B)$$

with $A,B \in \mathcal{O}$, are absolutely integrable in x. This means that the election density at large distances approaches the background density ρ_B faster than $|\vec{x}|^{-2}$.

This condition guarantees the existence of the correlation functions of the electric field

$$\vec{E}(x,t) = - \int d^3y \ \ \vec{\nabla} V(x-y)(\rho(y,t)-\rho_B) , \qquad (6.2)$$

i.e. the weak convergence of $\alpha_L^t(\vec{E})$, as $L \to \infty$. To get the group law for the dynamics in the limit $L \to \infty$ one should actually have the ultrastrong convergence of α_L^t. For example, for the two-point function of the electric field one should have the integrability in the x,y variables of the correlation functions

$$|\vec{x}|^{-2}|\vec{y}|^{-2} \ <(\rho(x,t)-\rho_B)(\rho(y,t)-\rho_B)>_o \ .$$

Thus, in the following we will slightly strengthen condition A on the infrared regularity of the physical states ϕ by requiring that

A')   The correlation functions

$$|\vec{x}|^{-2}_{\ \phi}(A(\rho(x,t)-\rho_B)B), \quad A,B \in \mathcal{O} ,$$

are absolutely integrable.

By exploiting this condition one can discuss the removal of the infrared cutoff at each order of the Taylor expansion in t [22].

The covariance group of the infrared cutoffed dynamics α_L^t is the Galilei group. At arbitrary times, in particular at each order in t of a Taylor expansion around t=0, the local generation of the Galilei group by local charges can be taken as a basic feature in the characterization of the algebra $\mathcal{O}_1$ of essentially localized observables. This can actually be used as an algebraic tool to identify elements of $\mathcal{O}_1$.

One can then show that variables at infinity occur in the time evolution of local observables [16,22]. More specifically one can show [22] that

$$\frac{d^2}{dt^2} \, j_i(x,t)\Big|_{t=0} = \; : \partial_k j_k E_i:(x) + \partial_k T_{ki}(x)$$

$$-\omega_p^2 : \rho(j_i^{long} - j_i^\infty):(x) \tag{6.3}$$

where the dots : : denote Wick ordering, T_{ik} is a local gauge invariant function of ψ, ψ^* and E_i, j_i^{long} describes the longitudinal current belonging to $\mathcal{Q}_1$, (since it satisfies the above criterium), and j_i^∞ is the variable at infinity corresponding to the ergodic mean of the current.

Furthermore, the effective dynamics α_π^t gives rise to the following equations of motion

$$\frac{d}{dt} \, \alpha_\pi^t E_i(x) = -4\pi e^2 \, \alpha_\pi^t(j_i^{long}(x)) \; , \tag{6.4}$$

$$m \, \frac{d}{dt} \, \alpha_\pi^t(j_i(x)) = \alpha_\pi^t(:\rho E_i:(x) + \partial_k S_{ki}(x)) \quad , \tag{6.5}$$

$$S_{ki} = S_{ik} = (4m)^{-1} \{\psi^* \partial_i \partial_k \psi + (\partial_i \partial_k \psi^*)\psi$$

$$- \partial_k \psi^* \partial_i \psi - \partial_i \psi^* \partial_k \psi\} \; . \tag{6.6}$$

Clearly, all the above equations involve only elements of $\mathcal{Q}_1$ and provide the equation of motions for the effective dynamics.

For the Coulomb Fermi gas with uniform background the non-vanishing background density implies that the Galilei boosts $\beta^{\vec{v}}$ are spontaneously broken. In fact, given a pure ground state ϕ_o, invariant under space translations we have

$$< \frac{d}{dv_i} \, \beta^{\vec{v}}(j_k)\Big|_{\vec{v}=0} >_o = \delta_{ki} \, \rho_B \neq 0 \; . \tag{6.7}$$

As a result of the generalized Goldstone theorem, the energy spectrum at $k \to 0$ of the generalized Goldstone bosons is given by the classical motion of the variables at infinity associated to the symmetry breaking order parameter. In the case of the breaking of the Galilei boosts we have to consider the classical motion of j_i^∞, induced by the effective dynamics α_π^t. This can be done by using the equations of motion (6.4) (6.5) and the equation of motion for j_i^{long} 22

$$m \frac{d}{dt} \alpha_\pi^t \, j_i^{long} = \alpha_\pi^t (\pi - \lim_{L \to \infty} \partial_i \partial_k V_L * (:\rho E_k: + \partial_n S_{nk})). \tag{6.8}$$

The important fact is that the equation of motion for the variables at infinity, which can be obtained by taking the ergodic means of the above equations, are much more simple. Since α_π^t commutes with the space translations one has $\alpha_\pi^t \partial_k A = \partial_k \alpha_\pi^t A$, for any $A \in \mathcal{Q}_1$, and the ergodic mean of a space derivative of an element of $\mathcal{Q}_1$, vanishes on ϕ_0 and on $\phi^{\vec{v}} = (\beta^{\vec{v}}) * \phi_0$ since both are translationally invariant as states on the gauge invariant algebra $\mathcal{Q}_1$.

Furthermore the Maxwell equations

$$\partial_i E_i(x) = 4\pi e^2 (\rho(x) - \rho_B) \tag{6.9}$$

give

$$4\pi e^2 :\rho E_i: = \partial_k :E_k E_i: - \tfrac{1}{2}\partial_i :E_k^2: + 4\pi e^2 \rho_B E_i \tag{6.10}$$

and therefore

$$(:\rho E_i:)^\infty = \rho_B E_i^\infty , \tag{6.11}$$

i.e. an automatic linearization takes place. Thus, eqs. (6.4)(6.5) yield

$$\frac{d}{dt} \alpha_\pi^t (E_i^\infty) = -4\pi e^2 \, \alpha_\pi^t \, (j_i^{long})^\infty , \tag{6.12}$$

$$\frac{d}{dt} \alpha_\pi^t (j_i) = (\rho_B/m) \, \alpha_\pi^t (E_i^\infty) . \tag{6.13}$$

Similarly, from eq. (6.8) one gets[22]

$$\frac{d}{dt} \alpha_\pi^t (j_i^{long})^\infty = (\rho_B/m) \, \alpha_\pi^t (E_i^\infty) . \tag{6.14}$$

The solution for j_i^∞ is a periodic motion with frequency ω_p:

$$j_i^\infty(t) = j_i^\infty \cos \omega_p t + (\rho_B/m) E_i^\infty \sin \omega_p t / \omega_p \tag{6.15}$$

This implies that

i)   the plasmons are the generalized Goldstone bosons associated to the spontaneous breaking of the Galilei boosts;

ii) the plasmon spectrum at $k \to 0$ can be calculated exactly and it consists of a single point $\omega = \omega_p$, i.e. the plasmons are excitations with a lifetime which goes to infinity when $k \to 0$, namely the plasmons are quasi particles;

iii) the plasmon energy gap is generated by a phenomenon of non-symmetric quantum bifurcation and it is related to the non trivial classical motion of the variable at infinity j_i^∞.

The plasmon energy spectrum at $k \to 0$ can also be derived, more generally (without using the equations of motion for the effective dynamics eqs. (6.4) (6.5) (6.8)). In fact, one can directly show[16] that

$$\lim_{R\to\infty} <[G_{i,R}(t), \, j_i(0)]>_o = \rho_B \cos\omega_p t \qquad (6.16)$$

where $G_{i,R}$ is the local charge generating the Galilei boosts, and, by the connection between charge density commutators and energy spectrum, one gets the plasmon spectrum at $k \to 0$. It is worthwhile to stress that again the proof of eq. (6.16) requires a careful handling of the limit $L \to \infty$, required to define the time evolution of the Galilei charge starting from the infrared cutoffed dynamics α_L^t, and the limit $R \to \infty$, to be performed last.

REFERENCES
1)   For a brief account of the basic structures of QM$_\infty$, with emphasis on their interdisciplinary interest, see e.g. F. Strocchi, Elements of Quantum Mechanics of Infinite Systems, World Scientific, Singapore 1985.
2)   I.E. Segal, Ann. Math. 48, 930 (1947).
3)   R. Haag, in Les Problèmes Mathematiques de la Théorie Quantique des Champs, Coll. Inter. CNRS, Paris 1959;
     R. Haag and D. Kastler, J. Math. Phys. 5, 848 (1964).
4)   O. Brattelli and D.W. Robinson, Operator Algebras and Quantum Statistical Mechanics, Vol. I Springer 1979; Vol. II, Springer 1981.
5)   R. Haag and D. Kastler, J. Math. Phys. 5, 848 (1964).
6)   R. Haag, N.M. Hugenholz and M. Winnink, Commun. Math. Phys. 5, 215 (1967); D.W. Robinson, ibid. 7, 337 (1968).
7)   G.L. Sewell, Quantum Theory of Collective Phenomena, (Monographs on the Physics and Chemistry of Materials), Oxford Claredon Press 1986.
8)   J. Swieca, "Goldstone Theorem and Related Topics", in Cargèse Lectures Vol. 4, D. Kastler ed., Gordon and Breach 1970.

9) G. Morchio and F. Strocchi, Commun. Math. Phys. 99, 153 (1985); J. Math. Phys. 28, 622 (1987).

10) G. Morchio and F. Strocchi, "Infrared Problem, Higgs Phenomenon and Long Range Interactions", in Fundamental Problems of Gauge Field Theory, Erice 1985, G. Velo and A.S. Wightman dirs., Plenum 1986.

11) See e.g. S. Nakajima, Y. Toyozawa and R. Abe, The Physics of Elementary Excitations, Springer 1980.

12) D.A. Dubin and G.L. Sewell, J. Math. Phys. 11, 2990 (1970);
    G.C. Emch and H.J. Knops, J. Math. Phys. 11, 3008 (1970);
    M.B. Ruskai, Commun. Math. Phys. 20, 193 (1971);
    O. Brattelli and D.W. Robinson, Commun. Math. Phys. 50, 133 (1976);
    C. Radin, Commun. Math. Phys. 54, 69 (1977);
    H. Narnhofer, Acta Phys. Austr. 49, 207 (1978);
    G.L. Sewell, Lett. Math. Phys. 6, 209 (1982).

13) These variables have been introduced, with different motivations, by O. Landord and D. Ruelle, Commun. Math. Phys. 13, 194 (1969).

14) R. Haag, R. Kadison and D. Kastler, Commun. Math. Phys. 16, 81 (1970).

15) G. Morchio and F. Strocchi, J. Math. Phys. 28, 1912 (1987); Commun. Math. Phys. 111, 593 (1987).

16) G. Morchio and F. Strocchi, Ann. Phys. (N.Y.) 170, 310 (1986).

17) J. Bardeen, L. Cooper and J.R. Schrieffer, Phys. Rev. 108, 1175 (1957);
    P. Anderson, Phys. Rev. 112, 1900 (1958).

18) C. Kittel, Quantum Theory of Solids, J. Wiley 1963, Ch. 8.

19) W. Thirring, Comm. Math. Phys. 7, 181 (1968); Lectures in: The Many-Body Problem, Int. School of Physics, Mallorca, 1969, Plenum 1969, and references therein.

20) O. Brattelli and D.W. Robinson, Operator Algebras and Quantum Statistical Mechancis, Vol. II, Springer 1981, Ch. 6.

21) J.L. Van Hemmen, Fortsh. d. Phys. 26, 397 (1978).

22) G. Morchio, to be published.

THEORETICAL CONCEPTS IN QUANTUM PROBABILITY; QUANTUM MARKOV PROCESSES

Hans Maassen
Institute for Theoretical Physics
University of Nijmegen, Toernooiveld
6525 ED Nijmegen, The Netherlands

ABSTRACT. An introduction is given to some fundamental concepts in quantum probability, such as (quantum) probability spaces and (quantum) stochastic processes. Recent results are described relating to the question, what transition probabilities for an n-level quantum system are theoretically possible in a quantum Markov process.

Quantum probability theory is an attempt to generalise ordinary mathematical probability theory in such a way that it becomes a tool with which to discuss quantum mechanics. The theory was developed when the study of operator algebras for statistical mechanics and for field theory, made the parallels between probability theory and quantum mechanics stand out. In their fundamental paper [AFL], Accardi, Frigerio and Lewis formulated the concept of a quantum stochastic process: a small quantummechanical system, embedded in a large quantummechanical environment which evolves in a reversible way. If one wishes to satisfy the "fluctuation regression hypothesis" for the small system, the coupling to the environment must be singular, and the environment has to carry away to infinity all influences it feels from the system. This "Markov property" can only be realized in an environment with an infinite number of degrees of freedom, and the only way to deal with that theoretically is to use the formalism of von Neumann algebras and normal states. It is then wise to use this formalism from the very beginning and to pay attention to the surprising structure discovered by Tomita. This decision pays off, so it turns out. Certain puzzling difficulties, such as the problem, how to impose an initial condition on the small system without effecting the environment, are better dealt with in this language. Once the concept of a Markov process has been established, the "dilation problem" can be posed: what irreversible evolutions (or "fluctuation regressions") can be realised in these Markov processes? This problem has been a focus of study over recent years [Küm 1], [Küm 2], [EvL], [EAE], [Maa], [FrG], [LiM]. The present paper is structured as follows. In sections 1-6 the basic theory of quantum probability spaces and random variables is introduced. In sections 7, 8 and 9 quantum stochastic processes are introduced, and some recent results are described in section 10.

A. Amann et al. (eds.),
Fractals, Quasicrystals, Chaos, Knots and Algebraic Quantum Mechanics, 287–302.
© 1988 by Kluwer Academic Publishers.

1. QUANTUM PROBABILITY SPACES

The fundamental concept in quantum probability theory is that of a quantum probability space. This is a pair

$$(A, \phi),$$

where

>A is a von Neumann algebra;
>ϕ is a normal state on A, i.e.:
>
>>$\phi : A \to \mathbb{C}$ linear and weak-* continuous;
>>$\phi(a^*a) \geq 0$ for all $a \in A$;
>
>$\phi(\mathbb{1}) = 1.$

If ϕ is faithful, i.e.

$$\phi(a^*a) = 0 \quad \text{only for } a = 0,$$

then the quantum probability space (A, ϕ) is called nondegenerate. We shall only consider nondegenerate quantum probability spaces in this paper.
This concept of a quantum probability space generalises the idea of an ordinary probability space (called classical in this context). As is well known, this is a triple

$$(\Omega, \Sigma, \mathbb{P}),$$

where

Ω is a set (to be thought of as a tableau on which Chance will choose a point).

Σ is a subdivision of Ω, a σ-algebra of subsets of Ω, called the events of the theory. At least two events are in Σ: the impossible event $\emptyset$ and the sure event Ω.

$\mathbb{P}$: $\Sigma \to [0,1]$ associates to every event a probability in a σ-additive way and such that $\mathbb{P}(\emptyset) = 0$ and $\mathbb{P}(\Omega) = 1$.

Because $\mathbb{P}$ is defined on Σ, the elements of Σ are also called the measurable subsets of Ω. A function $f : \Omega \to \mathbb{C}$ is called measurable if all the sets $\{\omega \in \Omega \,|\, a \leq f(\omega) \leq b\}$, $(a, b \in \mathbb{R})$ are measurable.

(A)   A classical probability space $(\Omega, \Sigma, \mathbb{P})$ yields an example of a quantum probability space (A, ϕ) if we put:
        $A = L^\infty(\Omega, \Sigma, \mathbb{P}).$
                This is the *-algebra of all bounded and measurable functions $f : \Omega \to \mathbb{C}$, two such functions being identified if they differ only on a set of measure zero.
      $\phi : A \to \mathbb{C} : \phi(f) = \int_\Omega f(\omega) \mathbb{P}(d\omega).$
                $\phi(f)$ is the expectation of f with respect to $\mathbb{P}$.

This abstract form (A,ϕ) still contains all the information essential for the structure $(\Omega,\Sigma,\mathbb{P})$: the events are regained from (A,ϕ) by considering all the elements p of A with $p = p^2$ $(= p^*)$ (the projections). Indeed, $p^2 = p \in L^\infty(\Omega,\Sigma,\mathbb{P})$ implies that $p = 1_S$ for some $S \in \Sigma$. The probability measure $\mathbb{P}$ is then found back by defining $\mathbb{P}(S) = \phi(1_S)$. Only the concrete representation of the points of Ω is lost in the transition from $(\Omega,\Sigma,\mathbb{P})$ to $(L^\infty(\Omega,\Sigma,\mathbb{P}), \int \cdot d\mathbb{P})$.

Note that $L^\infty(\Omega,\Sigma,\mathbb{P})$ is a commutative algebra. In the converse direction, every quantum probability space (A,ϕ) with A commutative can be written in the form $(L^\infty(\Omega,\Sigma,\mathbb{P}), \int \cdot d\mathbb{P})$ for some classical probability space $(\Omega,\Sigma,\mathbb{P})$. In this sense we may say that a quantum probability space (A,ϕ) is classical if and only if A is commutative. Thus the classical is a special case of the quantummechanical.

Our main interest, of course, will be the non-classical quantum probability spaces. These group into different types according to the type of the von Neumann algebra A.

We give two examples:

(B)   An n-level system, (type I_n).

$A = M_n$, (the n×n-matrices);

$\phi(a) = \mathrm{tr}(\rho a)$, $(\rho \in M_n, \rho > 0, \mathrm{tr}\rho = 1)$.

In order that (A,ϕ) be non-degenerate, ρ is only allowed to have strictly positive eigenvalues. In particular, vector states $\phi : a \to \langle\psi,a\psi\rangle$, $(\psi \in \mathbb{C}^n)$ are excluded.

(C)   A free Bose field, (type III).

We shall describe this example below.

2. THE STANDARD FORM

Quantum probability spaces can be cast in standard form using the so-called Gel´fand-Naimark-Segal cyclic representation:

(H,π,ξ).

Here

H   is a suitably chosen Hilbert space;

π : $A \to \mathcal{L}(H)$ is a representation of A as bounded operators on H.

ξ $\in$ H is a cyclic vector for π, inducing the state ϕ on A, i.e.

$\pi(A)\xi$ is dense in H;

$$\langle\xi,\pi(a)\xi\rangle = \phi(a), \quad (a \in A).$$

This is a standard form in the sense that any two cyclic representations of (A,ϕ) must be unitarily equivalent.
For a nondegenerate quantum probability space the standard form can be obtained in the following way. Since $\phi(a^*a) = 0$ only for $a = 0$, an inner product on A is defined by $\langle a,b\rangle := \phi(a^*b)$. Then let H be the completion of A in the norm $a \to \|a\| = \langle a,a\rangle^{1/2}$. Define $\pi(a)$ on H by continuous extension of multiplication from the left: $\pi(a)b = ab$, and put $\xi = \mathbb{1}$.
In our examples the standard forms are the following:

(A) A classical probability space:
$$H = L^2(\Omega,\Sigma,\mathbb{P});$$
$$\pi(a) : \psi \mapsto a\cdot\psi, \quad (a \in A, \ \psi \in H);$$
$$\xi = 1.$$

(B) The n-level system:
$$H = M_n \text{ with inner product } \langle a,b\rangle = \mathrm{tr}(\rho a^*b);$$
$$\pi(a)b = ab;$$
$$\xi = \mathbb{1}.$$

(B') Alternatively:
$$H = \mathbb{C}^n \otimes \mathbb{C}^n \text{ (with ordinary inner product)}$$
$$\pi(a) = a\otimes\mathbb{1}$$
$$\xi = \sum_{j=1}^{n} \rho_j^{1/2}\epsilon_j\otimes\epsilon_j \quad \text{if} \quad \rho = \sum_{j=1}^{n} \rho_j|\epsilon_j\rangle\langle\epsilon_j|.$$

(C) A free Bose field in a universally invariant state.

$$H = F(L^2(\mathbb{R}^3))\otimes F(L^2(\mathbb{R}^3))$$
$$\pi(A_f) = \lambda A_f\otimes\mathbb{1} + \mu \mathbb{1}\otimes A_{\bar{f}}^*, \quad (\lambda,\mu \neq 0).$$
$$\xi = \Omega\otimes\Omega.$$

Here, F(K) is the symmetric Fock space $\oplus_{n \in \mathbb{N}}(K^{\otimes n})_{\text{symmetric}}$ over the Hilbert space K, and A_f and A_f^* are the annihilation and creation operators of a particle with wave function $f \in L^2(\mathbb{R}^3)$. The Fock vacuum vector Ω is given by $\Omega = 1\otimes0\otimes0\otimes0\ldots$.

3. TOMITA-TAKESAKI THEORY

Instead of the representation π of multiplication from the left, together with the inner product $\langle a,b\rangle = \phi(a^*b)$ on A, one could have chosen to represent A by multiplication from the right by the adjoint:

ñ(a) : b → ba*. This map ñ becomes an (antilinear) *-homomorphism if one provides A with the inner product $\langle a,b \rangle^{\sim} = \phi(ba^*)$. By associativity, the two representations commute with one another:

$$\pi(a)\tilde{\pi}(b)c = a(cb^*) = (ac)b^* = \tilde{\pi}(b)\pi(a)c.$$

The presence of two commuting representations of A on itself and two natural inner products on A leads to the Tomita-Takesaki structure [ToT], which is always there on a nondegenerate quantum probability space in standard form. We give a brief sketch of this structure.

The two inner products are connected by a - generally unbounded - positive operator Δ on H(or A):

$$\langle a,b \rangle^{\sim} = \langle a, \Delta b \rangle,$$

or, in a technically more correct formulation:

$$\langle \pi(b^*)\xi, \pi(a^*)\xi \rangle = \langle \Delta^{1/2}\pi(a)\xi, \Delta^{1/2}\pi(b)\xi \rangle.$$

It was proved by Tomita that the automorphisms of $\mathcal{L}(H)$ given by $X \to \Delta^{-it}X\Delta^{it}$, ($t \in \mathbb{R}$) leave the algebra $\pi(A)$ invariant. They therefore define a group of automorphisms of A:

$$\sigma_t^{\phi} : A \to A : a \mapsto \pi^{-1}(\Delta^{-it}\pi(a)\Delta^{it}),$$

called the <u>modular group</u> of (A,ϕ).

For a classical probability space the modular group is trivial: $\sigma_t^{\phi} = id_A$, since the two inner products coincide.

For the n-level system on the other hand, we find that the choice $\rho = \exp(-h)$ with $h = h^* \in M_n$ leads to the modular group

$$\sigma_t^{\phi}(a) = e^{ith}ae^{-ith}.$$

So in this case the state $\phi : a \to tr(\rho a)$ and the modular group σ_t^{ϕ} are related as a Gibbs state to the time evolution for which it describes thermal equilibrium. We shall return to this point in section 8.

The modular group of a quantum probability space has a nice characterisation, known as the Kubo-Martin-Schwinger (KMS) boundary condition [HHW]. An automorphism group σ_t, ($t \in \mathbb{R}$) of (A,ϕ) is said to satisfy the KMS condition if for all a,b $\in$ A the functions $\mathbb{R} \to \mathbb{C} : t \mapsto \phi(a\sigma_t(b))$ and $\mathbb{R}+i \to \mathbb{C} : t+i \mapsto \phi(\sigma_t(b)a)$ can be interpolated by a continuous function on the strip $\mathbb{R}+i[0,1]$ which is analytic on its interior $\mathbb{R}+i(0,1)$.

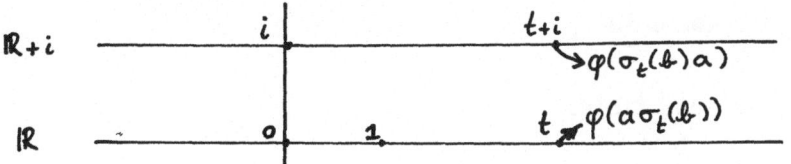

The other important element in the standard form (H, π, ξ) of (A, ϕ) is the presence of a second, antilinear representation π' of A on H, such that every operator $\pi'(a)$ commutes with all the operators $\pi(b)$. The representation π' is basically the representation $\tilde{\pi}$ of multiplication from the right, mentioned above. If S denotes the operator on H or A coming from the involution $a \to a^*$ on A, then $\tilde{\pi}$ can be written as

$$\tilde{\pi}(a) : b \mapsto ba^* = (ab^*)^* = S\pi(a)Sb,$$

so

$$\tilde{\pi}(a) = S\pi(a)S.$$

However, in the inner product $\langle \cdot, \cdot \rangle$ on H, the map $\tilde{\pi}$ is not a *-homomorphism, since $\tilde{\pi}(a)^* \neq \tilde{\pi}(a^*)$. In order to have the commutant representation π' in the same Hilbert space H as π, one must define instead:

$$\pi'(a) = J \pi(a) J,$$

where $J = \Delta^{1/2}S$, the star operator S adapted to the "wrong" inner product and made isometric.

It was proved by Tomita that not only does $\pi'(A)$ commute with $\pi(A)$, but it actually fills up its whole commutant:

$$\pi'(A) = \pi(A)' := \{X \in \mathcal{L}(H) | [X, \pi(a)] = 0 \, \forall a \in A\}.$$

Examples

(A) In classical probability J is complex conjugation and $\pi(A)$ equals its own commutant. $\sigma_t^\phi = \text{id}_A$ for all $t \in \mathbb{R}$.

(B) The n-level system $(H = M_n; \langle a, b \rangle = \text{tr}(\rho a^* b))$

$$\sigma_t^\phi(a) = \rho^{it}a\rho^{-it};$$

$$\pi(a)b = ab \quad ; \quad \pi'(a)b = b\rho^{1/2}a^*\rho^{-1/2};$$

$$J : a \mapsto \rho^{1/2}a^*\rho^{-1/2}.$$

(B') Alternatively $(H = \mathbb{C}^n \otimes \mathbb{C}^n$ with ordinary inner product)

$$\pi(a) = a \otimes \mathbb{1}; \quad \pi'(a) = \mathbb{1} \otimes a^*;$$

$$J : \psi \otimes \chi \to \bar{\chi} \otimes \bar{\psi}.$$

(C) The free Bose field.

$$\sigma_t^\phi(A_f) = (\mu/\lambda)^{2it} A_f;$$

$$\pi'(A_f) = \mu A_f^* \otimes \mathbb{1} + \lambda \mathbb{1} \otimes A_{\bar{f}}; \quad J : \psi \otimes \chi = \bar{\chi} \otimes \bar{\psi} .$$

4. MORPHISMS

In classical probability theory most objects of interest (random variables, conditional expectations, evolution in time) can ultimately be reduced to maps between probability spaces. The same can be done in this field. One considers the category of quantum probability spaces, and one studies morphisms between them [Küm 1]:

$$T : (A,\phi) \to (B,\psi).$$

This is a normal linear map $T : A \to B$ with the properties:
    T is completely positive;
        this means that for all $n \in \mathbb{N}$, T maps positive n×n
        matrices with entries in A to positive matrices with
        entries in B (when acting on the entries individually);
    $T(\mathbb{1}_A) = \mathbb{1}_B$;
    $\psi \circ T = \phi$.

These properties guarantee that the adjoint $T^* : B_* \to A_*$ maps states to states, in particular ψ to ϕ. (Complete positivity would not be needed for this, but turns out to be mathematically and physically natural to require.)

Examples.

A morphism whose inverse is also a morphism is an isomorphism. Isomorphisms preserve all the structure, and isomorphic probability spaces may be identified. An automorphism of (A,ϕ) is an isomorphism $(A,\phi) \to (A,\phi)$. The time evolution (in the Heisenberg picture) is a group of automorphisms. An embedding is a morphism $j : (A,\phi) \to (B,\psi)$ such that $j(ab) = j(a)j(b)$ and $\|j(a)\| = \|a\|$. If such an embedding is given, (A,ϕ) may be considered part of (B,ϕ). (However, in quantum probability, "the rest of (B,ψ)" is not always well-defined: systems cannot always be disentangled from their surroundings. See section 5). A conditional expectation is a morphism E which is also a projection: $E^2 = E$. See also section 5.
A transition operator on (A,ϕ) is any morphism $T : (A,\phi) \to (A,\phi)$. Such an operator has an adjoint which acts on states χ on A by

$$(T^*\chi)(a) = \chi(Ta).$$

If T is not an automorphism, then T^* "smears out" the probability distribution defined by χ, as the picture suggests.

5. QUANTUM RANDOM VARIABLES

By a <u>quantum</u> <u>random</u> <u>variable</u> we shall mean an embedding of one quantum probability space into another.

This is our generalisation of a classical random variable, which is a measurable function $X : \Omega \to \mathbb{R}$ on a probability space $(\Omega,\Sigma,\mathbb{P})$.

Such a function naturally determines an embedding
$j : L^\infty(\mathbb{R},\mu) \to L^\infty(\Omega,\mathbb{P})$ via

$$j(f) = f \circ X.$$

Here, μ is the induced measure $\mu = \mathbb{P} \circ X^{-1}$.
Generalising $L^\infty(\mathbb{R},\mu)$ to (A,ϕ) and $L^\infty(\Omega,\mathbb{P})$ to (B,ψ), one obtains the above <u>broad</u> concept of a quantum random variable. If one only makes $L^\infty(\Omega,\mathbb{P})$ non-commutative, changing it to a general (B,ψ), but keeping $L^\infty(\mathbb{R},\mu)$, one obtains a projection-valued measure P on $\mathbb{R}$ via

$$P(S) = j(1_S), \qquad (S \in \mathbb{R} \text{ measurable}).$$

This in its turn defines a self-adjoint operator X via

$$X = \int_{\mathbb{R}} \lambda P(d\lambda).$$

Conversely, every self-adjoint operator in the standard representation of (B,ψ) determines a projection-valued measure P on $\mathbb{R}$ via the spectral theorem. If all its projection operators $P(S)$ are in B, it determines an embedding of $L^\infty(\mathbb{R},\mu)$ into (B,ψ). Hence a self-adjoint operator, affiliated to B, determines a quantum random variable in a <u>narrow</u> sense.
To summarise: Position and momentum are quantum random variables in the narrow sense, but spin (in any direction) is a quantum random variable in the broad sense, since it is an embedding of the non-commutative space (M_2,ϕ) into some (B,ψ).

6. CONDITIONAL EXPECTATIONS

A quantum random variable $j : (A,\phi) \to (B,\psi)$ is said to be <u>conditionable</u> if there exists a morphism $E : (B,\psi) \to (A,\phi)$ such that $E \circ j = \text{id}_A$. In this case one has

$$(j \circ E)^2 = j \circ (E \circ j) \circ E = j \circ E,$$

so $j \circ E$ is a conditional expectation, projecting onto $j(A)$. (Identifying $j(A) \subset B$ with A itself, we also call E a conditional expectation). If $b = b^* \in B$ is a random variable (narrow sense), then $E(b) \in A$ is its expectation, given j.
In contrast to the classical situation, conditional expectations do not always exist:

<u>Theorem</u> (Takesaki ([Tak] 1972)). The random variable $j : (A,\phi) \to (B,\psi)$ is conditionable if and only if for all $t \in \mathbb{R}$:

$$\sigma_t^\psi \circ j(A) \subset j(A).$$

7. QUANTUM STOCHASTIC PROCESSES

A quantum stochastic process is a family of quantum random variables indexed by time:

$$j_t : (A,\phi) \to (B,\psi), \qquad (t \in T).$$

Here, the time T may be $\mathbb{R}$, $\mathbb{Z}$, $\mathbb{R}_+$ or $\mathbb{Z}_+$. Such a process determines correlation functions $F_n : T^n \times A^n \to \mathbb{C}$ of all orders $n \in \mathbb{N}$ by

$$F_n(t_1,\ldots,t_n;a_1,\ldots,a_n) = \psi(j_{t_1}(a_1)j_{t_2}(a_2)..j_{t_n}(a_n)).$$

The converse also holds:

<u>Reconstruction theorem</u> (Accardi, Frigerio, Lewis ([AFL], 1981)).

Given (A,ϕ) and all the correlation functions F_n, the quantum stochastic process $j_t : (A,\phi) \to (B,\psi)$ is determined up to isomorphism (provided that B is chosen in the minimal way).

A drawback on this result is the fact that in physical practice only the "pyramidal" correlation functions

$$\psi(j_{t_1}(a_1)\ldots j_{t_n}(a_n)j_{t_n}(b_n)\ldots j_{t_1}(b_1)),$$

with $t_1 \leq t_2 \leq \ldots \leq t_n$ and $a_1,\ldots,a_n$, $b_1,\ldots,b_n$ A can be actually measured.

8. STATIONARITY AND THERMAL EQUILIBRIUM

A quantum stochastic process $j_t : (A,\phi) \to (B,\psi)$ is called <u>stationary</u> if T is a group (i.e. $T = \mathbb{R}$ or $T = \mathbb{Z}$) and there is a representation $t \mapsto \tau_t$ of T in the automorphisms of (B,ψ) such that $j_t = \tau_t \circ j_o$ for all $t \in T$:

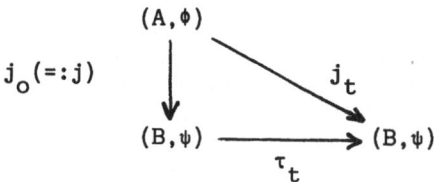

$$j_o(=:j)$$

Physically, the natural way to think of a stationary quantum stochastic process is to consider (A,ψ) as a small quantum system, embedded (by j) in an environment (B,ψ), on which a reversible dynamics τ_t is acting.

This image, and the interpretation of the modular group σ_t^ψ given in section 3, leads to the following definition: A stationary quantum stochastic process is said to occur in <u>thermal equilibrium</u> at inverse temperature β if

$$\tau_t = \sigma_{t/\beta}^\psi , \qquad (t \in T).$$

We note that this concept of thermal equilibrium is purely quantummechanical. Let us now consider such a process $j_t = \tau_t \circ j$ and suppose that it is conditionable: $E \circ j = id_A$ for some $E : (B,\psi) \to (A,\phi)$. It then follows from Takesaki's theorem that for all $t \in T$:

$$j_t(A) = \tau_t \circ j(A) = \sigma_{t/\beta}^\psi \circ j(A) = j(A).$$

Hence the minimal choice for (B,ψ) is just $(j(A),\phi \circ E)$, and the process collapses to a reversible evolution $E \circ \tau_t \circ j$ on (A,ϕ). Therefore, apart from this trivial possibility, the properties of being conditionable and occurring in thermal equilibrium exclude one another for stationary processes. In particular, (non-trivial) Markov processes, being conditionable by definition, cannot occur in thermal equilibrium. A disappointing conclusion which has been emphasised by several authors (e.g. [AFL], [SJG]). However, the situation for thermal equilibrium is even worse then this, as appears from the following theorem.

<u>Theorem</u> (Frigerio, Lindblad, Maassen (1986))

Stationary quantum stochastic processes in thermal equilibrium are <u>deterministic</u> in the sense that

$$A_{(-\infty,0]} = A_{(-\infty,\infty)}.$$

Here, for an interval $I \subset \mathbb{R}$, A_I denotes the von Neumann algebra generated by $\bigcup_{t \in I} j_t(A)$.

For deterministic processes, the future stochastic variables can be approximated by those from the past.

9. QUANTUM MARKOV PROCESSES

In what follows we shall only consider stationary conditionable processes. Barring reversible processes these are never in thermal equilibrium. We may now draw the diagram

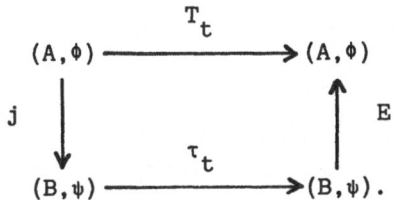

This diagram shows how the existence of E naturally leads to a family $T_t = E \circ \tau_t \circ j$, $(t \in T)$, of morphisms on (A, ϕ). These morphisms are not invertible in general, and hence may be said to form an <u>irreversible</u> evolution. In general there is no reason to expect even the <u>semigroup property</u>

$$T_{t+s} = T_t \circ T_s, \qquad (t, s \geq 0),$$

because of memory effects: the past behaviour of the quantum system (A, ϕ) may influence its future via perturbations of the environment. Such influences are excluded in the following definition: A stationary conditionable quantum stochastic process will be called a <u>quantum Markov process</u> if

$$E_{(-\infty, 0]} A_{[0, \infty)} = A_{\{0\}}.$$

(The existence of a conditional expectation E_I onto A_I follows from the existence of E.) This Markov property is quite the opposite of the deterministic property mentioned in section 8: the future $A_{[0, \infty)}$, far from being determined by the past, is independent of it, except for a correlation through the present. A consequence of this lack of memory is the above-mentioned semigroup property. We may then write

$$T_t = e^{tL}, \qquad (t \geq 0).$$

Such a semigroup of morphisms on a quantum probability space is called a <u>dynamical</u> <u>semigroup</u> in the literature. A markov process in which it occurs is called a (<u>Markov</u>) <u>dilation</u> of the dynamical semigroup [EvL], [EAE], [Küm 1].

10. MARKOV DILATIONS, RESULTS

In classical probability all dynamical semigroups possess a unique
Markov dilation. Due to this close correspondence, the term "Markov
process" is sometimes used to indicate the semigroup of transition
operators, as described by the Fokker-Planck equation, and sometimes
for the whole process, described by the Langevin equation. In quantum
probability these concept should be clearly distinguished, since Markov
dilations of dynamical semigroups need not be unique, neither do they
always exist. This is due to the fact that the semigroup T_t $(t \geq 0)$
only determines the pyramidal correlations:

$$\psi(j_{t_1}(a_1)..j_{t_n}(a_n)j_{t_n}(b_n)..j_{t_1}(b_1))$$

$$= \phi(a_1 T_{t_2-t_1}(a_2..T_{t_n-t_{n-1}}(a_n b_n)..b_2)b_1),$$

$$(t_1 \leq t_2 \leq .. \leq t_n \; ; \; a_1,...,a_n,b_1,...,b_n \in A).$$

In this section we shall take $(A,\phi) = (M_n, tr(\rho \cdot))$, as in example B of
section 1. By a theorem of Lindblad [Lin] every dynamical semigroup on
(A,ϕ) is of the form $T_t = e^{tL}$ with

$$L(x) = i[h,x] + \sum_{j=1}^{k} v_j^* x v_j - \frac{1}{2}\{v_j^* v_j, x\},$$

where $h \in M_n$ is self-adjoint and $v_j \in M_n$ arbitrary $(j = 1,...,k)$,
$(k \in \mathbb{N})$. We use the notation $\{a,b\} := ab+ba$.
   Which of these semigroups admit a Markov dilation? This question
(the dilation problem for M_n) is still open, but we shall see several
results below. A first requirement on L is, of course, that $\phi \circ L = 0$.
It can furthermore be shown [Küm 1] that one must have

$$L \circ \sigma_t^\phi = \sigma_t^\phi \circ L, \qquad (t \in \mathbb{R}).$$

All dilations in continuous time, known up to now, are of the form

$$
\begin{array}{ccc}
(M_n,\phi) & \xrightarrow{\ e^{tL}\ } & (M_n,\phi) \\
{\scriptstyle id \otimes \mathbb{1}} \downarrow & & \uparrow {\scriptstyle id \otimes \nu} \\
(M_n \otimes N, \phi \otimes \nu) & \xrightarrow[\ U_t^*(\mathbb{1} \otimes S_t(\cdot))U_t\ ]{} & (M_n \otimes N, \phi \otimes \nu)
\end{array}
$$

Here (N,ν) is a quantum white noise or noise, i.e. a quantum
probability space with independent subspaces $(N_{[s,t]}, \nu)$ for all time
intervals $[t,s]$, $(N_{(-\infty,\infty)} = N)$, and a group of automorphisms $S_t : N \to N$

mapping $N_{[s,u]}$ into $N_{[s+t,u+t]}$. The noise is used as input for a Langevin equation whose solution is a family U_t ($t \geq 0$) of unitary operators on the standard Hilbert space of $(M_n \otimes N, \phi \otimes v)$:

$$dU_t = (d(\text{noise}))U_t \quad ; \quad U_o = 1.$$

Physically, U_t is the evolution operator in the interaction picture between system and environment. The classical noises are white noise and the Poisson processes of different intensities (also called "shot noise" by physicists). In quantum probability theory the list has been extended to include the following noises. Bose noise or quantum Brownian motion [CoH] is generated by $A_t = A(1_{[0,t]})$, the integrated Bose field on the real line, as described in section 2. Fermi noise F_t is constructed in very much the same way, replacing commutation relations by anticommutation relations [ApH]. Clifford noise C_t is given by $C_t = F_t + F_t^*$ [BSW]. The quantum Poisson process P_t over an arbitrary measure space (M,μ) (unnormalized quantum probability space) is constructed on the Fock space over the standard representation (H,π,ξ) of $L^\infty(\mathbb{R},dt) \otimes (M,\mu)$ as follows [TrM]:

$$P_t(a) = W(\xi)^* d\Gamma(\pi(1_{[0,t]} \otimes a))W(\xi).$$

Here, $W(\eta)$ is the Weyl operator associated to the vector $\eta \in H$, and $d\Gamma$ is the second quantisation map. We now briefly list recent results as to what semigroups can be dilated using these noises.

Theorem 10.1 (Kümmerer, Maassen (1985) [KüM])
    A dynamical semigroup e^{tL} on (M_n, ϕ) admits a Markov dilation in a classical environment if and only if L is of the form

$$L(x) = i[h,x] + \sum_{j=1}^{k} (a_j x a_j - \frac{1}{2}\{a_j^2, x\}) + \sum_{i=1}^{\ell} \kappa_i(u_i^* x u_i - x),$$

    where $h, a_1, \ldots, a_k \in M_n$ are self-adjoint, $\kappa_1, \ldots, \kappa_\ell > 0$ and $u_1, \ldots, u_\ell \in M_n$ are unitary.

Note that $\text{tr}(L(x)) = 0$ for all x, so that we may always take $\phi = \frac{1}{n}\text{tr}$ in this classical case. The dilations employ Brownian motion B_t and Poisson processes P_t of various intensities. They can be found by solving the Langevin equations

$$dU_t = dP_t(u-1)U_t$$

for the u-terms, and

$$dU_t = (ia \otimes B_t - \frac{1}{2}a^2 \otimes \mathbb{1}dt)U_t$$

for the a-terms.

Theorem 10.2 (Frigerio, Gorini ([FrG], 1984)).

The semigroup e^{tL} on (M_n, ϕ) admits a Markov dilation using Bose noise if and only if L is of the form

$$L(x) = i[h,x] + \sum_{j=1}^{k} (1-e^{-\beta_j})^{-1}\Big((v_j^* x v_j - \frac{1}{2}\{v_j^* v_j, x\})$$

$$+ e^{-\beta_j}(v_j x v_j^* - \frac{1}{2}\{v_j v_j^*, x\})\Big),$$

with $h = h^* \in M_n$ such that $\sigma_t^\phi(h) = h$ and $v_j \in M_n$ such that $\sigma_t^\phi(v_j) = e^{i\beta_j t} v_j$, $(j = 1, \ldots, k)$.

For the v-terms one solves the Langevin equation

$$dU_t = (v \otimes dA_t^* - v^* \otimes dA_t - \frac{1}{2} \frac{v^* v + e^{-\beta} v v^*}{1-e^{-\beta}} \otimes \mathbb{1}dt) \, U_t.$$

Theorem 10.3 (Frigerio, Maassen (FrM] 1986)).

The semigroup e^{tL} on (M_n, ϕ) admits a Markov dilation using the quantum Poisson process if and only if L is of the form

$$L(x) = i[h,x] + (id \otimes \mu)(u^*(x \otimes \mathbb{1})u)$$

for some quantum measure space (M, μ) and some unitary $u \in M_n \otimes M$ satisfying $\sigma_t^{\phi \otimes \mu}(u) = u$.

The dilation is based on the Langevin equation

$$dU_t = dP_t(u - \mathbb{1})U_t.$$

We note that, although the semigroups to be dilated using Brownian motion or Bose noise, satisfy detailed balance, this is by no means a condition for dilatability. Indeed, the "Poisson terms" (containing u) in 10.1 and 10.3 do not satisfy the detailed balance condition. Moreover, as a word of warning, it should be said that the necessity to use any noise at all has never been proved. The final theorem which we mention does not presuppose the use of a noise (neither in fact did the first). It can be used to derive theorem 10.3 in a simple way, and it reduces the analysis of dilations in continuous time to that in discrete time. We cite its restriction to M_n:

Theorem 10.4. (Kümmerer ([Küm 2] 1986))

Let T_t ($t \geq 0$) denote a dynamical semigroup on (M_n, Φ). Then the following are equivalent.

(i) T_t admits a Markov dilation;

(ii)   For each $t_o > 0$ the discrete time semigroup $m \mapsto (T_{t_o})^m$ has a Markov dilation;

(iii) $T_t = \lim_{j \to \infty} e^{\lambda_j (T_j - id) t}$,

where for all $j \in \mathbb{N}$ the semigroup $m \to T_j^m$ admits a Markov dilation, and $\lambda_j > 0$.

ACKNOWLEDGEMENT

The author gratefully acknowledges support from the Netherlands Organisation for the Advancement of Pure Research ZWO.

REFERENCES

AFL   Accardi L., Frigerio A., Lewis J.T.: Quantum Stochastic Processes, Publ. RIMS, Kyoto, 18 (1982) 97-133.

ApH   Applebaum D., Hudson R.L.: Fermion Itô's formula and Stochastic Evolutions, Commun. Math. Phys. 96 (1984) 473.

BSW   Barnett C., Streater R.F., Wilde I.F.: The Itô-Clifford Integral, J. Func. Anal. 48 (1982) 172-212.

CoH   Cockroft A.M., Hudson R.L.: Quantum Mechanical Wiener Processes, J. Multiv. Anal. 7 (1977) 107-124.

EAE   Emch G.G., Albeverio A., Eckmann J.-P.: Quasi-free generalised K-flows, Rep. Math. Phys. 13 (1978) 73-85.

EvL   Evans D.E., Lewis J.T.: Dilations of dynamical semigroups, Commun Math. Phys. 50 (1976) 219-227.

FrG   Frigerio. A., Gorini V.: Markov Dilations and Quantum Detailed Balance, Commun. Math. Phys. 93 (1984) 517.

FrM   Frigerio. A., Maassen H.: Quantum Poisson processes and dilations of dynamical semigroups. Preprint, Nijmegen.

302

HHW   Haag R., Hugenholtz N.M., Winnink M.: On the equilibrium states in quantum statistical mechanics, Commun. Math. Phys. 5 (1967) 215-236.

Küm 1   Kümmerer B.: Markov Dilations on W*-algebras. J. Func. Anal. 63 (1985) 139-177.

Küm 2   Kümmerer B.: Markov Dilations and Non-Commutative Poisson Processes. Preprint, Tübingen.

KüM   Kümmerer B., Maassen H.: The essentially commutative dilations of dynamical semigroups on M_n, Commun. Math. Phys. 109 (1987) 1-22.

Lin   Lindblad G.: On the generators of quantum dynamical semigroups. Commun. Math. Phys. 48 (1976) 119-130.

Maa   Maassen H.: Quantum Markov Processes on Fock space described by integral kernels. In: "Quantum Probability and Applications II", Lecture Notes in Mathematics 1136, Springer 1985.

Tak   Takesaki M.: Conditional Expectations in von Neumann Algebras. J. Func. Anal. 9 (1972) 306-321.

ToT   Takesaki M.: Tomita's Theory of Modular Hilbert Algebras and its Applications, Lecture Notes in Mathematics 128, Springer 1970.

SJG   Schramm P., Jung R., Grabert H.: A closer look at the quantum Langevin equation: Fokker Planck equation and quasi-probabilities. Phys. Lett. 107A (1985) 385-389.

THE LARGE DEVIATION PRINCIPLE IN STATISTICAL MECHANICS

J.T. Lewis
School of Theoretical Physics
Dublin Institute for Advanced Studies
10 Burlington Road
Dublin 4
Ireland

The "method of the largest term" or "Laplace's method", as it is various-
ly called, is at the heart of statistical mechanics but making it completely convincing can be a tedious business. Varadhan's Large Deviation
Principle is a highly efficient way of organizing such arguments which
is attracting growing interest in the field of statistical mechanics.It
not only furnishes new, clearer proofs of old results but it provides
the means for solving problems which other methods have not touched.

A survey of applications to classical statistical mechanics is to
be found in the book of Ellis [5] together with a comprehensive treatment
of the Large Deviation Principle. The serious student cannot do better
than study §3 of Varadhan's original paper on the subject [10]. A short
expository account of the Large Deviation Principle is given in [6] and
a review of applications to the Boson gas in [2]. Varadhan's Theorem is
an essential ingredient in the treatment of the Huang-Yang-Luttinger
model of a boson gas interacting via a repulsive hard-core which is given
in [1]. It is proved in [2] that the grand canonical distribution of the
particle number density in the free boson gas satisfies the Large Deviation Principle; this proof makes use of the existence in the thermodynamic limit of the grand canonical pressure for the free boson gas. An alternative proof, based on the existence of the canonical free energy density, is given in [7].

In [3], a quantum system of n identical spins of magnitude j is
considered; we introduce an integrated density of states of definite
total spin angular momentum. The underlying sequence of probability
measures is proved to satisfy the Large Deviation Principle. The Berezin-
-Lieb inequalities are used to obtain upper and lower bounds for the
limiting specific free-energy of the spins interacting with a second
quantum system. The method is illustrated by applications to the strong-
-coupling BCS model and to the Dicke maser model. In [8], the method is
applied to a detailed study of the phase-transition in the Dicke maser
model. Duffield and Pulè [4] have supplemented the method with techniques
from convex analysis to obtain an expression for the free-energy density
of the full BCS model. Raggio [9] has used similar ideas to solve the
heterogeneous Dicke maser model.

A. Amann et al. (eds.),
Fractals, Quasicrystals, Chaos, Knots and Algebraic Quantum Mechanics, 303–304.
© 1988 by Kluwer Academic Publishers.

304

REFERENCES

[1]   M. van den Berg, J.T. Lewis, J.V. Pulè : 'Some Models of an
       Interacting Boson Gas'. *Commun.Math.Phys.* (in press).

[2]   M. van den Berg, J.T. Lewis, J.V. Pulè : 'Large Deviations
       and the Boson Gas'. Conference Proceedings : *Stochastic
       Mechanics and Stochastic Processes*, Swansea 1986,
       ed. I.M. Davies and A. Truman, *Springer Lecture Notes
       in Mathematics* (in press).

[3]   W. Cegła, J.T. Lewis, G.A. Raggio : 'The Free Energy of Quantum
       Spin Systems and Large Deviations'. *Commun.Math.Phys.*
       (in press).

[4]   N. Duffield, J.V. Pulè : 'Thermodynamics of the Full BCS Model
       through Large Deviations'. *Letters in Math.Phys.* $\underline{14}$ 329
       (1987).
       'A New Method for the Thermodynamics of the BCS Model'.
       (submitted to *Commun.Math.Phys.*).

[5]   R.S. Ellis : *Entropy, Large Deviations and Statistical
       Mechanics*. Springer, New York 1985.

[6]   J.T. Lewis : 'The Large Deviation Principle in Statistical
       Mechanics: An Expository Account'. Conference Proceedings:
       Stochastic Mechanics and Stochastic Processes, Swansea
       1986,
       ed. I.M. Davies and A. Truman, *Springer Lecture Notes
       in Mathematics* (in press).

[7]   J.T. Lewis, J.V. Pulè, V.A. Zagrebnov : 'The Large Deviation
       Principle for the Kac Distribution'. (submitted to *Helv.
       Phys.Acta*).

[8]   J.T. Lewis, G.A. Raggio : 'The Equilibrium Thermodynamics of
       a Spin-Boson Model'. *J.Stat.Phys.* (in press).

[9]   G.A. Raggio : 'The Free Energy of the Full Spin-Boson Model'.
       (in course of preparation).

[10]   S.R.S. Varadhan : 'Asymptotic Probabilities and Differential
       Equations'. *Comm.Pure Appl.Math.* $\underline{19}$ 261 – 286 (1966).

CHIRALITY AS A CLASSICAL OBSERVABLE IN ALGEBRAIC QUANTUM MECHANICS

Anton Amann
Laboratory of Physical Chemistry
ETH-Zentrum
CH-8092 Zürich
Switzerland

ABSTRACT. Ordinary Hilbert-space quantum mechanics leads to a wrong prediction for the ground state of chiral molecules such as alanine. This does not mean that quantum mechanics is incorrect but only that it is not applied properly. A detailed analysis shows that chirality corresponds to a *classical observable* (a superselection rule) which is generated by the *environment*, i.e. by the influence of an *infinite system*. For both, classical observables and infinite systems, Hilbert space quantum mechanics is inappropriate and has to be replaced by algebraic quantum mechanics.
    Two models for chirality are discussed :
  - The *spin-boson model*, where the single (eventually chiral) molecule is described by a two-level system. The infinitely many bosons of the model mimick the radiation field (the environment) which is inseparably coupled to the molecule.
  - The *Ising model with a transverse field*, which is built up of infinitely many spins representing, e.g., an infinite crystal.
Chiral KMS-states (thermodynamic states) arise only in the latter model. It is shown that this result fits nicely into a more subtle discussion of the different notions of states and their interpretation in algebraic quantum mechanics. For single individual molecules chirality may only be described on the level of *pure* states of the system. The possibility of a phase transition in the spin-boson model is discussed.

1. INTRODUCTION

In 1927, F. Hund ([1]) was the first to recognize that the existence of optical isomers (chiral molecules) is not easily reconciled with the first principles of quantum mechanics. Consider, for example, the molecule alanine with the chemical formula $CH_3CH(NH_2)COOH$. For this molecule the Hilbert-space formalism of traditional quantum mechanics predicts a *nondegenerate space-reflection invariant* ground state. This prediction contradicts all experimental results : Space-reflection invariant pure states for alanine do not exist. The experimentally observed states of lowest energy are energetically degenerated and chiral, that is, the

305

A. Amann et al. (eds.),
Fractals, Quasicrystals, Chaos, Knots and Algebraic Quantum Mechanics, 305–325.
© 1988 by Kluwer Academic Publishers.

molecules exist only in a left- or righthanded form:

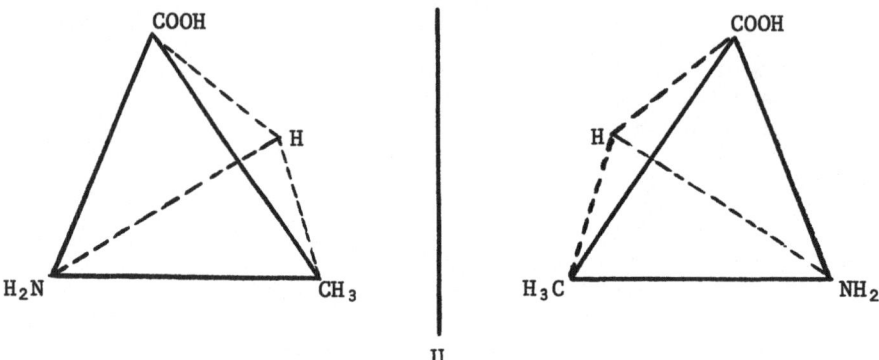

Figure 1: The left- and right-handed forms of alanine are transformed into one another by a space--reflection U.

If $\hat{H}_{Ala}$ is the Hamiltonian for alanine and Ψ_0 (Ψ_1) the state vector for the ground state (first excited state) with energy E_0 (E_1)

$$\hat{H}_{Ala}\Psi_0 = E_0\Psi_0$$
$$\hat{H}_{Ala}\Psi_1 = E_1\Psi_1 \quad , \tag{1}$$

handed state vectors (which are *not* eigenfunctions of $\hat{H}_{Ala}$) are easily constructed as follows :

$$\Psi_{\pm} \overset{\text{def}}{=} \frac{1}{\sqrt{2}} (\Psi_0 \pm \Psi_1). \tag{2}$$

If the unitary operator $\hat{U}$ represents space reflection, one has

$$\hat{H}_{Ala}\hat{U} = \hat{U}\,\hat{H}_{Ala}$$
$$\hat{U}\,\Psi_0 = \Psi_0 \tag{3}$$
$$\hat{U}\,\Psi_1 = -\Psi_1 \quad ,$$

whereas the handed state vectors Ψ_+ and Ψ_- are transformed into one another :

$$\hat{U}\,\Psi_+ = \Psi_- \quad , \quad \hat{U}\,\Psi_- = \Psi_+ \tag{4}$$

The 'paradox of optical isomers' (F. Hund) may then be posed in a different way : Supposing the handed state vectors $\Psi_{\pm}$ to exist, why is it impossible to generate the proper ground state vector Ψ_0 as a coherent superposition $\Psi_0 = \frac{1}{\sqrt{2}} (\Psi_+ + \Psi_-)$? The interesting aspect of chirality is thus not only the existence of handed states but also the breakdown of the universal validity of the superposition principle.

If the phenomenon of chirality disturbs the quantum physicist, the chemist on the other hand is disturbed by the fact that handed states may *not* arise even if a center of chirality is there. The typical example of this situation is NHDT, a derivate of ammonia NH_3 :

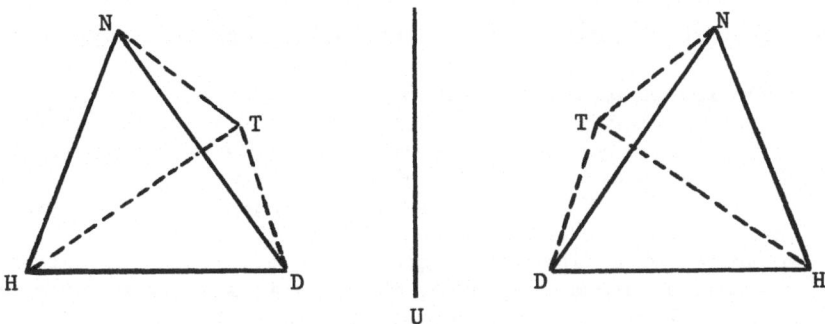

Figure 2: Handed forms of NHDT, which do
not exist as stationary states.

In NHDT (and NH_3) the experimentally observed ground state is space-reflection invariant and stationary handed states (Fig.2) do not exist.

The explanation of the phenomenon of chirality which was given by Hund may be paraphrased as follows : "*If* the molecule alanine is in a handed state, then it will remain in this state for a very long time." This explanation relies on the fact that the difference between the energies E_1 and E_0 of the first excited state and the ground state of $\hat{H}_{Ala}$ is very small, $E_1 - E_0 \cong 10^{-70}$ a.u. (atomic units, 1 a.u. $\cong$ 628 kcal/mol). Comparatively, the same difference for NH_3 is at about 10^{-6} a.u. ([2]).

Hund's explanation may be criticized for various reasons (cf.[2]):
- It is not clear, how handed configurations arise.
- Arbitrary small interactions between the molecule and its environment may perturb the handed states. A decay of $\Psi_\pm$ to the ground state Ψ_0 by radiation interaction may take place.
- There exist chiral molecules for which it is impossible to prepare reflection-invariant pure states. Why ?

The discussion of the 'paradox of optical isomers' in the last 60 years (cf. [2] for references) made clear the following aspects of the problem :
- Chirality of a single molecule corresponds to a *classical observable* (i.e., an observable which in every physical state has a dispersion-free value : +1 or -1 in our example).
- This classical observable is generated by the influence of the *environment*, i.e. by the influence of an infinite system.
- There is a sort of phase transition arising in the problem to the extent that the molecule is chiral if $(E_1 - E_0)$ exceeds a certain critical value.

For both, classical observables and infinite systems, ordinary Hilbertspace quantum mechanics is inappropriate (cf.[3]) and has to be replaced by the slightly generalized formalism of *algebraic quantum mechanics*.

2. ALGEBRAIC QUANTUM MECHANICS

2.1. The formalism

In W*-algebraic quantum mechanics (cf.[3,4]) a physical system is described by
- a W*-algebra of observables M
  [M may be viewed as a *-subalgebra of the *-algebra of all bounded
  linear operators $B(H)$ on a Hilbertspace H with the property :
  $M = (M')'$ where $S' = \{x \in B(H) | xy = yx, \forall y \in S\}$ is the commutant of
  a subset S of $B(H)$. A W*-algebra represented in this way on a
  Hilbertspace is called a von Neumann algebra]
- an action of a kinematical group G (e.g., space reflection $\mathbb{Z}_2$)
- a dynamical group $\{\alpha_t | t \in \mathbb{R}\}$, where α_t is an automorphism of M for
  every $t \in \mathbb{R}$ with $\alpha_{t_1} \circ \alpha_{t_2} = \alpha_{t_1 + t_2}$, $t_1, t_2 \in \mathbb{R}$.

Examples for W*-algebras are
- the algebras of $n \times n$-matrices, $n = 1, 2, \ldots$,
- algebras of essentially bounded measurable functions on a phase
  space (describing, e.g., a point particle of classical mechanics).

A C*-algebra is a *-algebra which is *-isomorphic to a *norm-closed*
*-subalgebra of operators on a Hilbertspace H. The norm of an operator
$x \in B(H)$ is given by $\|x\| = \sup_{\substack{\xi \in H \\ \|\xi\| = 1}} \{\|x\xi\|\}$. Every W*-algebra is a C*-algebra
but not conversely.

A *state* ϕ on the *-algebra M is a linear positive normalized
functional $\phi : M \longrightarrow \mathbb{C}$:

$$\text{(i)} \quad \phi(x + \lambda y) = \phi(x) + \lambda \phi(y)$$
$$\text{(ii)} \quad \phi(x^*x) \geq 0 \quad , \quad \forall x \in M \tag{5}$$
$$\text{(iii)} \quad \phi(1) = 1 \quad , \quad \text{where} \quad 1 \in M \quad \text{is the unit operator.}$$

A state ϕ is called *pure*, if every convex decomposition $\phi = \lambda \phi_1 + (1-\lambda)\phi_2$,
$0 < \lambda < 1$, into states ϕ_1 and ϕ_2 is trivial : $\phi = \phi_1 = \phi_2$. If $M \cong B(H)$ is
the algebra of all bounded linear operators on a Hilbertspace H, every
pure state ϕ on M is a vector state, i.e., given by $\phi(x) = \langle \xi | x \xi \rangle$, $x \in M$,
for a suitable vector $\xi \in H$.

Classical observables of the system described by a W*-algebra M
are nontrivial selfadjoint elements of the center $Z(M) \stackrel{\text{def}}{=} \{x \in M | xy = yx,$
$\forall y \in M\}$. For every pure state $\Psi : M \longrightarrow \mathbb{C}$ an arbitrary element $x = x^*$ (x^*
is the adjoint of $x \in M$) has a dispersion-free value $\Psi(x) : \Psi(x^2) = (\Psi(x))^2$.

2.2. Infinite systems in algebraic quantum mechanics

The algebra C which describes an infinite system is built up of 'local'
subalgebras C_n , $n \in I$, where I is an index set, e.g., $I = \{1,2,3,\ldots\}$.
The algebras C_n refer to 'finite' systems (e.g. systems of finitely many
degrees of freedom or of finite volume). The union $(\bigcup_{n \in I} C_n)$ contains all
'local' observables. The closure in norm

$$\left\{ \bigcup_{n \in I} C_n \right\} \longrightarrow = \overline{\left\{ \bigcup_{n \in I} C_n \right\}}^{\text{norm}} = C$$

can be done in a unique way. The C*-algebra C is not yet an observable
algebra (a W*-algebra). To find the relevant W*-algebra, say M, $C \subseteq M$,

one has to select a faithful representation $\pi : C \longrightarrow B(H)$ of C by bounded linear operators on a Hilbertspace H and takes M as the smallest W*-algebra containing $\pi(C)$, $M = \pi(C)''$. If C refers to a system of finitely many degrees of freedom there is a unique completion to a W*-algebra M (as asserted by the Stone – von Neumann theorem (cf.[5])). If C refers to a system of infinitely many degrees of freedom, there exist infinitely many disjoint representations $\pi_j : C \longrightarrow B(H_j)$, $j \in J$. Accordingly infinitely many different observable algebras can be found, corresponding to *different physical situations* (cf.[3]).

Two representations π_1 and π_2 with generated W*-algebras M_1 and M_2 are called *disjoint*, if the W*-algebra M generated by the direct sum representation $\pi_1 \oplus \pi_2 : C \ni x \longrightarrow \pi_1(x) \oplus \pi_2(x)$ on

$$H_1 \oplus H = \{(\xi,\eta) \,|\, \xi \in H_1, \ \eta \in H_2 \,; \ \langle(\xi_1,\eta_1)|(\xi_2,\eta_2)\rangle \overset{\text{def}}{=} \langle\xi_1|\xi_2\rangle_{H_1} + \langle\eta_1|\eta_2\rangle_{H_2}\}$$

(6)

is given as

$$M = M_1 \oplus M_2 = \{(x,y) \in B(H_1) \oplus B(H_2) \,|\, x \in M_1 , \ y \in M_2\}.$$

(7)

In particular, M then contains the central observables $c1_1 \oplus d1_2$, $c,d \in C$, where 1_j is the unit operator in M_j, $j = 1,2$. *Thus disjoint representations (arising only with systems of infinitely many degrees of freedom) offer the opportunity to introduce an observable algebra M with classical observables* ([6,7,8]). The above example just refers to chirality where the corresponding two-valued classical observable is of the form $1_1 \oplus (-1_2)$.

Suitable representations of a C*-algebra C are often constructed as Gelfand – Naimark – Segal (in short GNS-) representations $(\pi_\phi, H_\phi, \Omega_\phi)$ with respect to a certain state ϕ on C. In such a representation $\pi_\phi : C \longrightarrow B(H_\phi)$ the state ϕ is implemented by a vector Ω_ϕ:

$$\phi (x) = \langle\Omega_\phi|\pi_\phi(x)\Omega_\phi\rangle \quad , \quad \forall x \in C.$$

Furthermore the smallest sub-Hilbert space of H_ϕ containing the vectors $\pi_\phi(x)\Omega_\phi$, $x \in C$, coincides with H_ϕ. By these conditions $(\pi_\phi, H_\phi, \Omega_\phi)$ is fixed uniquely up to unitary equivalence.

Two states ϕ_1 and ϕ_2 are called disjoint, if the respective GNS-representations π_{ϕ_1} and π_{ϕ_2} are disjoint.

A typical example for a quasilocal C*-algebra is the CCR-algebra ('canonical commutation relations') $\Delta(H_0)$ describing a system of infinitely many bosons ([9]). Here H_0 is a pre-Hilbert space (i.e. not necessarily complete with respect to the norm of the scalar product) and $\Delta(H_0)$ is the C*-algebra generated by the operators $W(f)$, $f \in H_0$, fulfilling the relation

$$W(f_1)W(f_2) = e^{-i\text{Im}\langle f_1|f_2\rangle}W(f_1+f_2) \quad , \quad f_1,f_2 \in H_0 .$$

(8)

The Weyl operators $W(f)$, $f \in H_0$, are related to the annihilation and creation operators a_f and a_f^* by

$$W(f) = \exp\{i(a(f) + a(f)^*)\} \quad, \quad f \in H_0 \quad.$$

3. THE SPIN-BOSON MODEL

3.1. Description of the model

The spin-boson model ([2,10]) consists of a single two-level system (spin 1/2) coupled to a reservoir of infinitely many bosons. The spin is described by 2×2-matrices and mimicks the two lowest lying energy levels of the eventually chiral molecule. The bosons represent the radiation field. *The idea behind the spin-boson model is that the chirality of the molecule is generated by the radiation field which is inseparably coupled to it.* The Hamiltonian

$$H = \mu(\sigma_1 \otimes 1) + 1 \otimes \sum_{j=-\infty}^{+\infty} \omega_{k_j} a_{k_j}^* a_{k_j} + \sigma_3 \otimes \sum_{j=-\infty}^{+\infty} \lambda_{k_j} (a_{k_j}^* + a_{k_j}) \qquad (9)$$

$$\underbrace{}_{\text{molecule}} \quad \underbrace{}_{\text{radiation field}} \quad \underbrace{}_{\substack{\text{coupling between} \\ \text{the molecule and} \\ \text{the radiation field}}}$$

of the spin-boson model ($\hbar = 1$) acts on the infinite tensor product Hilbert space $K = H_2 \otimes H_\infty \otimes H_\infty \otimes \ldots \otimes H_\infty \otimes \ldots$ where H_2 is a two-dimensional and H_∞ a separable infinite-dimensional Hilbert space. The wave vectors $\{k_n = (2\pi/L)n \mid n \in \mathbb{Z}^3 \setminus \{0\}\}$ depend on the size L of the box in which the radiation field is described. For $L \longrightarrow \infty$ H will contain integrals of operators instead of sums. The ω_k's are the angular frequencies of the bosons with wave vectors k ($\omega_k = c|k|$ is proportional to $|k|$). The λ_k's are the coupling constants between the molecule and the mode with wave vector k of the radiation field. The operators a_k and a_k^* are annihilation and creation operators, respectively. The quasilocal C*-algebra B is here generated by the spin operators $\sigma_0 = \begin{pmatrix} 1 & 0 \\ 0 & 1 \end{pmatrix}$, $\sigma_1 = \begin{pmatrix} 0 & 1 \\ 1 & 0 \end{pmatrix}$, $\sigma_2 = \begin{pmatrix} 0 & -1 \\ 1 & 0 \end{pmatrix}$, $\sigma_3 = \begin{pmatrix} 1 & 0 \\ 0 & -1 \end{pmatrix}$ and operators of the type

$$\sigma_0 \otimes 1 \otimes 1 \otimes 1 \otimes \ldots 1 \otimes y \otimes 1 \otimes \ldots \otimes 1 \otimes 1 \quad, \quad y \in B(H_\infty). \text{ Define}$$

$$\Lambda \stackrel{\text{def}}{=} \sum_j \frac{|\lambda_{k_j}|^2}{\omega_{k_j}}$$

$$\Gamma \stackrel{\text{def}}{=} \sum_j \frac{|\lambda_{k_j}|^2}{\omega_{k_j}^2} \quad, \qquad (10)$$

where Λ is supposed to be finite and Γ is supposed to diverge (see [2]). *The infrared divergence $\Gamma = \infty$ will be the relevant condition to generate chirality!* This divergence will not arise for finite box size L but only in the limit $L \longrightarrow \infty$ (cf. the discussion in chapter 5).

3.2. A phase transition in the spin-boson model

The following (well founded) conjecture was made by P. Pfeifer in his
thesis ([2]) : The ground state of the spin-boson model with $\Gamma = \infty, \Lambda < \infty$
is non-degenerate and totally symmetric for $\mu \geq 2\Lambda$ and twofold degenerate
(with *disjoint* chiral state vectors Ψ_+ and Ψ_-) for $\mu < 2\Lambda$. In the latter
case Ψ_+ and Ψ_- have vanishing interference terms $\langle \Psi_+ | x \, \Psi_- \rangle = 0$, where
$x \in B$ (or $x \in B''$) is an arbitrary observable of the system. Ψ_+ and Ψ_- lie
in different sectors of the Hilbert space K and generate a classical
observable. A superposition $c_+ \Psi_+ + c_- \Psi_-$ of Ψ_+ and Ψ_- ($c_+, c_- \in C$) leads to
a statistical mixed state in the sense of

$$\langle c_+ \Psi_+ + c_- \Psi_- | x(c_+ \Psi_+ + c_- \Psi_-) \rangle = |c_+|^2 \langle \Psi_+ | x \Psi_+ \rangle + |c_-|^2 \langle \Psi_- | x \Psi_- \rangle \ , \ x \in B'' \ .$$

3.3 An algebraic version of the spin-boson model

The infinite tensor product Hilbert space K of the spin-boson model is
nonseparable and somewhat artificial. It is more convenient to introduce
a reasonable quasilocal algebra C for the joint system (molecule and radiation field) and to study the W*-closure of C in suitable representations. Recall that for a given representation π of C on a Hilbert space
H this W*-closure is given by $\{\pi(C)\}''$. The particular Hilbert space H is
not important. If π_j are representations of C on H_j, $j = 1, 2$, then π_1
and π_2 are called quasiequivalent if an *-isomorphism $\kappa : \pi_1(C)'' \longrightarrow \pi_2(C)''$
exists with $\kappa(\pi_1(x)) = \pi_2(x)$, $\forall x \in C$. Quasiequivalent representations
describe the same physical situation (with the same proper observable
algebra $\pi_1(C)'' \cong \pi_2(C)''$) and will be identified.

As quasilocal C*-algebra for the spin-boson model consider

$$A \stackrel{\text{def}}{=} M_2 \otimes \Delta(H_o)$$

$$H_o \stackrel{\text{def}}{=} \left\{ f \in L^2(\mathbb{R}) \mid \int_{\mathbb{R}} \frac{|f(k)|^2}{|k|} \, dk < \infty \right\} \tag{11}$$

where $L^2(\mathbb{R})$ is the usual Hilbert space of square integrable complex-
-valued functions on $\mathbb{R}$. The condition

$$\int_{\mathbb{R}} \frac{|f(k)|^2}{|k|} \, dk < \infty$$

is particularly adapted to the infrared divergence $\Gamma = \infty$. The dimension
of the configuration space is reduced from 3 to 1 ($\mathbb{R}^3$ is replaced by $\mathbb{R}$).
The algebra M_2 of 2×2-matrices mimicks the molecule whereas the CCR-
-algebra $\Delta(H_o)$ describes the radiation field. Every operator $\underline{x} \in A$ may
be written in the form

$$\underline{x} = \begin{pmatrix} x_1 x_2 \\ x_3 x_4 \end{pmatrix} \ , \ x_j \in \Delta(H_o) \ ,$$

of a 2×2-matrix with entries from $\Delta(H_o)$, the algebraic operations being done in the obvious way just as for matrices. By definition, A is generated by elements of the form

$$\underline{W(f)} \overset{\text{def}}{=} \begin{pmatrix} W(f_1) & W(f_2) \\ W(f_3) & W(f_4) \end{pmatrix} \quad , \quad f_j \in H_0 \ , \ j = 1,2,3,4.$$

The space-reflection symmetry τ is given by

$$\tau(\underline{W(f)}) \overset{\text{def}}{=} \begin{pmatrix} W(-f_4) & W(-f_3) \\ W(-f_2) & W(-f_1) \end{pmatrix} \quad , \quad f_j \in H_0 \ . \tag{12}$$

The dynamics $\{\alpha_t^o | t \in \mathbb{R}\}$ corresponding to the Hamiltonian

$$H_0 = 1 \otimes \int_{\mathbb{R}} |k| \ a_k^* a_k \ dk + \sigma_3 \otimes \int_{\mathbb{R}} \lambda(k) \ (a_k^* + a_k) \ dk \tag{13}$$

for $\mu = 0$ is given by ([10,11])

$$\alpha_t^o \{\underline{W(f)}\} \overset{\text{def}}{=}$$

$$\begin{vmatrix} e^{2i\text{Re}\langle \frac{\lambda}{|k|}(e^{-it|k|}-1)|f_1\rangle} W(e^{it|k|}f_1) & W(e^{it|k|}f_2 - 2i \frac{\lambda}{|k|}(e^{it|k|}-1)) \\ W(e^{it|k|}f_3 + 2i \frac{\lambda}{|k|}(e^{it|k|}-1)) & e^{-2i\text{Re}\langle \frac{\lambda}{|k|}(e^{-it|k|}-1)|f_4\rangle} W(e^{it|k|}f_4) \end{vmatrix}$$

$$f_j \in H_0 \ , \quad j = 1,\ldots,4 \ . \tag{14}$$

Here $e^{it|k|}f$, $f \in H_0$, stands for the element $\{\mathbb{R} \ni k \longrightarrow e^{it|k|}f(k)\}$ of H_0, whereas

$$\text{Re} \ \langle \frac{\lambda}{|k|}(e^{-it|k|}-1)|f\rangle \ , \quad f \in H_0 \ ,$$

is the real part of the complex number

$$\int_{\mathbb{R}} \frac{\lambda(k)}{|k|} (e^{+it|k|}-1) \ f(k) \ dk \ .$$

If γ^o denotes the quasifree dynamics

$$\gamma_t^o (W(f)) \overset{\text{def}}{=} W(e^{i|k|t}f) \ , \quad f \in H_0 \ , \tag{15}$$

on $\Delta(H_0)$, α^o is given by

$$\alpha_t^o (x) = S_t \{\text{Id} \otimes \gamma_t^o(\underline{x})\} S_t^* \ , \quad \underline{x} \in A = M_2 \otimes \Delta(H_0), t \in \mathbb{R} \tag{16}$$

where S_t , $t \in \mathbb{R}$, is defined to be the unitary operator

$$S_t \overset{\text{def}}{=} \begin{vmatrix} W(-i\, \frac{\lambda}{|k|}\, \{e^{+i|k|t}-1\}) & 0 \\ & \\ 0 & W(+i\, \frac{\lambda}{|k|}\, \{e^{+i|k|t}-1\}) \end{vmatrix} \,, \quad t \in \mathbb{R} \,.$$

Remark 1: Every state ϕ on A corresponds to four linear functionals ϕ_j, $j = 1,2,3,4$, on $\Delta(H_0)$ with

$$\phi(\underline{W(f)}) = \sum_{j=1}^{4} \phi_j(W(f_j)) \,.$$

The linear functionals ϕ_j will always be supposed to be *regular*, which means that for arbitrary $f, g \in H_0$ the map

$$\mathbb{R} \ni z \longrightarrow \phi_j(W(f+zg)) \,, \quad j = 1,\ldots,4,$$

is analytic. This condition implies the existence of correlation functions

$$\phi_j(a^{\#}(f_1)\ldots a^{\#}(f_n)) \,, \quad f_k \in H_0 \,,$$

where $a^{\#}$ stands for a or a^*. Furthermore the correlation functions will be supposed to be continuous with respect to the scalar product

$$\langle f|g \rangle_{\sim} \overset{\text{def}}{=} \int_{\mathbb{R}} (1 + \frac{1}{|k|})\, f(k)^* g(k)\, dk \,,$$

i.e. for $j = 1$ and $j = 4$ we suppose

$$|\phi_j(a^*(f_1)\ldots a^*(f_n)a(g_1)\ldots a(g_m))|^2 \le C^{n+m}\, n!m!\, \prod_{k=1}^{n} \|f_k\|_{\sim}^2 \prod_{\ell=1}^{m} \|g_\ell\|_{\sim}^2 \,, \quad (17)$$

$$f_k, g_\ell \in H_0 \,, \quad k = 1,\ldots,n \,;\, \ell = 1,\ldots,m.$$

3.4. Perturbation of dynamical systems (cf.[9])

Let $(B,\mathbb{R},\alpha)$ be a C*- or W*-system. Then a dynamics $\mathbb{R} \ni t \longrightarrow \beta_t$ is called a *local perturbation* of the dynamics $\mathbb{R} \ni t \longrightarrow \alpha_t$, if the derivation

$$\delta(x) \overset{\text{def}}{=} \frac{d(\alpha_t^{-1}\beta_t(x))}{dt}\bigg|_{t=0} \,, \quad x \in B$$

is of the form $\delta(x) = i[V,x]$, where $V = V^* \in B$. This corresponds to a change of the Hamiltonian H_0 which generates α to $H = H_0 + V$.

Perturbation theory permits to define a dynamics $\{\alpha_t^\mu | t \in \mathbb{R}\}$ for $\mu \neq 0$ in the spin-boson model :

$$H^\mu = \mu(\sigma_1 \otimes 1) + H_0 \,.$$

Remark 2 : Note that the dynamics α^o of the spin-boson model is not pointwise norm-continuous, that is, the mappings $\mathbb{R} \ni t \longrightarrow \alpha_t^o(x)$, are *not* continuous for all $x \in A$ with respect to the norm on A. As a consequence, the perturbation of α^o has to be done in an appropriate representation. A representation π is 'appropriate' if there exists a σ-weakly continuous automorphism group $\tilde{\alpha}^o$ of $\pi(A)''$ such that $\tilde{\alpha}_t^o(\pi(x)) = \pi(\alpha_t^o(x))$, $\forall x \in A$. The dynamics $\tilde{\alpha}^o$ on $\pi(A)'' \in B(H)$ is called σ-weakly continuous if the mappings $\mathbb{R} \ni t \longrightarrow \phi(\tilde{\alpha}^o(x))$, $x \in \pi(A)''$, are continuous for all normal states on $\pi(A)''$. This is equivalent to the continuity of the mappings

$$\mathbb{R} \ni t \longrightarrow \langle \xi | \tilde{\alpha}_t^o(x) \xi \rangle \in \mathbb{R} \quad , \quad \forall \xi \in H \ , \ \forall x \in \pi(A)'' \quad .$$

In simpler models (such as the transverse Ising model to follow) continuity of the C*-system is fulfilled.

3.5. β-KMS-states in algebraic quantum mechanics (cf.[9])

The β-KMS-states (with $\beta = \frac{1}{kT}$ the inverse temperature) generalize Gibbs states (given by a density operator $e^{-\beta H}/Tr\{e^{-\beta H}\}$) to general systems in algebraic quantum mechanics. The precise definition runs as follows : Let $\phi : C \longrightarrow \mathbb{C}$ be a state on a C*-algebra C. Consider a dynamical group $\{\gamma_t | t \in \mathbb{R}\}$ on C. Then ϕ is called a β-KMS-state on C if for every pair of elements $x, y \in C$ there is a function $G_{xy} : \mathbb{C} \longrightarrow \mathbb{C}$ which is holomorphic in the open strip $0 < Im(z) < \beta$, $z \in \mathbb{C}$, and bounded and continuous on the closed strip $0 \le Im(z) \le \beta$, such that

$$G_{xy}(t) = \phi \{x \alpha_t(y)\}$$

$$G_{xy}(t+i\beta) = \phi \{\alpha_t(y)x\} \quad , \quad \forall t \in \mathbb{R} \quad . \tag{18}$$

KMS-states are stable in the sense that for each small local perturbation of the dynamics there is another state close to the original one which is a KMS-state with respect to the perturbed dynamics.

4. WHEN DOES CHIRALITY ARISE ?

4.1. Chirality and thermodynamic states

Consider a *single* molecule described by the algebra M_2 of 2×2-matrices and coupled to an environment with C*-algebra C. The full (not necessarily quasilocal) C*-algebra of the system is given by the tensor product $B = M_2 \otimes C$. Let the space-reflection symmetry be given by an automorphism τ of B and the dynamics of the system by a (pointwise norm-continuous) automorphism group $\{\gamma_t^o | t \in \mathbb{R}\}$ of B with the properties

$$(i) \quad \tau \circ \gamma_t^o = \gamma_t^o \circ \tau$$

$$(ii) \quad \tau \{\sigma_1 \otimes 1\} = \sigma_1 \otimes 1 \ . \tag{19}$$

These properties were of course fulfilled in all our previous considerations : (i) corresponds to the invariance of the Hamiltonian $\hat{H}_{A1a}$ under space-reflection and (ii) corresponds to the invariance of the ground state $\langle\Psi_0|\cdot\Psi_0\rangle$ and the first-excited state $\langle\Psi_1|\cdot\Psi_1\rangle$ under space-reflections ($\hat{U}\Psi_0 = \Psi_0$, $\hat{U}\Psi_1 = -\Psi_1$).

Theorem 1
Consider a C*-algebra $B = M_2 \otimes C$, an automorphism τ of B and a dynamical group $\{\gamma_t^0|t \in \mathbb{R}\}$ fulfilling the conditions (i) and (ii). Let $\{\gamma_t^\mu|t \in \mathbb{R}\}$ be the perturbed dynamical groups with

$$\left.\frac{d}{dt} (\gamma_t^\mu\gamma_{-t}^0)(\cdot)\right|_{t=0} = i [\mu(\sigma_1\otimes 1),\cdot]$$

for $\mu \in \mathbb{R}$. Let β be a fixed inverse temperature, $0 < \beta < \infty$. Assume that all β-KMS-states with respect to γ^0 are τ-invariant. Then all β-KMS--states with respect to $\{\gamma_t^\mu|t \in \mathbb{R}\}$ are τ-invariant.

Proof
The proof is an immediate consequence of ([9] : Corollary 5.4.5).

   Of course, theorem 1 can be formulated on the W*-level if γ^0 does not fulfill continuity properties. In plain words, it says that *on the level of thermodynamic states* (β-KMS-states, $\beta < \infty$) a phase transition with respect to μ does not exist. The situation is all the same irrespective of the special value of μ! *Therefore the specific expectations explained in the introduction (μ small $\leftrightarrow$ chirality arises) cannot be fulfilled with any thermal environment (temperature $T \neq 0$) of the single individual molecule whatsoever!* In the spin-boson model, this result can even be sharpened:

Theorem 2 ([10])
Consider the full spin-boson model (A,α^μ,τ). For every $\mu \in \mathbb{R}$ and $0 < \beta < \infty$ there is exactly *one* β-KMS-state ω_β^μ with respect to the dynamics α^μ. In particular, $\omega_\beta^\mu \circ \tau = \omega_\beta^\mu$, i.e. the space reflection symmetry remains unbroken and chiral β-KMS-states do not exist.

4.2. A simple model with chiral β-KMS-states (the transverse Ising model)

The situation described in the theorems 1 and 2 is changed immediately, if one does not consider a single molecule but infinitely many molecules (a crystal, e.g.) instead. A simple model with chiral β-KMS-states is given by the 'Ising model with a transverse field' ([12,13,14,15]) : Let $\Xi \subset \mathbb{Z}^3$ be an arbitrary finite sublattice of the infinite lattice $\mathbb{Z}^3$ in 3-dimensional space and define the Hamiltonian

$$\hat{H}_\Xi^{\mu,\lambda} \overset{\text{def}}{=} \sum_{a\in\Xi} \mu\sigma_1^a + \sum_{\substack{|a-b|=1\\a,b\in\Xi}} \lambda\,\sigma_3^a\otimes\sigma_3^b \quad,\quad \mu,\lambda \in \mathbb{R} . \tag{20}$$

$\hat{H}_\Xi^{\mu,\lambda}$ is an element of the local C*-algebra

$$C_\Xi \overset{\text{def}}{=} \underset{a\in\Xi}{\otimes} M_2^a \quad,\quad M_2^a \cong M_2 \quad,\quad \forall\, a\in\Xi \quad. \tag{21}$$

The operators σ_j^a, $j = 1,2,3$, are the Pauli matrices in M_2^a, $a \in Z^3$. The Hamiltonians $\hat{H}_\Xi^{\mu,\lambda}$ take into account only nearest neighbor interactions. The terms $\mu\sigma_1^a$, $a \in \Xi$, stand for the level splitting between ground state and first excited state of the molecule sitting at $a \in Z^3$. The quasilocal algebra, the space-reflection symmetry and the dynamics of the model are defined by

$$C = \overline{\underset{\substack{\Xi \subset Z^3 \\ |\Xi| < \infty}}{U} C_\Xi}^{\text{norm}} ,$$

$$\tau(x) \overset{\text{def}}{=} \underset{|\Xi| \to \infty}{\lim} (\underset{a \in \Xi}{\otimes} \sigma_1^a) \; x \; (\underset{a \in \Xi}{\otimes} \sigma_1^a) \quad , \quad x \in C \tag{22}$$

$$\gamma_t^{\mu,\lambda}(x) \overset{\text{def}}{=} \underset{|\Xi| \to \infty}{\lim} (e^{+i\hat{H}_\Xi^{\mu,\lambda} t} \; x \; e^{-\hat{H}_\Xi^{\mu,\lambda} t}) \quad , \quad x \in B \quad , \quad t \in R .$$

Theorem 3
There is a constant $C > 0$, such that for $\mu/\lambda < C$ and large enough β chiral β-KMS-states with respect to $\gamma^{\mu,\lambda}$ do exist.

Proof
The proof is a consequence of the results in ([16]). The existence of a phase transition with respect to μ (i.e. non-existence of chiral states for μ large) is always assumed in papers with a more heuristic approach. The best exact result in this direction seems to be given in ([15]).

Existence of 'chiral' β-KMS-states simply means that there is a β-KMS-state ϕ^β with $\phi^\beta \circ \tau \neq \phi^\beta$. Due to the commutation properties $\tau \circ \gamma_t^{\mu,\lambda} = \gamma_t^{\mu,\lambda} \circ \tau$, $t \in R$, the state $\phi^\beta \circ \tau$ is again β-KMS with respect to $\gamma^{\mu,\lambda}$. Therefore $\phi^\beta \circ \tau$ and $\phi^\beta = (\phi^\beta \circ \tau) \circ \tau$ may be regarded as (left- and right-) handed states of the system.

The transverse Ising model gives an explanation for the chirality of large systems (an alanine-crystal, for example). Since the existence of *single* chiral molecules (e.g. pheromones) is well established, this is not fully satisfactory.

4.3. Ground states of the spin-boson model

Def.: A ground state ϕ of a C*-system (C,R,γ) is an ∞-KMS-state with respect to γ. This means that for every pair $x,y \in C$ there exists a function $G_{xy} : C \longrightarrow C$ which is continuous for Im $z \geq 0$ and analytic and bounded in Im $z > 0$ such that

$$G_{xy}(t) = \phi (x \gamma_t (y)) \quad , \quad \forall t \in R \quad .$$

An equivalent condition ([9] : Proposition 5.3.19) runs as follows : ϕ is γ-invariant, and if

$$e^{itH_\phi} \pi_\phi(x)\Omega_\phi \overset{\text{def}}{=} \pi_\phi (\gamma_t (x))\Omega_\phi$$

is the corresponding unitary representation of $\mathbb{R}$ on the GNS-Hilbert-space H_ϕ, then $H_\phi \geq 0$. *Here always this latter characterization will be used* (in fact slightly adapted due to reasons given in remark 2).

The set K_∞ of ground states of a C*-system $(C,\mathbb{R},\gamma)$ is a convex compact face ([9] : Theorem 5.3.37). Therefore K_∞ is generated by its extremal points, i.e. the elements ϕ of K_∞ which do not admit a non-trivial convex decomposition $\phi = \lambda\,\phi_1 + (1-\lambda)\phi_2$, $0 < \lambda < 1$, $\phi_1, \phi_2 \in K_\infty$. *The extremal ground states are pure states on C.*

Theorem 4 ([11])
Consider the spin-boson model $(A, \mathbb{R}, \alpha^0)$ for $\mu = 0$ and assume that

$$\Gamma \overset{\text{def}}{=} \frac{|\lambda(k)|^2}{k^2}\, dk$$

diverges (see section 3.1). Then there exist exactly two extremal (pure) ground states ω_+ and ω_- :

$$\omega_+ \begin{pmatrix} W(f_1) & W(f_2) \\ W(f_3) & W(f_4) \end{pmatrix} \overset{\text{def}}{=} e^{-2i\mathrm{Re}\langle \frac{\lambda}{|k|}|f_1\rangle}\, e^{-\frac{1}{2}\|f_1\|^2}$$

$$(23)$$

$$\omega_- \begin{pmatrix} W(f_1) & W(f_2) \\ W(f_3) & W(f_4) \end{pmatrix} \overset{\text{def}}{=} e^{+2i\mathrm{Re}\langle \frac{\lambda}{|k|}|f_4\rangle}\, e^{-\frac{1}{2}\|f_4\|^2} \quad , \quad f_j \in H_o.$$

The states ω_+ and ω_- are disjoint and chiral with $\omega_+ = \omega_- \circ \tau$ and $\omega_- = \omega_+ \circ \tau$. The limit state

$$\omega \overset{\text{def}}{=} \lim_{\beta \to \infty} \omega_\beta^0$$

of the β-KMS-states with respect to the dynamics α^0 is given by $\omega = \frac{1}{2}(\omega_+ + \omega_-)$.

Up to now, there is no contradiction at all between the conjecture of Pfeifer (see section 3.2) and the results of the theorems 1 and 2. Nevertheless, even if this conjecture (whose proof would be a generalization of theorem 4 to $\mu \neq 0$, which has not yet been done) is correct, a reasonable argument could be given which devalues this conjecture and any other results on ground states of the spin-boson model :
'Even if the absolute temperature of a real system is low, it is never zero. Consequently, one has to consider β-KMS-states with large β instead of ground states. Thus for all practical purposes the space-reflection symmetry remains unbroken and chiral states do not exist.'

Of course, this argument does not apply to the transverse Ising model since there the space-reflection symmetry is broken for large β (if $\mu/\lambda < C$) and therefore chiral ground states are an acceptable caricature to β-KMS-states with large β.

To clarify the situation of the spin-boson model and in fact all models describing a *single* molecule and its environment (see theorem 1), a closer look at the concept of a 'state' in algebraic quantum mechanics is necessary.

4.4. States in algebraic quantum mechanics

In algebraic quantum mechanics *individual systems* (such as single mole-
cules) are always described by pure states. Non-pure states (such as
β-KMS-states, $\beta \in \mathbb{R}$) may be used for different reasons :
  - The algebra under discussion does not describe an individual
    system but a statistical ensemble.
  - Pure states may be unstable under perturbations and, as a con-
    sequence, not be determinable experimentally. This forces the
    experimentalist to introduce statistical states which describe
    the measurable quantities properly but do not describe a single
    individual system in all details.

*Postulate : An individual system (even with infinitely many degrees of
freedom) is always in a pure state (cf.[3,17]). Therefore the question
of chirality of a single molecule has to be treated on this level.*

The problem of chirality may be also regarded from a more dynamical point of view. Pure states of the single molecule may be coupled to
pure states (e.g. the Fock vacuum ϕ_0 or, more generally, any coherent
state) of the radiation field. If this is done for two states, say $\binom{1}{0}$
and $\binom{0}{1}$ of the molecule, the states $\phi_+ = \binom{1}{0} \otimes \phi_0$ and $\phi_- = \binom{0}{1} \otimes \phi_0$ may
develop under the time evolution to states which finally get disjoint
in the limit $t \to \infty$ and generate different superselection sectors. In the
following case, these limit states are again ground states of the system
(which is not necessarily the case).

Theorem 5
Let ϕ_0 be the Fock state on $\Delta(H_0)$, $\phi_0(W(f)) \stackrel{\text{def}}{=} e^{-\frac{1}{2}\|f\|^2}$, $f \in H_0$. Consider
the pure product states

$$\phi_+ \stackrel{\text{def}}{=} \binom{1}{0} \otimes \phi_0 \quad \text{and} \quad \phi_- \stackrel{\text{def}}{=} \binom{0}{1} \otimes \phi_0$$

on the quasilocal C*-algebra $A = M_2 \otimes \Delta(H_0)$. Then the limit states

$$\lim_{t \to \infty} \phi_+ \circ \alpha_t^0 \quad \text{and} \quad \lim_{t \to \infty} \phi_- \circ \alpha_t^0$$

are the disjoint ground states ω_+ and ω_- of the system with respect to
the dynamics α^0 (see section 3.3). That is, they lie in different sectors and generate a classical chirality-observable.

Proof

$$\phi_+(\alpha_t^0(\underline{W}(\underline{f}))) = e^{-2i\operatorname{Re}\langle \frac{\lambda}{|k|} (e^{-it|k|}-1)|f_1\rangle} e^{-\frac{1}{2}\|f_1\|^2}.$$

$$\lim_{t \to \infty} \langle \frac{\lambda}{|k|} (e^{-it|k|}-1)|f_1\rangle = (\lim_{t \to \infty} \int_{\mathbb{R}} \frac{\lambda(k)}{|k|} e^{it|k|} f_1(k)dk) - \langle \frac{\lambda}{|k|}|f_1\rangle =$$

$$= - \langle \frac{\lambda}{|k|}|f_1\rangle \text{ by use of the Riemann-Lebesgue lemma ([18] : Theorem IX.7).}$$

Therefore $\lim_{t \to \infty} \phi_+ \circ \alpha_t^0 = \omega_+$. The proof for ϕ_- can be done similarly.

Remarks:
-   Theorem 5 holds as well if ϕ_0 is replaced by an arbitrary coherent state

$$\phi_{f_0}(W(f)) \overset{\text{def}}{=} e^{i(\langle f | f_0 \rangle - \langle f_0 | f \rangle)} e^{-\frac{\|f\|^2}{2}} , \quad f \in H_0$$

where $f_0 \in H_0$ is an arbitrary fixed vector.

-   For finite time t, the states $\phi_+ \circ \alpha_t^0$ and $\phi_- \circ \alpha_t^0$ get 'more and more disjoint' :
    Let $\xi_+ \in H$ and $\xi_- \in H$ be vector states implementing ϕ_+ and ϕ_- in a representation $\pi : A \longrightarrow B(H)$:

$$\phi_+(x) = \langle \xi_+ | \pi(x) \xi_+ \rangle$$

$$\phi_-(x) = \langle \xi_- | \pi(x) \xi_- \rangle , \quad x \in A .$$

Then the transition probabilities $\langle \xi_+ | \pi(\alpha_t^0(x)) \xi_- \rangle$ converge to zero for $t \to \infty$.

4.5. How to understand the uniqueness-result of Fannes et al.

Altering the chirality of a single molecule (e.g. in a chemical reaction) means that the radiation field coupled to this molecule is radically changed. If the experimentalist would indeed observe the system (that is, the molecule and its radiation field) on the level of individual (pure) states, he could detect such an alteration of chirality by just looking at the radiation field alone.

Of course, the determination of the precise individual pure state of a system with infinitely many degrees of freedom is impossible. Due to their extremely non-robust nature, pure states of infinite systems are never experimentally detectable. For the purpose of experiments, one has to change to a statistical (nonindividual) view and there (as the result of Fannes et al. asserts) the alteration of chirality should (and does) not change the superselection sector of the radiation field. This is perfectly in order and reflected in the counterpart to theorem 5 below. If experimentally accessible (e.g. thermodynamic) chiral states of the radiation field would exist in our model which change sign if the molecule's chirality is changed, we had to consider the model as being somewhat doubtful.

Nevertheless, looking at the molecule as a single individual entity enforces an individual interpretation of the whole system. This joint system then is in a pure state even if this state cannot be determined by experiments. The same phenomenon arises in classical mechanics where a point particle sits at a position $\vec{q} \in \mathbb{R}^3$ even if this position can never be determined : It is, for example, impossible to measure if $\vec{q}$ has rational or irrational coordinates. A measurement only gives some probability distribution on the phase space Ω.

Theorem 6
Consider product states $\binom{1}{0} \otimes \phi_\beta$ and $\binom{0}{1} \otimes \phi_\beta$ on $A = M_2 \otimes \Delta(H_o)$, where

$$\phi_\beta(W(f)) \overset{def}{=} \exp\{-\tfrac{1}{2}\langle f|\coth(\tfrac{\beta|k|}{2})f\rangle\} \quad , \quad f \in H_o \quad ,$$

is the β-KMS-state on $\Delta(H_o)$ with respect to the free evolution of the boson system. Then the limit states

$$\phi_+ \overset{def}{=} \lim_{t \to \infty} \left\{\binom{1}{0} \otimes \phi_\beta\right\} \circ \alpha_t^o$$

$$\phi_- \overset{def}{=} \lim_{t \to \infty} \left\{\binom{0}{1} \otimes \phi_\beta\right\} \circ \alpha_t^o$$

$$\phi_+(\underline{W(f)}) = e^{-2\mathrm{Re}\langle\frac{\lambda}{|k|}|f_1\rangle} \; e^{-\frac{1}{2}\langle f_1|\coth\frac{\beta|k|}{2}f_1\rangle}$$

$$\phi_-(\underline{W(f)}) = e^{+2\mathrm{Re}\langle\frac{\lambda}{|k|}|f_4\rangle} \; e^{-\frac{1}{2}\langle f_4|\coth\frac{\beta|k|}{2}f_4\rangle}$$

are quasiequivalent.

Proof
See ([10] : Lemma 3.3).

4.6. Outlook

The problem of generalizing theorems 4 and 5 to the case of $\mu \neq 0$ and the derivation of an eventually existing phase transition has not yet been solved. It seems to be more difficult than expected by P. Pfeifer in his thesis : There ([2] : Postulate 2, p.86; p.56) the restriction of all ground states (even for $\mu \neq 0$) to the two-level system are assumed to be *pure* states. This expectation is not unreasonable at all, but, unfortunately, it is only true for $\mu = 0$. For $\mu \neq 0$ the restriction of a ground state to the two-level system is always a *mixed* state.

Theorem 7
Let ϕ be a state on the spin-boson algebra $A = M_2 \otimes \Delta(H_o)$ and assume that the restriction ϕ_1 of ϕ to M_2
$$\phi_1 : M_2 \ni x \longrightarrow \phi(x \otimes 1)$$
is a *pure* state on M_2. Then ϕ cannot be invariant with respect to the dynamics α^μ, $\mu \neq 0$, on A.
    The proof of theorem 7 is given in section 4.7.
    Assume chiral ground states ϕ_+ and ϕ_- to exist for $\mu \neq 0$ and consider the algebra $\tilde{M}_2$ of the 'dressed' two-level system. The algebra $\tilde{M}_2$ is isomorphic to the algebra of all 2×2-matrices. The restrictions of the chiral states ϕ_+ and ϕ_- are implemented by the vectors $\binom{1}{0}$ and $\binom{0}{1}$ in appropriate coordinates. In fact $\tilde{M}_2$ is characterized by the latter property. Theorem 7 now states that $\tilde{M}_2$ and M_2 - though being isomorphic - do *not* coincide. Generally, $\tilde{M}_2$ cannot even be expected to be a subalgebra

of the spin-boson algebra A, but only a subalgebra of an observable (W*-) algebra generated by A.

Remark : Some work in the direction of nonvanishing level splitting has been done ([19,20,21]) : Leggett et al. ([19]) use a perturbation expansion up to second order (cf.[22]). Their results do *not* coincide with Pfeifer's conjecture and should certainly be used as a source of inspiration for an exact treatment. The approach ([20]) of Spohn et al. gives rigorous results and leads to even more fascinating conjectures (differing again from what Pfeifer proposed). Its connection with algebraic methods is not yet entirely clear to the present author.

4.7. Proof of theorem 7

Consider the dynamical groups α^+ and α^- on $\Delta(H_0)$ defined by

$$\alpha^+_t(W(f)) \stackrel{\text{def}}{=} e^{\pm 2i\text{Re}\langle \frac{\lambda}{|k|}(e^{-it|k|}-1)|f\rangle} W(e^{it|k|}f) \quad , \quad f \in H_0. \tag{24}$$

Note that α^+ and α^- enter into the definition of the dynamics α^0 (14) on the spin-boson algebra $A = M_2 \otimes \Delta(H_0)$.

Lemma 1
There is no α^+- and α^--invariant regular state on $\Delta(H_0)$.

Proof
Suppose the state ϕ on $\Delta(H_0)$ to be α^+- and α^--invariant.

$$\Rightarrow \quad \phi(W(e^{i|k|t}f)) \exp\left\{-2i\text{Re}\langle \frac{\lambda}{|k|}(e^{-i|k|t}-1)|f\rangle\right\} =$$

$$= \phi(W(e^{i|k|t}f)) \exp\left\{+2i\text{Re}\langle \frac{\lambda}{|k|}(e^{-i|k|t}-1)|f\rangle\right\} , \quad \forall f \in H_0.$$

Let $f_0 \in H_0$ be such that $\text{Re}\langle \frac{\lambda}{|k|}(e^{-i|k|t_0}-1)|f_0\rangle = d \neq 0$ for suitable $t_0 \in \mathbb{R}$. Then $\phi(W(c.e^{i|k|t_0}f_0))$ can be nonzero only for $c = \frac{n\pi}{2d}$, $n \in \mathbb{Z}$. Consequently the function $\mathbb{R} \ni c \longrightarrow \phi(W(c.e^{i|k|t_0}f_0))$ is not continuous and ϕ is not regular.   Q.E.D.

Lemma 2
Let ϕ be a state on $A = M_2 \otimes \Delta(H_0)$ with restrictions ϕ_1 and ϕ_2 to M_2 and $\Delta(H_0)$, respectively. Assume ϕ_1 to be a *pure* state on M_2. Then ϕ is of product form, $\phi = \phi_1 \otimes \phi_2$.

Proof

$$\phi_1(x) \stackrel{\text{def}}{=} \phi(x \otimes 1) \quad , \quad x \in M_2$$

$$\phi_2(y) \stackrel{\text{def}}{=} \phi(1 \otimes y) \quad , \quad y \in \Delta(H_0) .$$

Let p be the support projection ([23] : III.3.6) of the *pure* state ϕ_1. Then p is an atomic projection and $pxp = \phi_1(x).p$, $x \in M_2$, holds.

$$\Rightarrow \quad \phi(x \otimes y) \overset{(\#)}{=} \phi\{(p \otimes 1)(x \otimes y)(p \otimes 1)\} = \phi\{(pxp) \otimes y\} =$$

$$= \phi\{(\phi_1(x)p) \otimes y\} = \phi_1(x) \cdot \phi\{p \otimes y\} \overset{(\#)}{=} \phi_1(x) \cdot \phi\{1 \otimes y\} =$$

$$= \phi_1(x) \cdot \phi_2(y) \quad , \quad x \in M_2 \quad , \quad y \in \Delta(H_0) \quad .$$

The Cauchy-Schwarz inequality is used at (#).   Q.E.D.

Proof of theorem 7
Taking into account lemma 2, it has to be proven that a product state
$\phi = \phi_1 \otimes \phi_2$ on $A = M_2 \otimes \Delta(H_0)$ (ϕ_1 a pure state on M_2, ϕ_2 regular on $\Delta(H_0)$)
cannot be invariant under the dynamics α^μ, $\mu \neq 0$. Assume to the contrary
that $\phi_1 \otimes \phi_2$ is α^μ-invariant (α^μ exists only on the W*-level!).
Let $(H_{\phi_2}, \pi_{\phi_2}, \Omega_{\phi_2})$ be the GNS-representation of $\Delta(H_0)$ with respect to
ϕ_2. Let (H, π, Ω) be the GNS-representation of $M_2 \otimes \Delta(H_0)$ with respect to
$\phi = \phi_1 \otimes \phi_2$. Since ϕ is assumed to be α^μ-invariant, there exists a unitary
one-parameter group $\mathbb{R} \ni t \longrightarrow U_t^\mu \in B(H)$ implementing α^μ by

$$\alpha_t^\mu(y) = U_t^\mu y (U_t^\mu)^* \quad , \quad t \in \mathbb{R} \quad , \quad y \in \pi(A)'',$$

and with $U_t^\mu \Omega = \Omega, \forall t \in \mathbb{R}$. Due to ([9] : Corollary 5.4.2) α^0 is implemented by a unitary group, too : $\pi(\alpha_t^0(x)) = U_t^0 \pi(x)(U_t^0)^*$, $t \in \mathbb{R}$. If H^0 is the
infinitesimal generator of U^0 and H is the infinitesimal generator of
U^μ, the following holds :
$$H = H^0 + \mu \pi (\sigma_1 \otimes 1) \quad .$$

Let ϕ_1 be implemented by the vector $\binom{c_1}{c_2} \in H_2$. Then the following relations hold (unitary equivalence) :

$$H \cong H_{\phi_2} \oplus H_{\phi_2} \quad ,$$

$$\pi \begin{pmatrix} x_1 & x_2 \\ x_3 & x_4 \end{pmatrix} \cong \begin{pmatrix} \pi_{\phi_2}(x_1) & \pi_{\phi_2}(x_2) \\ \pi_{\phi_2}(x_3) & \pi_{\phi_2}(x_4) \end{pmatrix} \quad , \quad x_j \in \Delta(H_0) \quad , \quad j = 1,2,3,4,$$

$$\Omega \cong c_1 \Omega_{\phi_2} \oplus c_2 \Omega_{\phi_2} \quad .$$

H^0 commutes with $\pi(\sigma_3 \otimes 1) = \begin{pmatrix} 1 & 0 \\ 0 & -1 \end{pmatrix}$, and therefore has the form

$$H^0 = \begin{pmatrix} H_+ & 0 \\ 0 & H_- \end{pmatrix}$$

where H_+ implements $\{\alpha_t^+ | t \in \mathbb{R}\}$ and H_- implements $\{\alpha_t^- | t \in \mathbb{R}\}$ via π_{ϕ_2}.

$$H \Omega = 0 \quad \Rightarrow \quad \begin{pmatrix} H_+ & \mu \\ \mu & H_- \end{pmatrix} \begin{pmatrix} c_1 \Omega_{\phi_2} \\ c_2 \Omega_{\phi_2} \end{pmatrix} = \begin{pmatrix} 0 \\ 0 \end{pmatrix}$$

$$\Rightarrow \quad H_+ \Omega_{\phi_2} = -\frac{c_2}{c_1} \mu \, \Omega_{\phi_2}$$

$$H_- \Omega_{\phi_2} = -\frac{c_1}{c_2} \mu \, \Omega_{\phi_2} \quad .$$

[If $c_1 = 0$, one has $\mu c_2 \Omega_{\phi_2} = 0$ and consequently $c_2 = 0$ which is a contradiction. For $c_2 = 0$ one can argue similarly.]

$$\Rightarrow \quad e^{itH_+}\Omega_{\phi_2} = e^{-i \frac{c_2}{c_1} \mu t} \Omega_{\phi_2}$$

$$e^{itH_-}\Omega_{\phi_2} = e^{-i \frac{c_1}{c_2} \mu t} \Omega_{\phi_2} \quad .$$

$\Rightarrow \quad \phi_2$ is α^+- and α^--invariant and regular which contradicts lemma 1.   Q.E.D.

5.   DO DISJOINT STATES EXIST ?

The infrared divergence

$$\Gamma = \int_{\mathbb{R}} \frac{|\lambda(k)|^2}{|k|^2} \, dk = \infty$$

does only arise in the limit $L \longrightarrow \infty$, where L is the size of the box of the radiation field (see section 3.1). Since $\Gamma = \infty$ is a necessary condition for the disjointness of the ground states in theorem 4, it might be doubtful if disjoint states really exist. A cosmological argument supporting this doubt is the "finiteness of the universe".
     Some principal objections can be made to such an argumentation : If the size L has to be kept finite, the 'problem of optical isomers' depends strongly on L and cannot be given a clear solution. *The limit L $\longrightarrow \infty$, on the contrary, may be considered as a regularization of this problem.* Furthermore the ground states ω_+ and ω_- (see section 4.3) get more and more disjoint if L goes to infinity. Approximate disjointness may be measured by the vanishing of the transition probabilities between ω_+ and ω_- or by the vanishing of a Hellinger type integral. *Thus disjointness of ω_+ and ω_- for L = ∞ is a reasonable caricature for the situation where L is large but finite.* The argumentation here is similar to that one needed for finite time t instead of the limit $t \longrightarrow \infty$ in the remarks after theorem 5.

Acknowledgment
My thanks go to Prof. H. Primas who patiently explained to me almost everything written up in this paper. Furthermore, I am indebted to Dr. G. Raggio for communicating unpublished results, to Prof. J. Fröhlich for some fruitful discussions and to Dr. E. Zass who provided me with an extensive literature search on the transverse Ising model.

324

References

[1]   F. Hund : 'Zur Deutung der Molekelspektren III'. Z.Phys. 43,
       805 – 826 (1927).

[2]   P. Pfeifer : 'Chiral Molecules – a Superselection Rule Induced
       by the Radiation Field'. Thesis, Eidgenössische Technische
       Hochschule, Diss. ETH No. 6551 (1980).

[3]   H. Primas : Chemistry, Quantum Mechanics, and Reductionism.
       Perspectives in Theoretical Chemistry. 2nd corrected
       edition. Springer, Berlin (1983).

[4]   A. Amann : 'Observables in W*–Algebraic Quantum Mechanics'.
       Fortschr. Physik 34, 167 – 215 (1986).

[5]   R. Jost : Quantenmechanik I. Verlag des Vereins der Mathematiker
       und Physiker an der ETH–Zürich, Zürich (1969).

[6]   A. Amann, U. Müller-Herold : 'Momentum operators for large
       systems'. Helv.Phys.Acta 59, 1311 – 1320 (1986).

[7]   A. Amann : 'Broken symmetries and the generation of classical
       observables in large systems'. Helv.Phys.Acta 60, 384 – 393
       (1987).

[8]   J. Pöttinger : 'Global quantities in algebraic quantum mechanics
       of infinite systems : Classical observables or parameters ?'.
       Preprint, ETH–Zürich (1987).

[9]   O. Bratteli, D.W. Robinson : Operator Algebras and Quantum Statis-
       tical Mechanics III. Springer, New York (1981).

[10]   M. Fannes, B. Nachtergaele, A. Verbeure : 'The Equilibrium States
       of the Spin-Boson Model'. To be published in Commun.Math.
       Phys..

[11]   G. Raggio, personal communication.

[12]   R.J. Elliott, P. Pfeuty, C. Wood : 'Ising model with a transverse
       field'. Phys.Rev.Lett. 25, 443 – 446 (1970).

[13]   J. Oitmaa, M. Plischke : 'Critical behaviour of the Ising model
       with a transverse field'. J.Phys.C 9, 2093 – 2102 (1976).

[14]   P. Pfeuty : 'Quantum classical crossover critical behavior of
       the Ising model in a transverse field'. Physica 86 – 88 B,
       579 – 580 (1977).

[15]   L.E. Thomas, Z. Yin : 'Low Temperature Expansions for the Gibbs
       States of Weakly Interacting Quantum Ising Lattice Systems'.

<u>Commun.Math.Phys.</u> <u>91</u>, 405 – 417 (1983).

[16]   F.J. Dyson, E.H. Lieb, B. Simon : 'Phase Transitions in Quantum
       Spin Systems with Isotropic and Nonisotropic Interactions'.
       <u>J.Stat.Phys.</u> <u>18</u>, 335 – 383 (1978).

[17]   A. Amann : 'Jauch–Piron states in W*–algebraic quantum mechanics'.
       <u>J.Math.Phys.</u> <u>28</u>, 2384 – 2389 (1987).

[18]   M. Reed, B. Simon : <u>Methods of Modern Mathematical Physics II.</u>
       <u>Fourier Analysis, Self–Adjointness.</u> Academic Press, New York
       (1975).

[19]   A.J. Leggett, S. Chakravarty, A.T. Dorsey, M.P.A. Fisher, A. Garg,
       W. Zwerger : 'Dynamics of the dissipative two–state system'.
       <u>Rev. Mod. Phys.</u> <u>59</u>, 1 – 85 (1987).

[20]   H. Spohn, R. Dümcke : 'Quantum Tunneling with Dissipation and the
       Ising Model over $\mathbb{R}$'. <u>J.Stat.Phys.</u> <u>41</u>, 389 – 423 (1985).

[21]   H. Spohn : 'Models of Statistical Mechanics in One Dimension
       Originating From Quantum Ground States'. In : <u>Statistical</u>
       <u>Mechanics and Field Theory : Mathematical Aspects. Lecture</u>
       <u>Notes in Physics</u> <u>257</u>. Springer, Berlin (1986).

[22]   C. Aslangul, N. Pottier, D. Saint-James : 'Spin-boson systems :
       equivalence between the dilute-blip and the Born approxima-
       tions'. <u>J.Physique</u> <u>47</u>, 1657 – 1661 (1986).

[23]   M. Takesaki : <u>Theory of Operator Algebras I.</u> Springer, New York
       (1979).